CHEMISTRY RESEARCH AND APPLICATIONS

NEW TRENDS IN CHEMICAL PHYSICS RESEARCH

CHEMISTRY RESEARCH AND APPLICATIONS

Additional books in this series can be found on Nova's website under the Series tab.

Additional E-books in this series can be found on Nova's website under the E-books tab.

PHYSICS RESEARCH AND TECHNOLOGY

Additional books in this series can be found on Nova's website under the Series tab.

Additional E-books in this series can be found on Nova's website under the E-books tab.

CHEMISTRY RESEARCH AND APPLICATIONS

NEW TRENDS IN CHEMICAL PHYSICS RESEARCH

JAMES N. LAPHERTY
EDITOR

Nova Science Publishers, Inc.
New York

LIBRARY OF CONGRESS CATALOGING-IN-PUBLICATION DATA
New trends in chemical physics research / editor, James N. Lapherty.
p. cm.
Includes index.
ISBN 978-1-61668-535-5 (hardcover)
1. Chemistry, Physical and theoretical. I. Lapherty, James N.
QD453.3.N49 2010
541--dc22
2010036440

Published by Nova Science Publishers, Inc. † New York

CONTENTS

PREFACE

This book presents and discusses recent research developments in the field of chemical physics. Topics discussed herein include quantum dynamics in base pair DNA systems; chemical physical properties of bioprotectant-biomolecule systems; chemical physics of phonons; optimization of polar solar cells; computer molecular modelling; wood pyrolysis; and phonon dynamics of fe-based glassy alloys.

Chapter 1 - Single particle mass spectrometry (SPMS), that is capable of sizing and chemically analyzing single nanoparticles, has offered an opportunity for better understandings of mechanisms of aerosol-based nanoparticle synthesis, origin of environmental aerosol particles, and combustion of energetic nanoparticles. Also as ultrafine and accumulation-mode aerosols in the size range of 30 - 300nm can penetrate deeply into human lung and cause serious diseases, the need for advanced SPMS has been rapidly growing. In this article, the authors therefore offer a broad review of single particle mass spectrometry based on their previous publications, from its working principle and several concerns for interpretation of the data, to the applications for probing nanoscale phenomena,. This article also describes ways to develop and improve the SPMS for better quantification of data and more efficient detection, so as to provide simple design guidelines for SPMS developers.

Chapter 2 - The interfacial region with molecular organization and non-cento symmetry between two phases in contact has in variably different dielectric properties from those of the bulk phases. The response of polar solvent molecules to the changing electric fields is certainly different at different points in the interaction zone as the polar solvent molecules near the electrode surface are rigidly oriented and ordered in comparison to the unoriented or poorly oriented polar molecules towards the bulk in electrolyte. It has been observed that none of the existing model is able to explain the multiple dielectric relaxations exhibited by the polar solvent in the interfacial region. The significant change in dielectric behavior of the polar solvent near the electrode's surface has been mainly attributed to the rearrangement and segregation of polar molecules in the electric double layer and from the continuous rupture and reformation of hydrogen bond due to surface effect. As a result of rearrangement of water molecule dipole, the permittivity of water does not only vary from $\varepsilon=4$ on its surface to $\varepsilon=80$ in the bulk about 10nm away from surface but also shows its oscillatory character due to oscillatory electric field prevailing up to $10A^0$ from the solid surface due to surface effect. The fluctuating nature of permittivity would certainly affect the electromechanical and electrochemical properties of the interface.

Chapter 3 - In the DNA, the base pairs are involved in two qualitatively different mutual interaction types, hydrogen bonding and base stacking, and it is well known that these two major factors contribute to the stability of DNA double helix. The role of hydrogen bonds is well known in providing structure and directionality in the DNA base pairs. Base stacking interaction is one of the driving forces responsible for the stabilisation of the three-dimensional structure of DNA and RNA molecules. A very few studies are present in literature to study with a quantum approach these two processes and only the author has considered the time-dependent hydrogen transfer between the base pairs in a quantum way.

Chapter 4 - During its historical evolution, hip arthroplasty have encountered and still occasionally encounters enormous obstacles, which mainly deal with the quality of the used prosthetic materials. Since the first attempts, which used prostheses made of primitive materials at the end of the nineteenth century, many different materials have been experimented and improved surgical approaches undertaken. Hip prostheses have been conceived according to different design criteria, but they all have been following the wide underlying principle of low-friction arthroplasty. Such a basic concept yet represents the standard in total hip arthroplasty. In this paper, an analytical overview is given from a materials science perspective of the major problems encountered in the recent history of polyethylene, the most used material in hip arthroplasty. An outline is also given of an emerging Raman spectroscopic technology aimed at non-destructively monitoring the structural and environmental reliability of advanced biomaterials.

Chapter 5 - Ground state isomerism and dual fluorescence of 2-methylharmine anhydronium base (7-methoxy-1, 2-dimethyl-2H-pyrido[3,4-b]indole), HIA, have been studied in 2-methylbutane and other aprotic solvents. Absorption, steady state and time resolved fluorescence measurements reveal the existence of a ground state equilibrium between two HIA isomeric species, namely Q and Z, of presumed quinoid and zwitterionic structures, respectively. Upon light absorption, the Q isomer emits fluorescence at short wavelengths from the locally excited state, LE, resulting from a vertical Franck-Condon transition, with a lifetime around 4 ns. The Z species, once in its LE excited state, quickly relaxes to an internal charge transfer state, ICT, emitting long wavelength fluorescence with a lifetime around 11 ns.

Chapter 6 - The present work is addressed to investigate the effects of some bioprotectant systems on an enzyme, the ornithine carbamoyltransferase (OCTase), and a protein, lysozyme.

The stabilizing effects of a class of bioprotectant systems, such as trehalose, PEG, NaCl, ornithine, sodium phosphate and ammonium sulphate, on OCTase and lysozyme structure and thermal inactivation are analysed by using small angle neutron scattering (SANS), circular dichroism and UV spectroscopy.

In particular, the behaviour of residual activity of OCTase has been determined as a function of time, while changes in secondary and tertiary enzyme structure are monitored by circular dichroism and UV spectroscopy. Finally SANS allow to study the conformational properties of lysozyme as a function of temperature in absence and in presence of trehalose and PEG.

Chapter 7 - Several potential heuristic angles are explored with regard to the parametrization of the superconducting transition temperature, T(c), and its relationship to the chemical physics of bonding. Fruitful angles explored in this paper include the relationship of the gas of Cooper pairs in a superconductor to a van der Waals equation of state for such a fluid and how it may be controlled and exploited through the consideration of such variables as pressure, P, and volume, V, of the fluid of Cooper pairs resident in a superconductor. Other angles explored, in an attempt to get a heuristic handle on the superconducting transition temperature, T(c), for superconducting compositions across the Periodic Table, include the introduction of a Morse's anharmonic chemical bond potential for the Debeye frequency of the phonons as it emerges from the central result for T(c) in superconductors from the Bardeen-Cooper-Schreifer (BCS) formulation of superconductivity. Yet another angle explored in this heuristic reasoning, is the use of a Badger's relationship between chemical bond force constant, k, and equilibrium internuclear distance, r(e), to try to gain some understanding of the nature of the electron-phonon coupling mechanism in superconductors. A relationship between r(e) and A(e) emerges as a result of this line of thought, given as r(e)·A(e) = "critical temperature constant", from which examination of spectroscopic data across the Periodic Table may yield a compositional solution of the appropriate chemical bonding, with the assumption of the proper electron-phonon coupling constant, C, and density of states, N(0), that ultimately may yield more desirable superconductivity transition temperatures.

Chapter 8 - Naturally occurring triglycerides, oils and wax esters have been used as lubricants for thousands of years and continue to play an important role in today's lubricants industry. Brassicas (Rapeseed-Mustard) is the second most important edible oil seed crop in India. The major fatty acid present in this oil is Erucic acid (40-45%). High Erucic acid Rapeseed oil has immense potential for industrial utilization. Erucic acid 13-docosenoic acid ($C_{22}H_{42}O_2$) (Mol.wt 338) is the precursor to various other chemicals, of which Erucamide is very important, because erucic acid is mono unsaturated, physical properties such as melting point and titre are comparable to oleic acid.

Chapter 9 - Metal-mediated coal pyrolysis chemistry precedes catalytic steam gasification, which may produce hydrogen-enriched gases with CO_2 capture, and remove sulphur, nitrogen, and ash pollutants. Catalytic species are formed during the pyrolysis, which are derived from aqua-inorganic species added to low-rank coal; the addition of these inorganic species must be done with a careful adjustment of the pH, otherwise a mixture of various inorganic compounds are physically added to the coal, which may not form catalytically active species. Inorganic compounds (Fe, Co, Ni, Na, K, Ca, Mg, Ba, Al, Cu, Pb, and Zn) have often been added to low-rank coals using various methods, resulting in coal samples containing poorly characterized inorganic mixtures. Metal-mediated pyrolysis chemistry is extremely complex and studies at a molecular level require both experimental and high-level molecular modelling computations. In this paper, molecular models of low-rank coal containing inorganic species have been constructed and optimised to ground state using semi-empirical quantum mechanics computations (SE). These models have been used to develop char models containing metal species; SE and *ab initio* density functional theory computations are used to examine metal-mediated pyrolysis chemistry and reaction routes to CO_2, H_2O, CO and H_2 formation. Insights from experimental and molecular modelling studies include: the nature of aqua-metal complexes added to coal, transformations of metal species in char molecules, and reaction mechanisms, particularly for the formation of CO_2, CO and H_2. Reactions leading to the formation of H_2 and CO from steam and char containing metal clusters and the nature of active sites have been examined using computational molecular modelling. Experimental and modelling data indicate that catalytic steam gasification of low-rank coals containing Fe and Ni significantly increases the yield of H_2.

Chapter 10 - Submonolayer to multilayer coadsorption of lanthanum with boron and gadolinium with boron on the surface of Mo(110) has been studied by means of Auger electron spectroscopy (AES), electron energy loss spectroscopy (EELS) and work function (ϕ) measurements. The equilibrium state of double adsorbate systems achieved either by adsorption of rare-earth metal (REM) on boron precovered Mo(110) surface held at room temperature or after moderate annealing of the system with opposite order of adsorption (B on REM films) is the layer which is the inhomogeneous mixture of boron and REM atoms with preferential concentration of boron in the surface area of the mixed film.

The work function of such films even at REM to boron concentration ratio much higher than 1/6 are very close to the values of corresponding bulk LaB_6 and GdB_6, favoring assumption of surface rearrangement as the dominant reason of high electron emission efficiency of hexaborides.

Almost total similarity of the results for La-B and Gd-B systems can be viewed as the consequence of weak participation of Gd f-electrons in determining the thermionic properties of corresponding double layers.

Chapter 11 - A promising approach to decrease the cost of photovoltaic devices, mainly made with inorganic materials, is the fabrication of solar cells based on organic semiconductors, like semiconducting polymers. During the last years new polymers have been prepared and tested as donor materials in solar cells with the aim to improve the light absorption and the efficiency of the corresponding device. Actually, power conversion efficiencies surpassing 5% have been demonstrated for polymer-fullerene solar cells, but to use these devices in commercial applications this value need to be increased further; in addition other aspects, like the large area processing and the lifetime need also to be improved. The power conversion efficiency for a solar cell is directly proportional to the short circuit current (I_{sc}) and to the open circuit voltage (V_{oc}). The maximum V_{oc} for polymer solar cells is related to the difference between the highest occupied molecular orbital (HOMO) of the electron donor and the lowest unoccupied molecular orbital (LUMO) of the electron acceptor. In addition, even the relative position of donor LUMO and acceptor LUMO is fundamental for an efficient charge transfer from the donor polymer to the acceptor fullerene.

In this paper the authors will briefly introduce the basic working principles of polymer solar cells (materials, device architecture, device physics, efficiency limits). Next they focused the reader's attention on the properties of the most used electron acceptors, namely the fullerenes, and the authors implemented a theoretical approach to design fullerene derivatives to be used as electron acceptor in polymer solar cells. The authors calculated the LUMO levels of fullerene derivatives successfully used as electron acceptors and they correlated the obtained values with the V_{oc} of the corresponding device. Furthermore, the authors investigated the possibility to tune the LUMO level varying the side group attached to the fullerene cage.

Chapter 12 - As in recent decades world petroleum costs are increasing the searching for new alternative energy sources was developed in many countries. Biomass pyrolysis is an important route for charcoal production, a fuel with several advantages in its use compared to biomass or coal combustion. Charcoal has a higher heating value and lesser water content than biomass, is more stable in the presence of microbial agents or plagues and it hardly ever contains sulfur.

Maximum pyrolysis temperature, raw material heating rate up to reaching pyrolysis temperature and pyrolysis time are major parameters in the determination of the resulting charcoal properties. The control of these parameters is hardly performed in industrial charcoal furnaces, but a well controlled pyrolysis process can be carried out at a laboratory level.

In the present work the influence of the maximum pyrolysis temperature and heating rate on the pyrolysis process and charcoal properties was studied. *Eucalyptus grandis* wood was used as the biomass raw material; pyrolysis was performed in a temperature programmable horizontal tubular laboratory furnace in high purity nitrogen atmosphere, with maximum pyrolysis temperatures selected in the range 300 – 650 °C and heating rates of 4 and 16 °C min^{-1}. Wood and charcoals characterization was performed through elemental and proximate analysis and determination of high and low heating values by using the calorimetric bomb method. Wood pyrolysis was also studied by TG, at the same conditions used in the laboratory furnace. For those charcoals obtained at 4 °C min^{-1} their combustion in air atmosphere were studied by DSC and TG/DTA; combustion apparent activation energies were determined for charcoals obtained at 4 and 16 °C min^{-1}. Results show that for both heating rates the heating values increase as maximum pyrolysis temperature increases, but there is a different behaviour between 300 and 350°C, where heating value decreases. This seems to be due to a different charcoal composition. In addition, combustion activation energy for the 300°C charcoal is higher than the 350°C one.

Charcoal yield is higher for the lower heating rate being the effect more remarkable for higher maximum pyrolysis temperatures. TG results show that charcoal yield for a 2 hs, 450°C wood pyrolysis is the same that for wood pyrolysis up to 600°C. A similar behaviour is found for a 2 hs, 300°C and wood pyrolysis up to 350°C. DSC and DTA for wood and charcoal combustion show two exothermic peaks due to cellulose, lignin and their pyrolysis products.

Chapter 13 - *Background,* the effects of composition of mixed nitric and sulphuric acids on the nitration of cellulose and the effects of average molar mass, degree of nitration and unperturbed polymer dimension of cellulose nitrate on bonding strength and tack of joints of selected substrates have been investigated. Nitration of cellulose was non-stoichiometric, the degree of nitration decreasing from 14 to 7 percent as the mole fraction of water in mixed acid increased. Degree of nitration increased non-linearly with average molar mass. Bond strength increased to a maximum and then decreased for the three substrates as average molar mass, degree of nitration and unperturbed polymer dimension increased.

Results indicate that maximum bond strength was obtained at a degree of nitration of about 12 percent and unperturbed polymer dimension of 1.20×10^{-12} cm^2 for the three substrates; at average molar mass of 6.87×10^3 for wood and leather and of 6.87×10^3 to 8.25×10^3 for paper. The tack of cellulose nitrate joints showed visco elastic response to applied stress. Onset of visco elasticity occurred at degree of nitration of about 13 percent, unperturbed polymer dimension of about 1.40×10^{-12} cm^2 and average molar mass of about 7×10^3 for the three substrates.

It is concluded that bond strengths of these joints for paper, wood and leather are influenced by degree of nitration, average molar mass and unperturbed polymer dimension of cellulose nitrate.

Chapter 14 - The well recognized model potential is used to investigate the vibrational properties of four Fe-based binary glassy alloys viz. $Fe_{90}Zr_{10}$, $Fe_{80}B_{20}$, $Fe_{83}B_{17}$ and $Fe_{80}P_{20}$. The thermodynamic and elastic properties are also computed from the elastic limits of the PDC. Three theoretical approaches given by Hubbard-Beeby (HB), Takeno-Goda (TG) and Bhatia-Singh (BS) are used in the present study to compute the phonon dispersion curves

(PDC). Five local field correction functions proposed by Hartree (H), Taylor (T), Ichimaru-Utsumi (IU), Farid *et al.* (F) and Sarkar *et al.* (S) are employed to see the effect of exchange and correlation in the aforesaid properties.

Chapter 15 - The phonon dynamics of some Zr-Ni bulk metallic glasses has been studied at room temperature in terms of phonon eigen frequencies of longitudinal and transverse modes employing three different approaches proposed by Hubbard-Beeby (HB), Takeno-Goda (TG) and Bhatia-Singh (BS). The well recognized model potential is employed successfully to explain electron-ion interaction in the metallic glass. The present findings of phonon dispersion curve are found in fair agreement with available theoretical as well as experimental data. The thermodynamic properties obtained by HB and TG approaches are found very lower than those obtained by BS approach.

Chapter 16 - This chapter is devoted to consideration of the method of the transition from prebiotic organic microsystems to the primordial communities of prokaryotes. During the initial transformations prebiotic organic microsystems acquired the unstable bifurcate state, with its following relative stabilization by means of balanced oscillations around the bifurcation point. Then microsystems of the type converted into the primary living units – probionts, which were characterized with the ability to concentrate free energy and information already. Through the synergy effect probionts were organized into the communities that represented themselves minimum self-sufficient unit of life.

The appeared communities of probionts evolved to more complex communities of progenotes (characterized by C.Woese) and then prokaryotes. This earliest biological evolution proceeded through development of the initial nucleoprotein interaction, formation of RNA and then DNA, etc. Emergence of species had led to conversion of the initial communities of probionts and progenotes ("embryos" of ecosystems) into the real ecological systems, which were hierarchically organized of different microbial species.

Chapter 17 - Glassy materials are crucial to a number of major technological devices notably in domains of electronics, energy storage, and converters and environmental monitoring such as batteries, fuel cells and sensor technology. In this paper the structure of vitreous phosphate and borate glasses have been investigated. Also, effects of addition of alkali ions on the structure of vitreous phosphate and borate have been reviewed. It is observed that the structure of glassy phosphoric oxide is based on tetrahedral building block which termed as ultraphosphate. By Addition of metal oxides to phosphoric oxide, other structures are possible such as meta, pyro, and ortho phosphate which can be made depending on the [O]/[P] ratio. Investigations have been demonstrated that the structure of vitreous B_2O_3 is formed by boroxol rings which joined together by a common oxygen atom. Addition of alkali ions to B_2O_3 glass converts two BO_3 (triangle) units into two BO_4 (tetrahedron) units.

Chapter 18 - This article presents a systematic exploration of the direct relation existing between electrical properties (permittivity and conductivity) and free methanol content (MC) in biodiesel (BD), as a function of temperature, and also between flash point (FP) and MC.

Precisely known amounts of methanol (up to 2.5% V/V) were added to BD, and electrical properties were determined between 20Hz and 2 MHz from room temperature to 81°C. Measurements show a clear dependence between MC and the permittivity and conductivity of the samples. Particularly, the value of the real part of the permittivity extrapolated to 0°C shows a linear dependence on MC. Moreover, since MC is closely related to FP of biodiesel, the correlation between both properties was also determined. An excellent fitting to a potential function was found.

The results presented in this article show that electrical properties (EP) and flash point measurements are an interesting alternative for the verification of methanol content in biodiesel.

Versions of these chapters were also published in *Chemical Physics Research Journal,* Volume 3, Numbers 1-4, edited by Frank Columbus, published by Nova Science Publishers, Inc. They were submitted for appropriate modifications in an effort to encourage wider dissemination of research.

In: New Trends in Chemical Physics Research
Editor: James N. Lapherty
ISBN: 978-1-61668-535-5

Chapter 1

UNDERSTANDING NANOSCALE PHENOMENA USING SINGLE PARTICLE MASS SPECTROMETRY AND IMPROVEMENT OF ITS PERFORMANCE: A REVIEW

***Donggeun Lee*[*†a], *Kwang-Sung Lee*[a‡], *Sung-Woo Cho*[a§], *Lei Zhou*[b**], *Arthur Miller*[c††], *Kihong Park*[d‡‡] and *Michael R. Zachariah*[b§§]**

[a] Pusan National University, Geumjeong, Busan, Korea
[b]University of Maryland, MD
[c]NIOSH/Spokane Research Lab, WA
[d]Gwangju Institute of Science and Technology, Gwangju Gwangju, Korea

ABSTRACT

Single particle mass spectrometry (SPMS), that is capable of sizing and chemically analyzing single nanoparticles, has offered an opportunity for better understandings of

[*] To whom correspondence should be addressed:Tel: +82-51-510-2365, Fax: +82-51-512-5236, Email: donglee@pusan.ac.kr

[†] School of Mechanical Engineering, RIMT, Pusan National University, 30 Jangjeon, Geumjeong, Busan 609-735, Korea

[‡] School of Mechanical Engineering, RIMT, Pusan National University, 30 Jangjeon, Geumjeong, Busan 609-735, Korea

[§] School of Mechanical Engineering, RIMT, Pusan National University, 30 Jangjeon, Geumjeong, Busan 609-735, Korea

[**] Department of Mechanical Engineering, University of Maryland, MD 20742, USA

[††] NIOSH/Spokane Research Lab, WA 99207, USA

[‡‡] Department of Environmental Science and Engineering, Gwangju Institute of Science and Technology, Gwangju Gwangju 500-712, Korea

[§§] Department of Mechanical Engineering and Department of Chemistry and Biochemistry, University of Maryland, MD 20742, USA mechanisms of aerosol-based nanoparticle synthesis, origin of environmental aerosol particles, and combustion of energetic nanoparticles. Also as ultrafine and accumulation-mode aerosols in the size range of 30 - 300nm can penetrate deeply into human lung and cause serious diseases, the need for advanced SPMS has been rapidly growing. In this article, we therefore offer a broad review of single particle mass spectrometry based on our previous publications, from its working principle and several concerns for interpretation of the data, to the applications for probing nanoscale phenomena,. This article also describes ways to develop and improve the SPMS for better quantification of data and more efficient detection, so as to provide simple design guidelines for SPMS developers.

Keywords: Single Particle Mass Spectrometry, Nano-scale Phenomena, Aerosol, Health Effect, Quantification, Efficient Detection

1. Introduction

Aerosols play an important role as both a source of environmental/health impact and building blocks to advanced materials. In the latter case, aerosols are synthesized from either gas or liquid phase precursors with the objective of producing particles with desired size (often sub-micrometers or smaller), morphology, and composition. While on-line characterization of size is a well developed method and morphology characterization based on light scattering and electron microscopy has received considerable attention in the last decade, chemical composition determination is still in its infancy. [1] Many such materials are produced by spray-pyrolysis techniques in which liquid droplets containing precursors are injected into a thermal reactor, with resulting evaporation of solvent and reaction of the precursors to form product particles. To have better control in such industrial particle production processes, there is a need to understand aerosol-phase reaction kinetics that govern the final characteristics of product particles. [2]

Metal nanoparticles, for example, aluminum nanoparticles are receiving considerable interest lately as a possible approach to enhance the energy release from energetic materials. It is well known that fine metal particles are highly reactive and in many cases pyrophoric, however, quantification of their reactivity has not been extensively explored. [3,4] Most measurements of condensed-phase reaction kinetics for aluminum particles have been carried out by dynamic thermal techniques such as thermogravimetry (TG) and differential thermal analysis (DTA) that use bulk samples of aluminum nanoparticles. It is well known that those techniques suffer from sampling artifacts associated with mass and heat transfer effects. [5,6] Ideally one would like to explore the intrinsic reactivity of nanoparticles in the absence of other rate limiting kinetic effects associated with heat and mass transfer to clarify what a mechanism is responsible for the reaction kinetics. This would be best achieved by in-situ kinetic measurement of aerosol-phase reaction of a single nanoparticle in air.

Diesel Engines emit a tri-modal size distribution of Diesel particulate matter (DPM) at concentrations as high as 10^9 particles/cm^3. [7] Ultrafine particles i.e. the nuclei- and accumulation-mode particles are known to be the most harmful because of their deep penetration and easy deposition into the human lung. Also coarse-mode DPM often contains alkali and transition metal species such as Ca, Al, Na, Ba, Zn, Fe, Se, Mg and so on. [8] For an efficient removal of such metal-containing particles prior to their emission, one must have the knowledge of contents and size distributions of metal species, and a mechanism of metal formation as well.

Here, we would like to introduce the concept of single particle analysis to address three different research projects of interest that seem to be not easily examined with any conventional characterization tools. If we have an ideal tool capable of sizing and chemically analyzing any single nanoparticle in the aerosol state, the tool may be able to provide key information for understanding the aforementioned physical and chemical nanoscale phenomena. Aerosol time-of-flight mass spectroscopy (ATOFMS or AMS) might be the closest approximation to such a tool. The Prather group reported the simultaneous online measurement of size and molecular composition of single environmental particle using the ATOFMS. [8,9] Recently, we successfully developed an improved Single Particle Mass Spectrometer (SPMS) based on similar principles with the specific goal of addressing such nanoscale phenomena. [6,10-13] During the course of previous research, we have developed and evaluated novel designs of two key components to greatly improve the performance of the SPMS. [15-17]. We also determined how the size of a particle was extracted directly from a mass spectrum signal of the particle. [14] With the help of the design improvements to our SPMS, the three above-mentioned phenomena were successfully investigated.

In this review article, we would like to address the development of the improved SPMS and its applications in regards to our recent research. It will be started in Section II with the brief introduction of the SPMS including a working principle and experimental setup. Section III deals with the three applications of the SPMS as well as several new findings in a view of nanoscale physical chemistry. Section IV is devoted to introduce two designing points to improve the performance of the SPMS.

2. Single Particle Mass Spectrometry (SPMS)

2.1. Working Principle of the Spms

A schematic diagram of the single-particle mass spectrometer (SPMS) used in our study is shown in Fig. 1. The primary components of this system consist of an aerodynamic lens inlet, a two-stage differential pumping system, a free-firing dissociation/ionization laser and optics system, a linear time-of-flight tube assembled with an ion optics and a detector, and data acquisition system composed of a fast, digital storage oscilloscope and a PC. The aerodynamic lens inlet is employed to separate particles from the carrier gas and collimate them to form an aerosol beam. They can be therefore delivered with high efficiency through differentially pumped chambers into the ionization region. The pressure in the chamber which houses both the flight tube and the ionization region is ~ $6x10^{-7}$ torr when the aerosol inlet is open.

A frequency-doubled pulsed Nd:YAG laser operating at 10 Hz in the internal Q-switch mode is used for ionization of a particle arriving at the ionization region. A 38 mm focal lens mounted internal to the 2nd vacuum chamber focuses the laser beam which intersects the particle beam. The estimated power density with a laser beam diameter of ~0.3 mm and the laser pulse duration of 5 ns is approximately ~ $3x10^{10}$ W/cm^2 at the focal point. The laser energy is maintained above ~100 mJ/pulse to ensure complete fragmentation and ionization of a single particle as determined by threshold ionization [6]. As configured, positive ions formed by laser ionization are accelerated along ~1 m long linear TOF tube and detected with a microchannel plate (MCP) detector. More complete details of the SPMS are described elsewhere [6,14-17].

Basic instrumentation of the conventional ATOFMS appears to be similar to that of the SPMS. Indeed, the ATOFMS works similarly to some extent. This system however has two significant constraints that are overcome by the current SPMS. First, the ATOFMS employs light scattering techniques for sizing particles as well as triggering the ionizing laser. This approach enables the system to know the time when a particle arrives at the focus of the ionizing laser. Thus, the ATOFMS can provide both size and composition of a single aerosol particle at low concentration. For this reason it is true that the ATOFMS has been a particularly exciting development in the field of environmental research. [8,9,18-20] On the other hand, this benefit is offset by the fact that the measurable size of particles is thus often limited to those larger than 200 nm due to the light scattering limitation. [20] This hinders the use of the ATOFMS to analyze ultrafine particles, which are becoming increasingly more important in various fields such as nanotechnology. Second, the ATOFMS instrument gives the mass spectra comprised of both positive and negative molecular ions. The different polarity of such ions unfortunately facilitates the rapid recombination and charge transfer between the ions. [6,21] This yields some bias in the measured composition of the particle. For this reason, the majority of ATOFMS-based research has focused on reporting qualitative assignment of molecular species in environmental aerosols.

Reents and Ge [21] suggested two simple but very useful approaches to solve the constraints of the ATOFMS. First, they reported that the use of a much stronger laser pulse might solve that problem, because if the beam energy is strong enough to tear a single particle to its constituent atoms and to extract electrons out of those atoms even with a high electron affinity, only positive atomic ions not likely to collide with each other should be produced. As such, the absence of the negative ions would present an advantage i.e. reduced recombination and transfer between ions. They also demonstrated that the measured composition of aerosols was closer to their theoretical value especially in the presence of multiply-charged atomic ions.

Second, they studied the potential of using the ion signal itself from a laser ionization time-of-flight mass spectrometer to estimate particle volume equivalent size. In their work they showed that the total peak area in a single particle mass spectrum can be used to estimate the original particle diameter fairly well within 10 %, even though the number of constituent atoms comprising the particle scatters a lot up to ~34 % depending on particle densities (see Table 1). This shows a good potential accuracy for the estimation of particle size that is not material-dependent.

2.3. Determination of Chemical Composition of a Particle Using Spms

The first version of our SPMS seen in Fig. 1 was developed by adopting their ideas. For characterization of our instrument, we generate NaCl nanoparticles by spray pyrolysis [2,6,23] as a standard material and put them into the SPMS.

Table 1. Size estimation from atomic densities for selected materials with various densities.

Molecular Formula	Molecular weight	Density [g/cc]	No. of atoms in a molecule	No. of atoms / 100 nm sphere	Estimated size [nm]
NaCl	58	2.165	2	2.35×10^7	97
Au(s)	197	19.3	1	3.08×10^7	107
Pb(s)	207	11.34	1	1.72×10^7	88
$(NH_4)_2SO_4$	132	1.769	7	2.95×10^7	105
$(NH_4)HSO_4$	115	1.78	6	2.92×10^7	105
$C_{20}H_{42}$	282	0.789	20	1.76×10^7	88
SiO_2	60	2.65	3	4.17×10^7	118
Al(s)	27	2.7	1	3.15×10^7	107
Al_2O_3(s)	102	4	5	6.17×10^7	134
C_8H_8(s)	104	1.05	8	$2.54 \times 10^7$7	100
Fe(s)	56	7.86	1	4.42×10^7	120
			Average[a]	$(3.2 \pm 1.3) \times 10^7$	106±14

[a]The smaller deviation for the average particle size comes from the calculation of the diameter using the cube root of the number of atoms.

An example of a typical mass spectrum for a single sodium chloride particle of nominal diameter ~70 nm is shown in Fig. 2. With the laser fluencies used we never observe any multi-atomic species, while multiply-charged atomic ion peaks are usually observed, and indicate that ionization efficiency is high. In addition to the signals for Na and Cl, we also see H and O presumably from our inability to completely dry the aerosols. In our prior work we demonstrated that under these conditions, relative ion concentration could be determined to within 15% for a variety of different aerosols [6].

The formation of multiply-charged positive ions is not surprising when one considers a simple estimation of the minimum threshold energy required for disintegration/ionization of a NaCl nanoparticle. The minimum energy required to convert a 100 nm NaCl particle into Na+ and Cl+ is 6×10^{-14} J. For the laser power density discussed above (100 mJ/pulse) the nominal energy a 100 nm diameter particle is exposed to is 1.1×10^{-8} J (see also table 3 in Reents and Schabel [22]), or > 5×10^6 times higher than the minimum threshold energy.

Elemental stoichiometry ratios evaluated for 168 sodium chloride particles are indicated in Fig. 3. The mode for the atomic sodium fraction (Na / (Na+Cl)) is at 0.516, which is very close to the theoretical value of 0.50. A further indication of the ionization efficiency is the invariance in the signal intensity, for 70 nm NaCl particles and 85 nm Al particles (DMA selected) with increasing laser power [14]. Above ~50 mJ/pulse no increase in ion intensity is observed indicating that the ionization process has been saturated. The results presented in Figures 2 and 3 are highly indicative that the ionization process, under the experimental conditions described here, is highly efficient.

2.4. Size Estimation of a Particle Size from its Mass Spectrum

Particle size estimation suggested by Reents and Ge [21] can be justified when two necessary conditions are satisfied. The first condition is "complete ionization", where all constituent atoms are converted to ions. As explained in the previous section, this condition is presumably satisfied with the use of extremely high laser fluencies [21,22] The second condition is that all ions generated through the complete ionization should be delivered from the laser ablation/ionization region to a detector with no traveling loss. When these two conditions are met, the ion signals in a mass spectrum will reflect the total number of atoms comprising the parent particle (∝ volume of particle) and the integrated signal intensities should have a linear relation to the particle volume. However, the cube root of the integrated ion current for various particle diameters (50, 70, 120, 150 nm) selected from the DMA is not linear as seen in Fig. 4. The possible reason for the nonlinearity will be explained in the next section. Instead, here we would like to emphasize that the correlation, regardless of the nonlinearity, is not material dependent. This implies that this correlation can still be used for proper estimation of particle size with unknown composition.

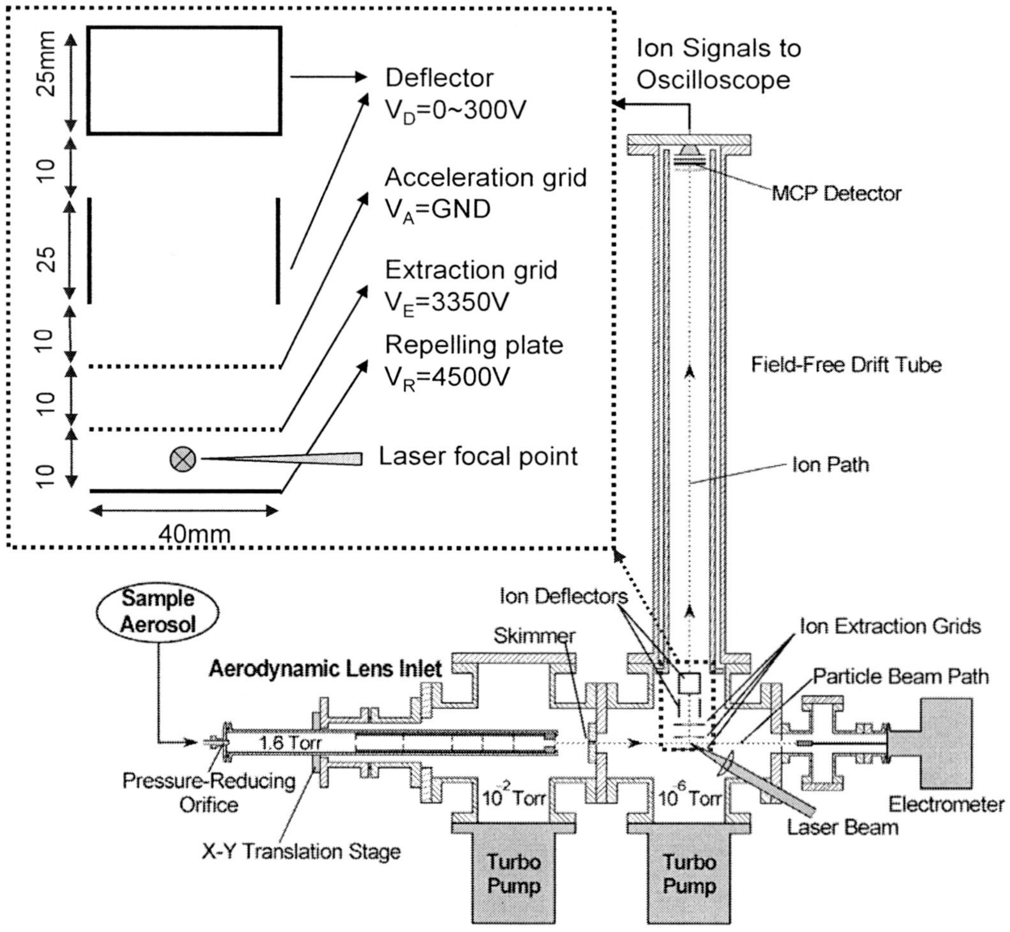

Figure 1. A schematic of the single particle mass spectrometer and a detail configuration of the ion optics.

Using the relationship in Fig. 2, we obtain a size distribution of size-polydisperse NaCl particles from more than 1000 mass spectra. For comparison, a commercial scanning mobility particle sizer (SMPS) is used to measure the size distribution of the same aerosols. Fig. 5 shows good agreement between the very different approaches to obtain particle size. For presentation purposes, the height of the SMPS size distribution is rescaled to match the SPMS-estimated size distribution. The result indicates that now we are ready to use the SPMS to obtain quantitative measurements of size and atomic composition of any aerosol particles.

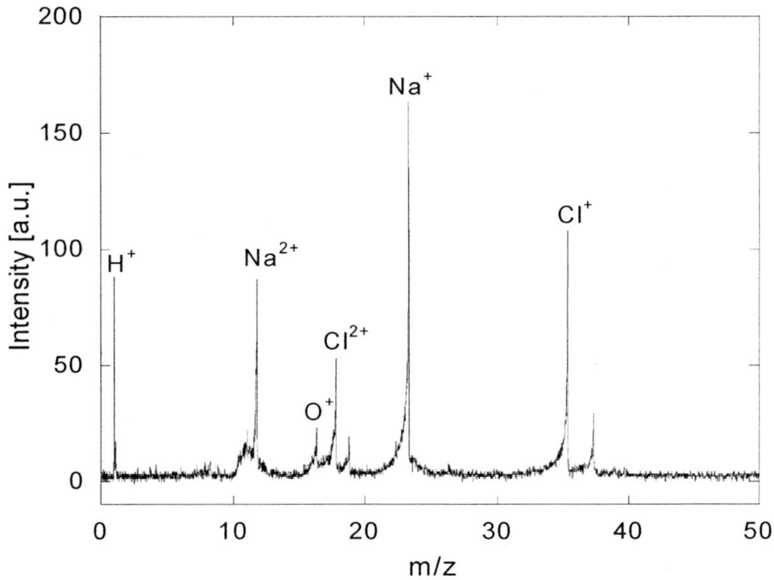

Figure 2. Typical mass spectrum of a NaCl nanoparticle (~70 nm).

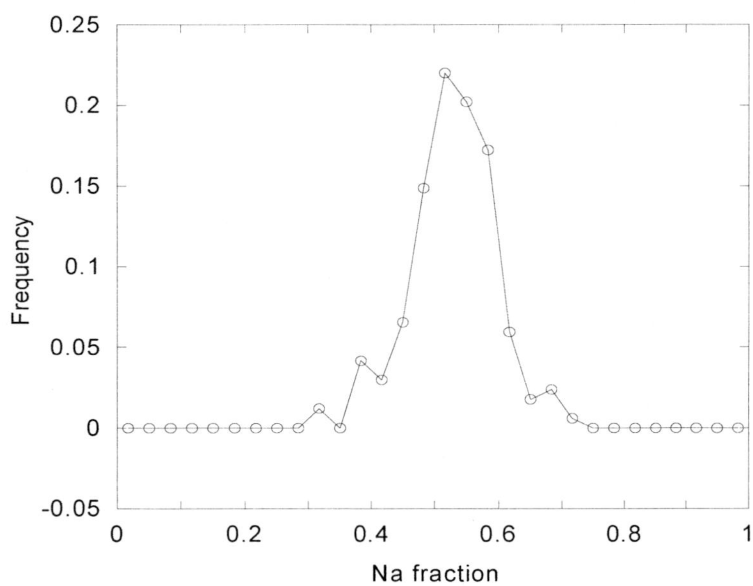

Figure 3. Histogram of Na fraction in polydisperse NaCl nanoparticles.

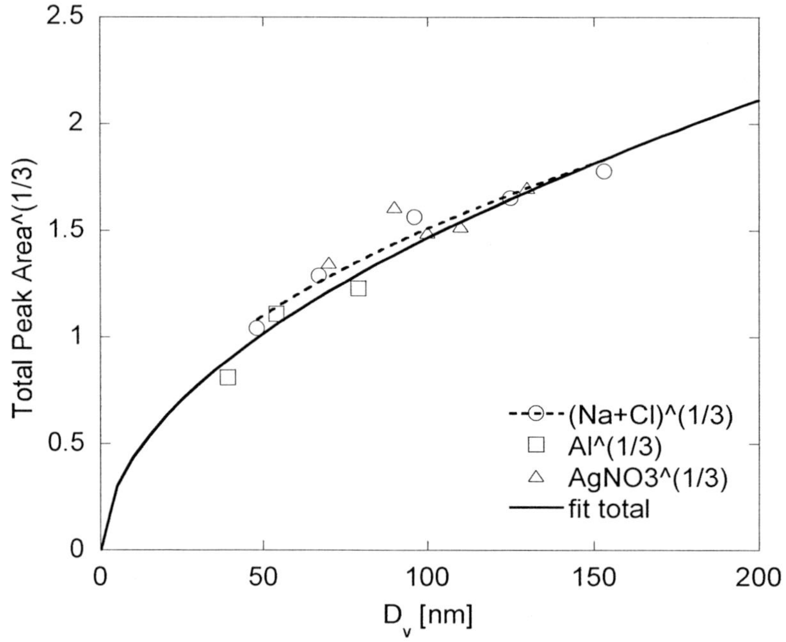

Figure 4. Non-linear relationship between the cube root of integrated ion signal intensity and particle size.

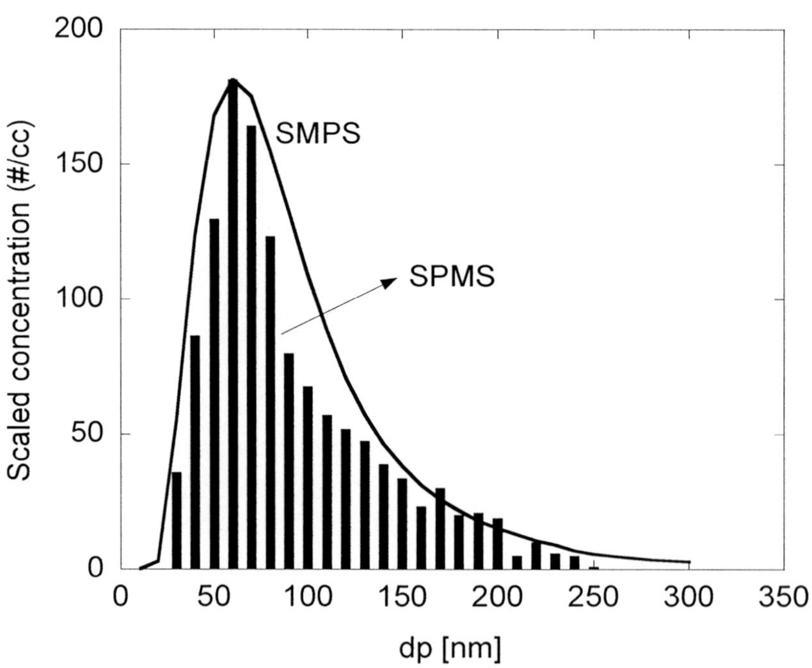

Figure 5. Comparison of size distributions measured by SPMS and Scanning Mobility Particle Sizer.

2.5. Transport Losses of Energetic Ions

Section of II-2 and II-3 described that such a strong laser pulse generates positive ions only. Under the voltage setup in Fig. 1, negative ions and electrons even if any exists can not be detected. To test this, inverse negative voltages were supplied to the repelling plate and extraction grid so as to ensure the non-existence of the negative ions. The experiment revealed that electrons are the only negatively-charged species. Taking this into account, the following simple model is developed to describe Fig. 4.

We first assume that the laser–particle interaction converts a particle into a spherical ion cloud of positive ions with a uniform charge density. The positive ions then undergo expansion due to hydrodynamic and Coulomb repulsion [24]. Ions initially located at the surface will be repelled by all inner charges and internal pressures from the developed plasma, with the outer atoms having the maximum possible kinetic energy (KE), and with the largest population as compared to the inner ions. When particles of different sizes are converted into their respective ion clouds, under the assumption of a constant cloud density, the cloud size is to first order proportional to original particle size. [24,25] Based on this, the KE of outer-surface ions dominates the mean KE and one might assume the KE may relate indirectly to the parent particle surface area (KE $\propto d_p^2$). In other words, larger particles might generate the more energetic ions.

At this moment it is interesting to investigate the origin of the nonlinearity. Following the energetic ion creation, some ions will arrive at the detector, while others can be lost at the TOF wall. The relative loss of total ions is first explored as a function of ion kinetic energy. The upward-ejected ions in Fig. 1 will be detected regardless of their initial kinetic energy and they will have the shortest time of flight. Backward-ejected ions decelerate and turn back to the MCP due to the electric field in the extraction region. Thus, the backward-ejected ions continue to be detected and have the longest time of flight. In contrast, horizontally ejected ions are most likely to collide with the inner wall of TOF tube and never make it to the detector. As the ejection angle with respect to the TOF upward axis increases, it becomes more difficult to detect the ions. Thus, it is possible to build a model to determine a certain range of the detectable angle at which ions depart. The two maximum detectable angles are in the proximity of both 0° and 180°. As all ions between these two maximum angles are lost, we would say that "a cone of silence" exists.

The trajectories of ions are simulated by using the commercial software SIMION. Fig. 6 shows the trajectories of Na^+ ions when they are ejected in all directions at four different kinetic energies. The voltages V_R, V_E and V_A are set for space focusing with values of 4500, 3350 and 0V, respectively as noted in Fig. 1. In the conventional plate-type ion optics, it is clear that a higher KE leads to greater ion loss. Fig. 7 shows a more quantitative explanation of this effect. Detection efficiency (DE), defined as the ratio of the number of ions arriving at the detector to the total number of ions emitted, is estimated at each specified KE. DE is highly KE-dependent, inverse-proportional to KE over a wide range (DE $\propto$ KE^{-1}). Substitution of the aforementioned KE $\propto d_p^2$ into the relationship of DE $\propto$ KE^{-1} results in the reduced form of DE $\propto d_p^{-2}$. Hence, the bigger the particle size, the greater the ion loss is. This explains the size-dependent deviation from a linear line representing 100% detection of ions.

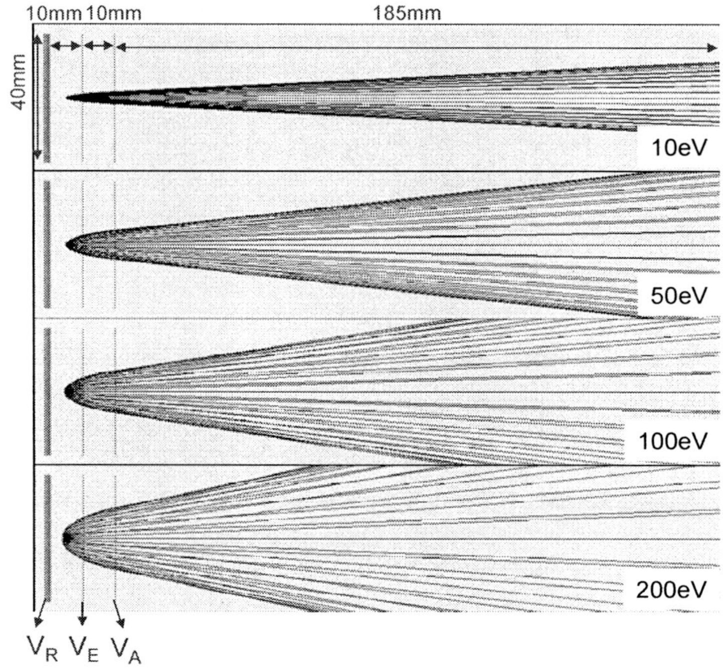

Figure 6. Trajectories of Na^+ ions launched at four different KEs in the three different E-field regions of extraction, acceleration, and field-free TOF regions.

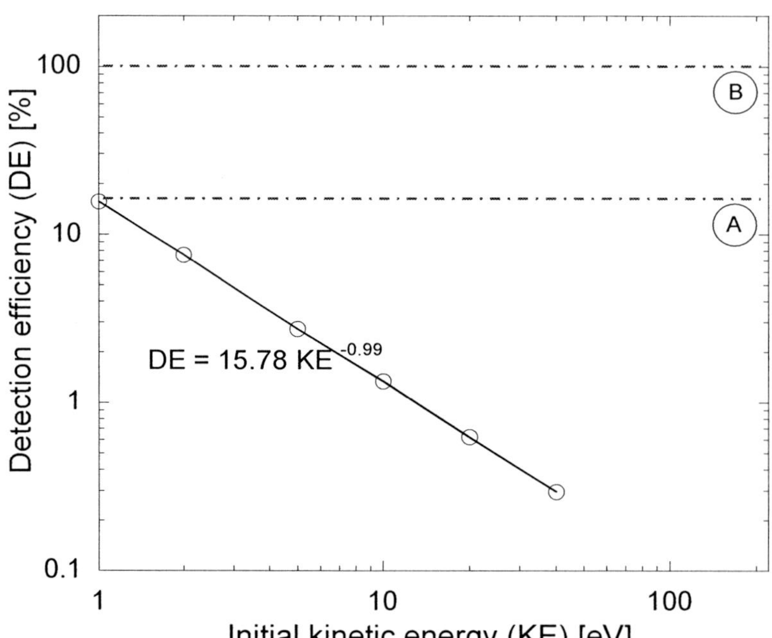

Figure 7. Profile of detection efficiency against the initial kinetic energy of ions in the conventional ion optics.

2.6. Nanosecond Laser-Induced Energetic Ion Formation

In this section, we will discuss in more detail the size-dependent energetic ion formation. A molecular dynamic simulation is used to investigate this. We postulate that positive ions generated by a nanosecond high-power laser pulse form a spherical ion cloud, which has often been adopted for the breathing sphere model [26]. The spherically symmetric system makes the calculation much simpler.

It has been observed that very energetic positive ions up to 1 MeV are formed when a femto-second laser pulse deposits a gas cluster [26,27]. The dynamics and maximum kinetic energy of the ions were predicted qualitatively well by Ditmire et al.'s nanoplasma model [26]. Note that the femto-second laser pulse has an extremely large intensity (> 10^{15} W/cm^2), which implies that even at the beginning of the pulse, enough electrons exist for collision ionization in a spherical plasma cloud. Collision ionization increases the electron density rapidly, often up to more than the critical value. The highly dense electrons in a nanoplasma ball are heated resonantly by the surrounding laser field, leading to a very rapid expansion of the plasma ball. A theory including tunneling, collision-induced ionization, and subsequent nanoplasma resonance with the pulse seem very likely to be accepted by the research community.

In contrast, the majority of nanosecond lasers used for the SPMS have a much lower intensity, typically less than 10^{10} W/cm^2. This intensity seems to be insufficient to generate electrons to such a high density. Therefore, resonant heating of the plasma ball may not occur, in turn, the above-mentioned hydrodynamic expansion may not be the case. Indeed, single-photon or multi-photon ionization of clusters or gas molecules with nanosecond laser pulse at low-to-moderate intensities has been studied, and the photofragmentation [28] and Coulomb explosion [29] dynamics have been proposed as the corresponding mechanism. The kinetic energy (KE) of the ions at those intensities was revealed to be much less than in the former case.

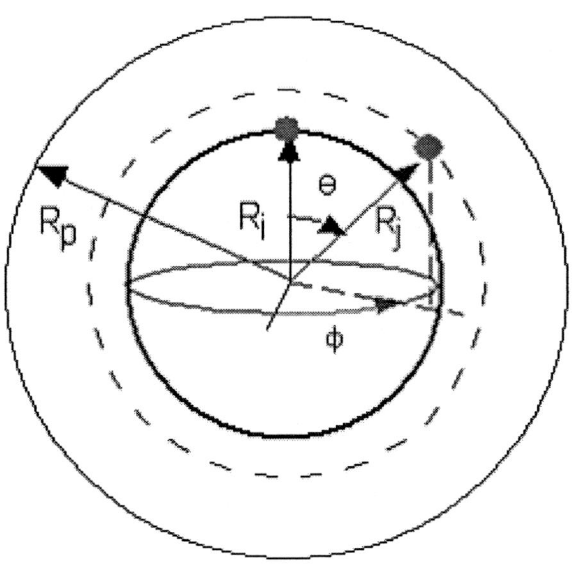

Figure 8. Spherical coordinate system for an ion cloud.

Together with contradictions in the mechanism, whether the cluster can be treated as plasma or not has given rise to a controversy. For example, classical dynamics simulations including the Coulomb field of the ions indicate that the electrons are quickly removed at the beginning of the interaction even from large clusters. [25] That makes the nanoplasma model [26] questionable. A thing to note is that all the above discussions are only for gas clusters, not for nanoparticles. There are no available experimental data for nanoparticle ionization. Are we between a nanoplasma ball and a positive ion cloud, or hydrodynamic and Coulomb expansion? Unfortunately, we are not currently in a position to answer that question. Regarding our conditions, for ionization of a nanoparticle by a longer (5 ns) much weaker ($\sim 10^{10}$ W/cm^2) laser pulse, the condition should lie between the above two limiting cases.

Therefore, we first choose the simplest model the main assumption of which is as follows. Particles are completely ionized to yield positive ions which form a spherical cloud. The interaction of each pair of ions is depicted by Coulomb repulsion. Fig. 8 shows the coordinate system, where R_i and R_j denote the radial position vectors of the i_{th} and the j_{th} ions as

$$\vec{r}_{ij} = r_j \sin\theta_j \cos\phi_j \hat{i} + r_j \sin\theta_j \sin\phi_j \hat{j} + (r_j \cos\theta_j - r_i)\hat{k} \qquad (1)$$

First, we derive the differential force vector d^3F_{ij} exerted on the ith ion by the jth ion by differentiating the corresponding Coulomb potential U_{ij}

$$U_{ij} = \frac{1}{4\pi\varepsilon_0}\frac{q_i q_j}{r_{ij}}, \quad d^3\vec{F}_{ij} = -\nabla U_{ij} = \frac{1}{4\pi\varepsilon_0}\frac{q_i q_j}{r_{ij}^{\ 3}}\vec{r}_{ij} \qquad (2)$$

where q_i and q_j are ionic charges of the ith and j^{th} ions, respectively and ε_0 is vacuum permittivity. Net force on the ith ion by all surrounding ions is directed upward due to the symmetry and is obtained by integrating the above d^3F_{ij} as

$$\vec{F}_i = \iiint d^3\vec{F}_{ij} \cdot \hat{k} = \frac{\rho^2 V_0}{3\varepsilon_0} r_i \qquad (3)$$

where ρ and V0 are constant charge density of ions in the ion cloud and volume of the cloud, respectively. Thus the ionic charge is given by qi = ρdVi. The potential energy on the ith ion is also obtained by integrating the above Uij over all jth ions as

$$U_i = \frac{1}{4\pi\varepsilon_0}\iiint \frac{\rho^2 V_0}{r_{ij}} r_j^{\ 2} \sin\theta_j d\theta_j d\phi_j dr_j = \frac{\rho^2 V_0}{12\varepsilon_0}(3R^2 - r_i^{\ 2})) \qquad (4)$$

where R is outer diameter of the spherical ion cloud. The total potential energy for the whole ion cloud U is given by integrating Ui against the volume as

$$U = \iiint U_i 4\pi r_i^2 dr_i = \frac{4}{15}\frac{\pi}{\varepsilon_0}\rho^2 R^5 \quad (5)$$

Mean total potential energy is obtained by dividing the U by total number of ions Ns as

$$\overline{U} = \frac{U}{N_s} = \frac{3Ue}{\rho 4\pi R^3} = \frac{\rho R^2}{5\varepsilon_0} \quad (6)$$

This is the first explicit expression for the size dependency of the potential energy of ions. The potential energy is completely converted into kinetic energy after sufficient time. Thus, Eq. 6 obviously shows that a larger particle or ion cloud leads to a higher potential and kinetic energy. As predicted in Section II-5, the mean kinetic energy of the ions in Eq. 6 is proportional to the surface area of the original particle. Interestingly, the previous experiment showed a 1.6th power for the R dependence [14], which is fairly close to the second power of R in Eq. 6. That proximity supports that the energetic ions may form through Coulomb repulsion.

Despite the fairly good agreement, it should be noted that the Coulomb repulsion model is derived under the assumption that the positive ions begin to expand in the absence of any electrons. All electrons escape from the ion cloud prior to the expansion. It is partly true in particular at the early stage of multi-photon ionization where electrons undergo less attraction. As more electrons depart from the cloud, the net charge of the ion cloud increases, forcing electrons to stay around positive ions. This seems to say the system is in the proximity to the nanoplasma, i.e. no electron escape. The caged electrons are then heated presumably by coherent electron motion [30] or resonant laser-electron coupling [26,31,32] either of which was proposed for the system of femto-second laser ionization of gas clusters. Such laser-driven hot electrons collide with atoms and/or ions, yielding multi-charge ionization, i.e. inverse Bremsstarhlung and at the same time expand rapidly with drawing out the surrounding ions. As both electrons and ions expand far enough, electrons exceeding Coulomb potential begin to escape, transforming the plasma sphere into the positive ion cloud. This is the time for Coulomb expansion, but generating less energetic ions as compared to the case of the pure Coulomb expansion model. Hence, the nanoplasma would be another limiting case of current system.

Here, we employed the 1-D hydrodynamic code based on a one-fluid two temperature (a single component particle; electron and ion temperature), which includes thermal conduction, a collision-radiative model for the ionization dynamics, the complete details for which are presented elsewhere [31]. From our prior simulations [32], we found the energetic ions are produced through two successive processes: 1) soft heating regime where single Al particle is evaporated to form a spherical fume 2) transition to a resonant absorption and hydrodynamic plasma expansion. During the expansion, electrons are highly non-uniformly distributed in radial direction, so that the surface electrons are first resonantly interacted with and heated very rapidly by the laser pulse. Interestingly, kinetic energy profile as a function of radial position R is pretty similar to Eq. 4 (see Fig. 5(a) in reference [32]). Moreover, we found that as particle size increases, the average kinetic energy as well as the maximum kinetic energy increases, following a similar power law relationship (KE ~ dp1.43). Totally different theories for the two limiting cases give similar results.

3. Applications of Single Particle Mass Spectrometry

In this section, we would like to demonstrate that the ability of SPMS goes beyond just measurement of size and composition of particles when it is applied to studies on the three aforementioned nanoscale phenomena on the basis of our previous researches [6,10-12].

3.1. Kinetic Measurement of Aerosol-Phase Spray Pyrolysis Reaction

Spray Pyrolysis Reaction

Fig. 9 shows the setup for production of particles by spray pyrolysis. A solution of nitrate, typically of 1 wt %, is prepared by mixing a kind of commercially available metal nitrate (> 99 % in purity) with filtered deionized water and atomized in air to generate aerosol droplets of the solution. The droplet-laden air is then passed through an aerosol dryer filled with dry silica gel to remove most of the water content in the droplets (say drying). After the solvent evaporates out of the droplet, the remaining salute forms a solid particle the size of which is controlled by the initial concentration of solute in the solution. The metal nitrate particles are delivered into a tube heated in a tube furnace. The heating decomposes metal nitrate particle into solid metal oxide and nitric oxide gas leaving the solid (say pyrolysis) according to

$$Al(NO_3)_3 \rightarrow \alpha Al_2O_3 + (1-\alpha)Al(NO_3)_3 \rightarrow Al_2O_3 + 3NO_2(g) \quad (7)$$

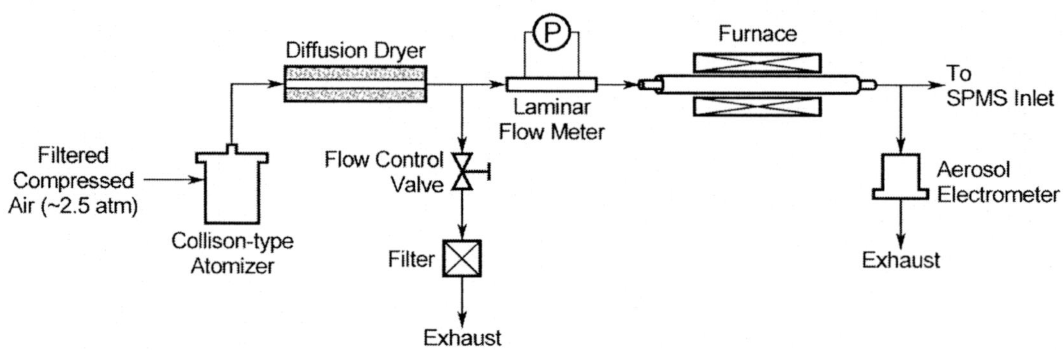

Figure 9. Schematic of the aerosol generator and thermal reactor setup

This is why this reaction is called spray pyrolysis.

The output aerosol from the furnace is split into two streams, one which is delivered to the SPMS and the other for monitoring the aerosol concentration with an aerosol electrometer. By analyzing the relative peak area of each element of the aerosol, the mass spectrometer can provide the stoichiometry ratios of O/Al, N/Al, and N/O at varied temperatures. At 100 °C as seen in Fig. 10, the ratio of nitrogen to aluminum is about 2.7 (expected stoichiometry of 3:1) while oxygen to aluminum ratio was ~8.3, and reflect that reactant particles are just starting to decompose. For calculation of the O/Al and N/O ratios at low temperatures, the oxygen contribution due to the presence of water, which was described earlier, is taken into account. Reduction of the nitrogen fraction, found in both N/Al and N/O, reveals progress of nitrate decomposition at higher temperatures. At 800 °C, the reaction seems to have completed, and the O/Al ratio reaches ~1.5 which corresponds to the theoretical stoichiometry of aluminum oxide, while the nitrogen fraction approaches a negligible level.

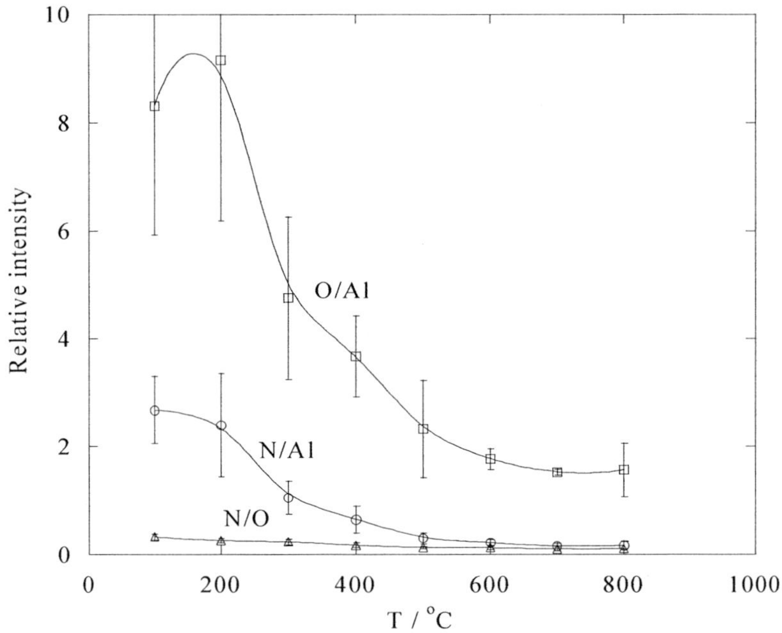

Figure 10. Change in relative intensities of nitrogen, and oxygen, and aluminum ion peaks during thermal decomposition of aluminum nitrate ($Al(NO_3)_3$).

Measurement of Kinetic Data of the Spray Pyrolysis Reaction

The extent of conversion, $\alpha \equiv (w_0 - w)/(w_0 - w_f)$, where w is the sample mass and the subscripts 0 and f represent the initial and final states of the reaction, respectively, could be calculated using the atomic fraction of nitrogen in particles <N> under a certain condition

$$\langle N \rangle = \frac{N}{Al + N + O} = \frac{1}{[Al/N] + 1 + [O/N]} \tag{8}$$

The extent of mass conversion α is found to be the same as the extent of mole conversion in Eq. 7. Fig. 11 provides a plot of the extents of conversion obtained for four metal nitrates (aluminum, calcium, silver, and strontium) at different reaction temperatures. Silver and aluminum nitrates exhibit lower decomposition temperatures (~200 °C for $\alpha = 0.5$) whereas the 50% conversion is observed at about 650 °C for calcium and strontium nitrates.

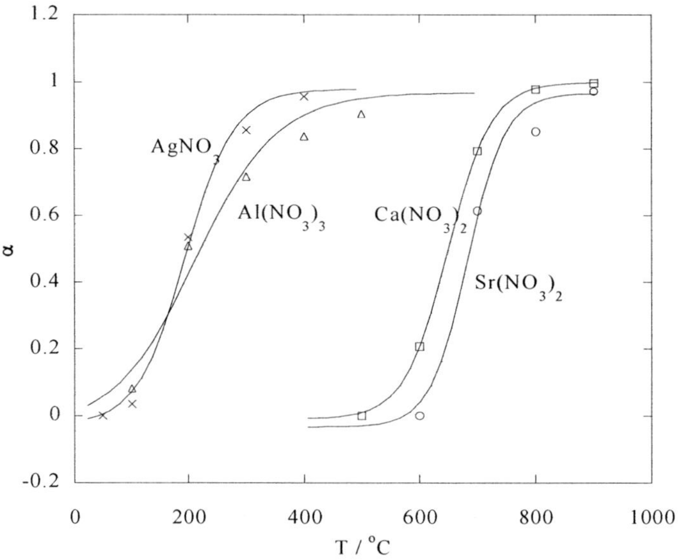

Figure 11. Temperature dependence of the extent of conversion obtained by SPMS for four metal nitrates.

The mathematical expression used for the kinetic parameters [6] is addressed as follows. Most condensed-phase reactions studied by traditional thermal analyses have been described using the following equation

$$\frac{d\alpha}{dt} = k(T)f(\alpha) \tag{9}$$

where k(T) is the reaction rate constant that is temperature-dependent, and f(α) is a mathematical representation of the reaction model. We assume a single-step first-order reaction for thermal decomposition of metal nitrates throughout the study, in which case f(α) is equal to 1-α. [33]

Eq. 9 can be written in an integrated form as

$$g(\alpha) \equiv \int_0^{\alpha} \frac{d\alpha'}{f(\alpha')} = \int_0^t k(T)dt' \tag{10}$$

where g(α) is equal to -ln(1-α) under the first-order reaction assumption [33]. For the SPMS experiment, we assume an isothermal reaction, and thus k(T) is constant with respect to the time. For our analysis, a plug-flow tube reactor model is used. Thus, the reaction duration time t is estimated by dividing the 30cm-long reaction zone volume by the volume flow rate which is 1.0 l/min measured at room temperature. The time t is on the order of a second at room temperature. At higher temperatures, the volume flow rate change is considered assuming the ideal gas law, and the reaction duration time is adjusted for analysis. By applying the Arrhenius relationship for k(T), Eq. 10 yields

$$\ln(g(\alpha)/t) = \ln k(T) = \ln A - \frac{E}{R}T^{-1} \tag{11}$$

Therefore, by repeating the measurement at various temperatures and determining corresponding extents of conversion, the activation energy and pre-exponential factor are obtained from an Arrhenius plot. The mathematical expression is essentially the same as used in conventional thermogravimetric analysis (TGA) [33,34]. Eq. 11 shows the relationship among α, k(T), and the Arrhenius parameters, i.e. the activation energy (E) and pre-exponential factor (A). The decomposition reaction is assumed to be single-step and first-order, and the heating process is approximated to be isothermal. Results are shown in Fig. 12. Each nitrate exhibited good linearity, which implies the single-step first-order assumption and use of the Arrhenius relationship are appropriate, although the number of points in the plots is limited.

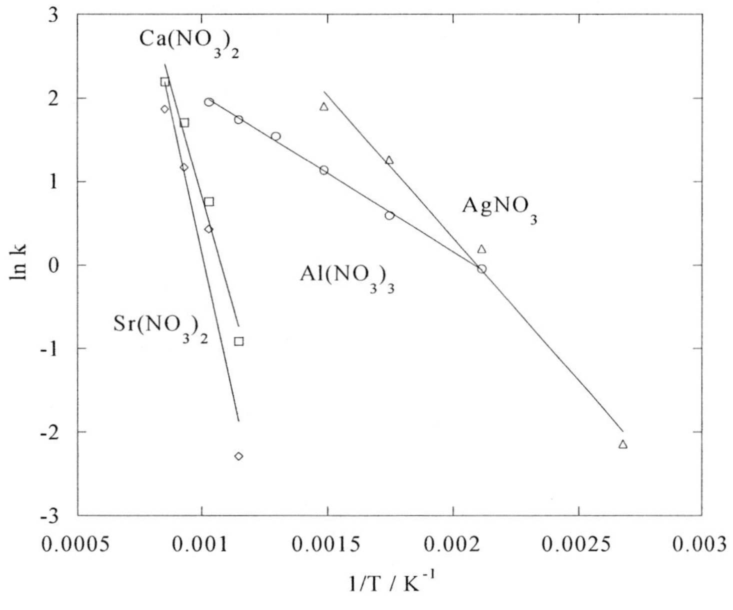

Figure 12. Arrehenius plot of for thermal decomposition of metal nitrates produced from obtained by SPMS data.

Change in Reaction Kinetics with Decreasing Initial Sample Mass

An important observation is to note that the reaction in the aerosol phase is taking place much faster than in the bulk powder state such as in a TGA dish. In Fig. 13, the Arrhenius plots obtained by SPMS and by dynamic TGA with the varied initial sample masses are shown together for calcium nitrate decomposition. The reaction rate constants for the TGA results are calculated by

$$\frac{d\alpha}{dT} = \frac{k(T)}{\beta} f(\alpha) \tag{12}$$

With an assumption that the term dα/dt can be expressed as the product of dα/dT and dT/dt, and defining a constant β be equal to dT/dt (the heating rate). We used this transformation which has been widely used in kinetics studies with conventional dynamic thermal techniques, although it has been criticized. [33] The curves for TGA in Fig. 13 show that the extent of conversion ranges from 0.1 to 0.9. The temperature ranges in which the rate constants are obtained by the two techniques do not overlap with each other. This is because there is a difference between the two techniques in the dynamic range for reaction rate measurement for a given temperature. The increase of the apparent reaction rate in TGA for decreasing mass is clearly demonstrated in this figure. As the mass decreases, it is interesting to note that the reaction rate constant determined by TGA moves up toward that derived using the SPMS. However, even with the smallest sample mass used in TGA (1.33 mg), there is a significant gap between TGA and SPMS, and the difference is approximately an order of magnitude at 600 °C. The initial sample size also affected the curve shape in the Arrhenius plot. For example, the Arrhenius curve has a straight line shape for the cases of the initial masses equal to or smaller than 2.29 mg. This usually indicates that the reaction model (single-step first-order) and Arrhenius relationship work quite well altogether for this reaction system. On the other hand, the Arrhenius curve has a non-linear structure when the initial sample mass was 7.45 mg or greater, and the same processing technique appears to have failed. This is another example of the initial mass effect that plagues reaction kinetics studies using TGA.

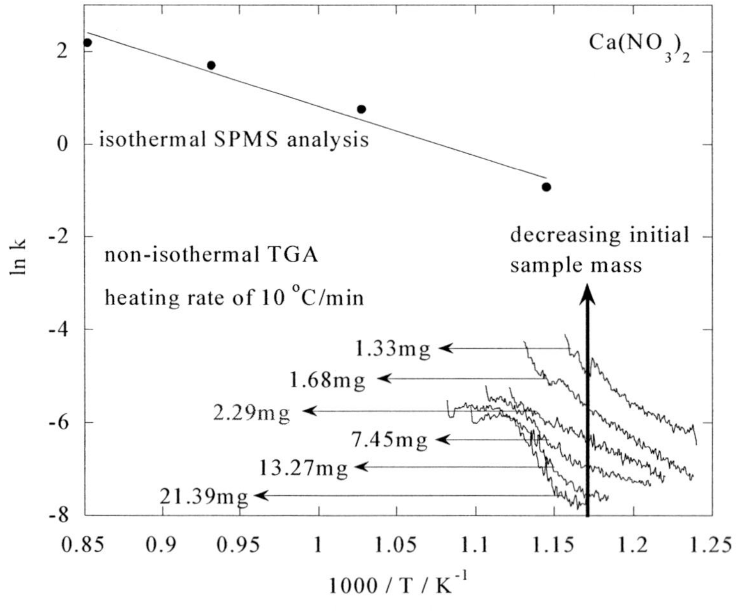

Figure 13. Comparison of the reaction rates of calcium nitrate obtained by SPMS and by conventional TGA. Note the upward shift of the reaction rate in the TGA results for smaller initial sample masses.

Discrimination of the causes of mass-loading on the observed kinetics in TGA is very difficult if not impossible to assess. The variety of potential mass-loading effects can range from simple heat capacity effects resulting in different actual heating rates for different sample sizes, to more complex phenomena. These might include longer diffusion paths for products of reaction for bulk samples. In the case of interest here, these might be evolution of NO or NO_2, where the observed kinetics would result from a slower mass loss rate, or the fact that the lowered loss rate of the NO_x results in a shift in the equilibrium constant as discussed by Ettarh and Galwey [35]. In general, however, the observed increase in the apparent rate constant as the sample size decreases implies that in the limit of very small sample sizes we should be observing rate behavior that is uncorrupted by these other transport related effects. Nevertheless, we feel confident therefore that our chemistry is not affected by any new phenomena that occurs at the nanoscale.

3.2. Size-Resolved Kinetic Measurement of Aluminum Nanoparticle Oxidation

Aluminum nanoparticle aerosols of various characteristics are generated by three different methods: DC arc discharge, laser ablation, and aerosolization of pre-made commercial aluminum powder. The resulting aerosols are fed into a tube furnace maintained at varied temperatures so as to react with air and are then injected into the SPMS as described in the previous section. Details of the aerosol generation are described elsewhere [10]. In this section, we will describe how the SPMS is used to obtain the size resolved oxidation rate of aluminum nanoparticles in air and to describe what the mechanism for the combustion is.

Kinetic Measurement of Aluminum Nanoparticle Oxidation

We first obtain isothermal kinetic data for DC arc-generated aluminum nanoparticle oxidation by passing particles through the tube furnace for a given residence time (~1 s) and temperature (25 °C ~ 1100 °C), and subsequently sampling with the SPMS. The SPMS-determined atomic fractions of aluminum (Al/(Al+O)) for 200~300 particles at a given temperature are used to estimate the extent of conversion α as defined by Levenspiel [36], i.e.

$$\alpha = \text{volume of converted material / total volume of particle.} \quad (13)$$

Thus, if the α equals one, a particle is completely oxidized to aluminum oxide (Al_2O_3), while the value of zero indicates the particle is pure aluminum. Values between 0 and 1 indicate a partially oxidized particle having the internal structure of an unreacted core (Al) surrounded by an aluminum oxide shell. Using the atomic fraction and size of an aluminum particle and size estimated from single measurement by SPMS, we can calculate the extent of conversion as a function of particle size at each temperature. The same mathematical expressions described in Section III-1 are used again to obtain the kinetic data, but as a function of particle size. In order to obtain a kinetic rate constant from the data, however, one must first propose a suitable model for $f(\alpha)$ and/or $g(\alpha)$ in Eqs 9-11. We choose the well known idealized "shrinking core" model for spherical particles of unchanging size [36] to describe aluminum nanoparticle oxidation. This process is repeated for three other sorts of aluminum particles, i.e. generated by the laser ablation and by the aerosolization of two commercially available powders.

Table 2. Summary of kinetic parameters of four sorts of aluminum nanoparticle oxidations.

Generation Method	Mean primary particle size[a]	Peak mobility size[b] (mode diameter)	Temperature	Fraction of completely oxidized particles
DC Arc discharge	~ 19 nm	~ 65 nm	900 °C	~ 68%
Laser ablation	~ 23 nm	~ 75 nm	1100 °C	~ 57%
Nanopowder (Aveka)	~ 29 nm	~ 70 nm	1100 °C	~ 13%
Nanopowder (Nanotechnologies)	~ 50 nm	~ 160 nm	1100 °C	~ 4%

[a]The mean primary particle size was measured using the TEM images.
[b]The peak mobility size was measured with the scanning mobility particle sizer (SMPS).

Table 2 summarizes the results of characterization and oxidations of the four different aluminum particles. For example, as temperature increased, the fraction of particles with α = 1 (i.e., Al_2O_3) increases, and at 900 °C, ~68 % of the DC arc-generated Al nanoparticles are completely oxidized to Al_2O_3. It is notable that the smaller particles oxidized first (i.e. for any given temperature, smaller particles have a larger extent of conversion). For the two kinds of Al particles with smallest primary particles (highest reactivity), i.e. DC Arc-generated and Laser ablated Al particles, each data set is binned by three size ranges (< 50 nm, 50-100 nm, and 100-150 nm). For these two particle types, Arrhenius parameters (activation energy and pre-exponential factor) are obtained using the size-resolved reaction rate constants as a function of temperature from Arrehenius plots (ln (k) versus 1/T), similar to the approach demonstrated in Figs. 12 and 13. We should note that with decreasing particle size, the rate constant increases and the activation energy decreases as summarized in Table 3. It is consistently observed that the activation energy decreases with decreasing particle size (agglomerate size). We hypothesize that the decreased reactivity with increasing the agglomerate size is caused by decreased particle surface area-to-volume ratio as the mobility size increases. Since the aluminum particles are agglomerates with some extent of "necking" or coalescing spheres (refer to TEM images in Fig. 10 of the reference [10]), the surface area-to-volume ratio should decrease as the agglomerate size increases. Only agglomerates with a point contact among primary particles have the constant surface area-to-volume ratio, independent of agglomerate size.

A Mechanism of Aluminum Nanoparticle Oxidation

The shrinking core model involves a reaction front moving radially inward and separating an unreacted core from a completely reacted ash outer layer. The overall reaction is believed to occur in the following three successive steps:

Step 1: Diffusion of gaseous reactant (oxygen) through the film surrounding the particle to particle surface (film resistance)

Step 2: Diffusion of gaseous reactant (oxygen) through the ash (oxide shell) to the surface of unreacted core (aluminum) (ash resistance)

Step 3: Chemical reaction of gaseous reactant (oxygen) with solid (aluminum) at unreacted core surface (chemical reaction resistance)

Since a solid ash (oxide shell) forms very rapidly when the particle is exposed to air, the first step (film resistance) can be ignored as a rate limiting step (i.e., the ash resistance is much greater than the film resistance). The time for complete conversion in each limiting step can be described by [10,36]

Table 3. Summary for Arrehenius parameters for aluminum nanoparticle oxidations.

Generation Method	Particle mobility size	Temperature range	Activation Energy (kJ/mol)	Pre-exponential factor (1/sec)
DC Arc discharge	< 50 nm	600~900 °C	31.8	$2.4*10^1$
	50-100 nm		56.9	$1.8*10^2$
	100-150 nm		174.6	$5.4*10^7$
Laser ablation	< 50 nm	800~1100 °C	24.6	$3.5*10^0$
	50-100 nm		104.8	$5.0*10^3$
	100-150 nm		120.2	$4.4*10^3$

$$\frac{t}{\tau_{diffusion}} = 1 - 3(1-\alpha)^{2/3} + 2(1-\alpha), \quad \tau_{diffusion} = \frac{\rho_{Al} D_p^{\ 2}}{32 D_e C_{O_2,\infty}}$$: Diffusion controlled step

$$\frac{t}{\tau_{reaction}} = 1 - (1-\alpha)^{1/3}, \quad \tau_{reaction} = \frac{\rho_{Al} D_p}{8/3 k_s C_{O_2,\infty}}$$: Chemical reaction controlled step (14)

where t is the reaction time, ρ_{Al} is the molar density of aluminum, D_e is the effective diffusion coefficient of oxygen through the ash layer, $C_{O2,\infty}$ is the oxygen molar concentration in gas, D_p is the particle diameter, and k_s is the first-order rate constant for the surface reaction.

To determine the rate limiting step, we also measure the extent of conversion by adjusting the reaction time (t) (i.e., the residence time in the furnace reactor) for a given temperature. By fitting each model (Eq. 14) to our experimental data (t vs α), we can infer the likely limiting step. Fig. 14 plots our experimentally determined reaction time (t) and extent of conversion at 1100 °C for aluminum nanoparticles (Aveka Inc.), and a best fit to each reaction model. The results presented indicate that our experimental data are better fitted to the diffusion controlled model. Thus, we argue that two stages of reaction take place for aluminum nanoparticle oxidation: one where the shell of the aluminum oxide layer is quickly formed followed by the condensed phase diffusion limited reaction. Once the very thin surface layer is formed, the kinetics would then be controlled by diffusion.

These results enable us to calculate the effective diffusion coefficients of oxygen through the ash layer using the rate constant as shown in Eq. 14, to give $D_e = 1.07*10^{-9}$ (600 °C) ~ $1.31*10^{-8}$ (900 °C) cm^2/s for DC arc generated aluminum nanoparticles of ~50 nm (i.e., $D_e = D_0 \exp(-E_a / RT)$ where $D_0 = 1.72*10^{-5}$ cm^2/s and $E_a = 69.5$ kJ/mol over 600 ~ 900 °C). To provide a more physical picture of the reactivity of nanoparticles, we compare the measured rate with the oxygen gas-collision rate [1] (i.e., the fastest possible oxidation rate).

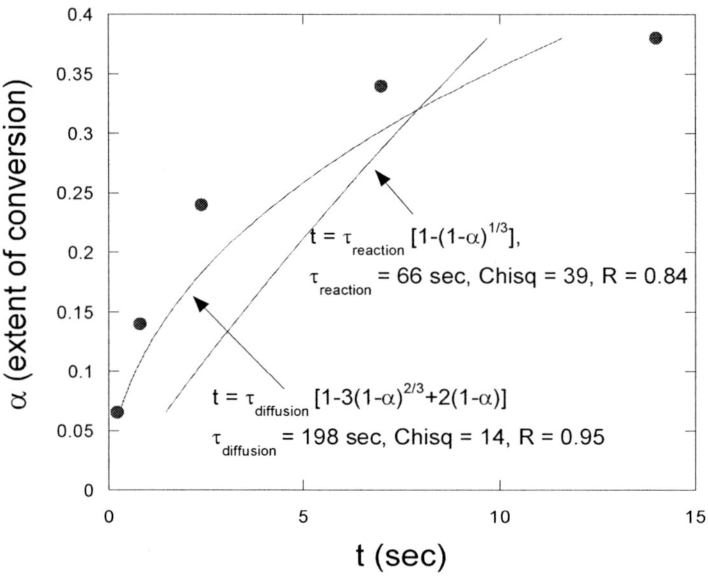

Figure 14. The reaction time (t) and extent of conversion at 1100 °C for aluminum nanoparticles (Aveka Inc.) with the fitted lines determined by diffusion-controlled and chemical reaction-controlled shrinking core models.

Theoretical collision rate (molecules/sec) = $\dfrac{P_{O_2}\pi D_p^2}{(2\pi m k_b T)^{1/2}}$

Measured reaction rate (molecules/sec) = $-\dfrac{dN_{O_2}}{dt} = \dfrac{4\pi D_e}{\left(\dfrac{1}{r_c} - \dfrac{1}{R}\right)} C_{O_2}$ (15)

Where PO2 is the partial pressure of oxygen in gas, m is the molecular mass of the gas molecules, kb is the Boltzmann's constant, R is the particle size, and rc is the radius of unreacted core. The theoretical collision rate at 900 oC for 50 nm particle is ~3.0*1012 molecules/sec, which is much greater than our experimental value (~7.4*105 molecules/sec).

3.3. Characterization of Metal-Containing Diesel-Emitted Particles

Classification of Particles and Size Distributions for their Classes

In this section, we would like to address the final application of single particle mass spectrometry to obtain the size- and composition- resolved statistics needed to gain insight into the formation mechanisms of metal-bearing diesel particles, for the full size range of diesel particulate matters (DPM). [12,15] For the experiment, a Isuzu diesel engine is operated at high or low engine load and with or without adding ppm levels of the ferrocene to the diesel fuel as a source of metal (Fe), the emitted DPM is sampled approximately isokinetically from the tailpipe and then fed into the SPMS as usual. Also DPM samples are taken for TEM/EDS analysis using a low-pressure cascade impactor (LPI). [37] Together with TEM/EDS observation, the particle mass spectra are sorted into the following basic classes or particle types.

1. "Elemental Carbon (EC)" particles, are defined as those where $H/C < 1$. These are particles containing greater relative amounts of elemental carbon (typically agglomerates of primary carbon spherules).
2. "Organic Carbon (OC)" particles are defined as those where $H/C > 1$. They are typically nanoparticles containing fuel or oil residues rich in organic carbon i.e. rich in hydrogen.
3. "Pure iron" particles, are those where $Fe/C > 10$. These are the nanoparticles of iron (presumably self nucleated in the engine).

The values of C, H and Fe are molar ratios and obtained from the corresponding peak areas in the individual mass spectra. Thousands of particle mass spectra are collected and analyzed to divide the particles into the classes. DPM in each class is then sorted and binned according to size. Before looking at detailed particle stoichiometry, we present for these three particle types the frequency distribution of each type as a function of particle size. As an example, Fig. 15 shows the number-weighted mobility size distributions for each of the three particle types at two different engine conditions and high (200 ppm) doping rates of iron. The EC particles are widely distributed in size, with a peak at around 70 nm, and constitute the bulk of the particulate material, which coincides with the size distribution of agglomerate particles measured by SMPS. The figure also shows that the OC particles are distributed in the lower size range (< 50 nm). At low engine load, we see virtually no "pure iron" particles. This is probably due to higher carbon generation at low load (characteristic of this particular engine), preferentially facilitating Fe adsorption onto carbon particles rather than nucleation of Fe.

At high engine load, the increase in Fe throughput combined with inherent soot reduction yields a higher Fe/C ratio, so the pure iron particles have increased in number and grown in size to a point where the aerodynamic lens is able to transmit them more efficiently to the ionization laser in the SPMS. Fig 15(b) shows that about 9% of the detected particles (by number) are pure iron nanoparticles. Note that in this figure the size distribution of the elemental carbon particles has not changed significantly, which is consistent with the SMPS measurement.

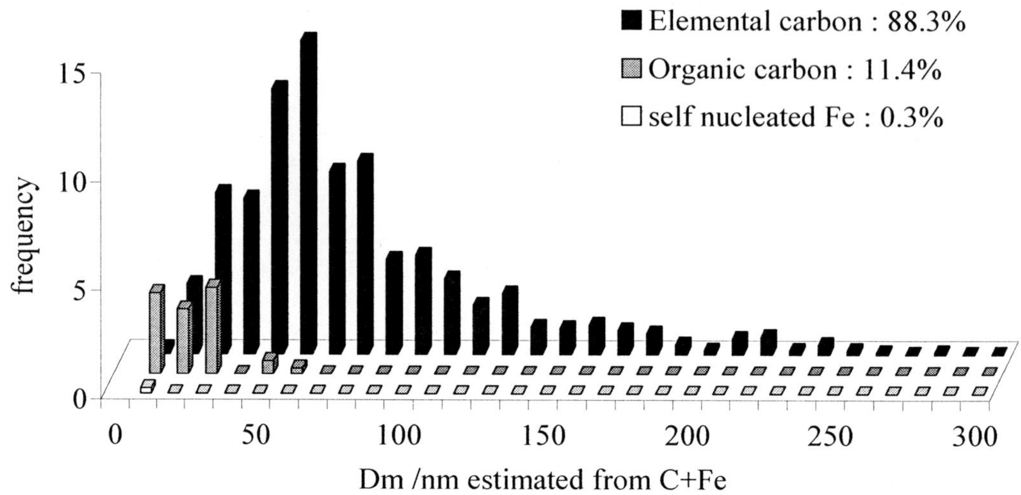

a)

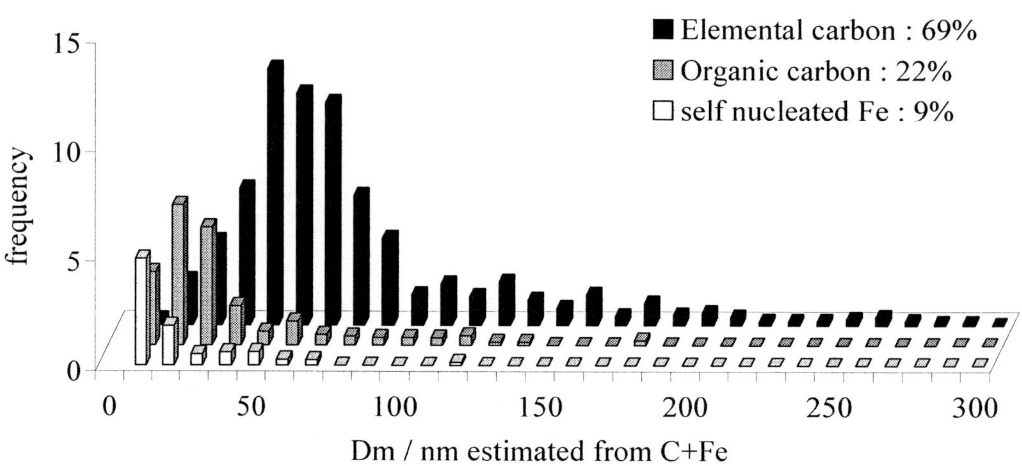

b)

Figure 15. Size distributions for three classes of diesel exhaust particles with high-Fe doping: (a) 0kW load, (b) 6kW load. Note that percent distributions refer to frequency of occurrence.

Mechanism of Particle Formation

We observe in Fig. 16 that the H/C ratios are clearly greater for smaller particles, but the scatter in the data demonstrates how elusive it would be to define a "typical" H/C ratio for diesel particles. The variation in H/C may reflect the locally varied temperature history that the particles underwent, which results in local variation in the (partial) burning of fuel and lube oil, as well as variation in the vapor deposition history. The solid line in Fig. 16 is drawn to indicate that the amount of hydrocarbon deposited to the preexisting iron-bearing carbon particles scales as D_m^2, i.e. approximately to the surface area of the accumulation mode PM. This might imply that a significant portion of the volatile hydrocarbon condenses near the tail pipe where exhaust gas containing volatile hydrocarbon is cooled. Under those conditions the very small sized PM would undergo free-molecular condensation, i.e. where condensation rate is proportional to the surface area of particle cores. Note that the horizontal axis in the figures corresponds to mobility size of composite particles estimated from the sum of carbon and iron peak areas.

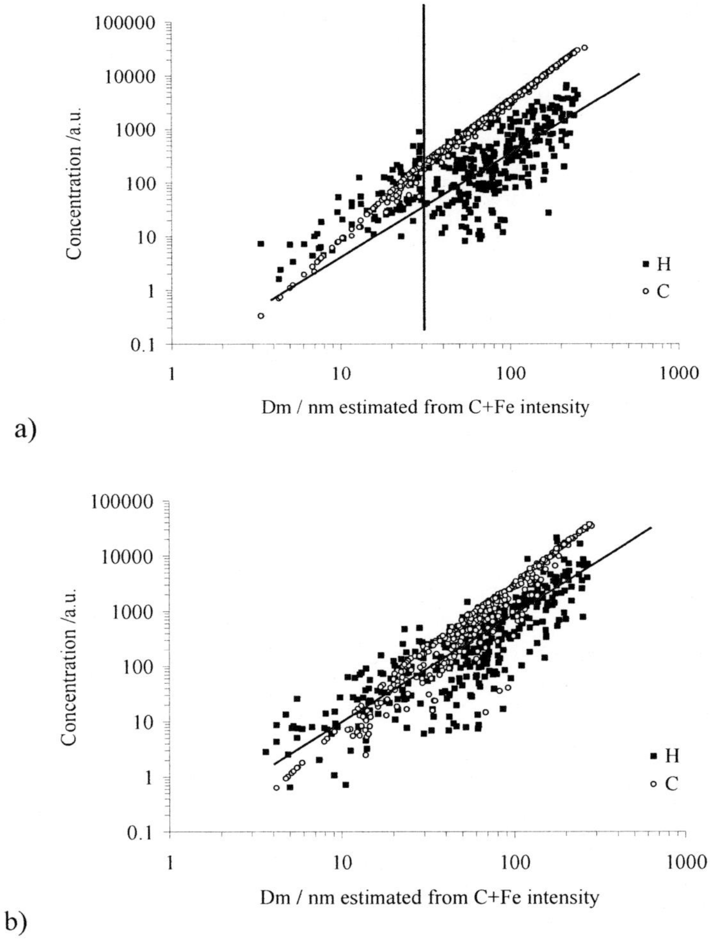

Figure 16. Size dependence of carbon and hydrogen in diesel exhaust particles for the case of high-Fe doping: (a) 0W load, (b) 6kW load. (Note: solid line through data shows $D_{m,c}^2$ dependence of elemental concentration).

Next, we are going to talk about the formation of iron species in DPM. For our experiment, high engine load conditions result in increased fuel flow rate i.e. to run the engine at constant speed as engine load increases. The quantity of Fe more than doubles due to this increased fuel flow rate, while the soot concentration decreases, due to improved fuel-air ratio at high load (for this engine) and to the soot suppression effect of metals [38-40]. The combined effect heightens the Fe/C ratio, effectively increasing the super-saturation ratio of iron vapor in the combustion chamber, in turn causing the nuclei mode particles to grow larger and more numerous. This is best seen in Fig. 17 (a-b), which show that for the 200 ppm doping rate, the higher Fe throughput at high engine load results in increased iron levels for the especially small particles. Note that at the low engine load (Fig. 17a) the data shows a tightly banded distribution of iron and a relatively low iron fraction over much of the particle size range i.e. lower engine loads produce less iron relative to carbon because the engine produces more carbon and less fuel is consumed to provide iron. In comparison, Fig. 17b shows that at higher engine load, many of the smaller particles lie above the carbon curve, suggesting they are "pure iron" particles, while many of the larger particles have iron content similar to the low load case. Overall, the relative amount of Fe is also higher and more variable over the entire size range for the high load case. The latter can be explained by the dominance of mutual coagulation i.e. over nucleation and condensation, between the three types of particles.

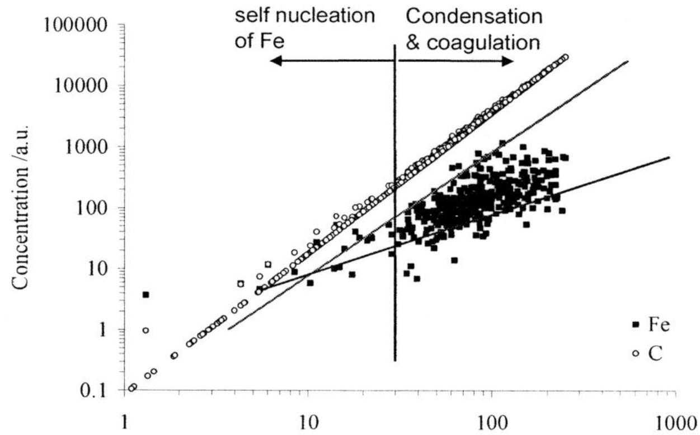

a)

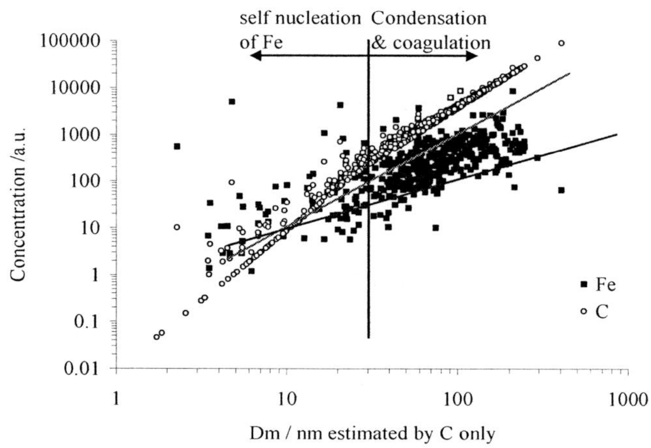

b)

Figure 17. Size dependence of iron and carbon fractions in diesel exhaust particles for the case of high-Fe doping: (a) 0W load, (b) 6kW load. (Note that upper line shows $D_{m,c}^{2}$ while lower line shows $D_{m,c}^{1}$).

Note that in Fig. 17a-b, the mobility size corresponds to carbon core mobility diameter that is estimated from the carbon peak area only. We postulated that the iron is condensed to the preexisting carbon core or self-nucleate inside the engine cylinder due to its very low vapor pressure. If this is the case, the behavior of particles may lie in the continuum regime, for which the deposition rate of Fe is proportional to diameter. [1] The slope of the plotted data Fe vs $D_{m,c}$ lies indeed between lower and upper solid lines representing continuum and free molecular regime, respectively, which is obviously smaller than the slope of H vs D_m in

Fig. 16. This is not surprising because inside the engine the pressure is very high, moving the particle behavior toward continuum conditions, where the deposition rate of Fe is proportional to $D_{m,c}$.

4. Development of Single Particle Mass Spectrometry

In this section we describe two new developments which are aimed at further improving the SPMS performance. In Sections II-5, we reported that significant particle size-dependent ion loss occurs prior to detection in conventional single particles mass spectrometry. This leads to increasing the lowest limit of particle mass measurement and more seriously to significant errors in the measured chemical composition of core-shell structured nanoparticles. In order to confirm the structural effect, similar method was applied for multi-component particles with two distinct structures such as light absorbing Al or Al_2O_3 particle coated with transparent NaCl and a mixture particle of those species. [41], We found that the core Al fraction of the core-shell particle is unlike the well mixed composite particles or pure salt. To prevent it, we need to develop a new type of ion optics in order to make the detection efficiency (DE) defined in Section II-5 to be KE invariant. In other words, the sloping DE versus KE profile should be flattened as suggested by lines A or B in Fig. 7. The second new development addresses the limitations of entraining and detecting ultrafine particles. To investigate ultrafine environmental aerosols one needs a well-designed aerodynamic lens to tightly focus and deliver the particles into the ionization laser focus. The narrower the focused aerosol beam, the more chance the laser has of hitting the particles, which becomes very important for ultrafine particles. In this regard, we acknowledge that two key elements, i.e. improved aerodynamic lens and ion optics characterize the uniqueness of our current SPMS. In this section, we would like to describe our efforts to progressively develop the best designs for those features.

4.1. Design of a New Ion Optics

Before addressing the design, we would like to mention that every ion optics system has a certain KE window of energetic ions to be measured. The objective of our design approach is to find a way to minimize ion losses by assessing and optimizing the KE of the ions. We first measured the maximum KE of the ions generated in the SPMS seen in Fig. 1. The upper limit is determined to be 200eV by estimating the initial KE of an ion from the peak width at the baseline of the mass spectrum. On the basis of the ion trajectories in the conventional ion optics as seen in Fig. 6, we used ion tracing software to evaluate several design trials including the curvature of the ion repelling plate, the addition of an Einzel lens, the use of a tubular electrode, and combinations of these.

Trial 1: Curved Repelling Plate

Observing the ion trajectories in Fig. 6, we are intuitively aware that diverging ion trajectories are compressed more if the configuration of the repelling plate is bent like a concave plane. As the first trial, a new repelling plate with various parabolic curvatures is applied (mathematical equation for parabola $y^2 = 4px$). A parabolic plate with greater curvature, such as a concave mirror, deflects ions more effectively than a less parabolic plate including a flat plate as seen in Fig. 18. This is because the curvature generates an electrical repulsion force that is expected to collapse the conical ion beam to the centerline by parallelizing equipotential lines to the curved electrode. The origin of the conical ion beam is located in the middle of the vertex (x=0) which coincides with center of the extraction grid where the ionization laser is fired. At a fixed repelling voltage, greater curvature of the plate clearly improves the ion focusing, but not enough to completely parallelize the beam (the meaning of this will be explained later). Furthermore, at p = 10, a number of ions are slightly overfocused, as highlighted with a dotted oval. When the curvature is increased further, the overfocusing becomes greater, leading to less detection efficiency. Also it is not possible to use such a repelling plate for the SPMS without drilling holes through which both laser and aerosol beams penetrate.

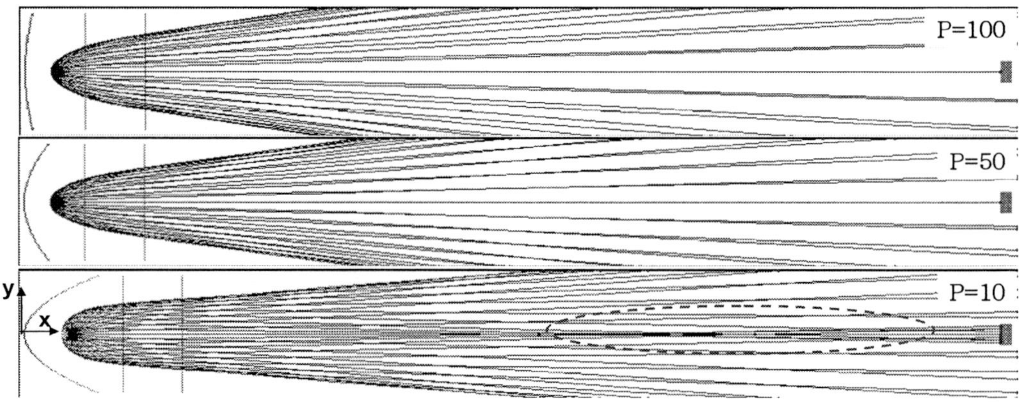

Figure 18. Curvature effect of the repeller plate on the ion trajectories at KE=100eV; the rectangle at the right end represents a MCP detector with an aperture of 18mm. The voltages for the three electrodes are V_R=4500V, V_E=3350V, and V_R=0V from the left.

Trial 2: Einzel Lens

Our next attempt to reduce ion losses entails employing an Einzel lens consisting of three cylindrical tubes. The first and the third tubes are electrically grounded while the central tube, which is equidistant from the first and the third ones, is held at a positive voltage as seen in Fig. 19. The incident ions are first decelerated and then accelerated. At the same time, they begin to move to the centerline of the TOF tube. Thus, the ion beam can be focused by using this lens system without imparting any net acceleration to the ions. For parallel-incident ions, the higher the KE of the ions, the longer the focal length of the lens becomes. Even with higher electrode voltage, confocal focusing of incident ions with different kinetic energies has never been achieved, though the focal length is shortened for all kinetic energies.

Taking a closer look at Fig. 19, it is observed that ions approaching far from the centerline of the lens are overfocused quicker than those closer to the centerline. In contrast, an Einzel lens both ends of which are bonded with fine metallic nets certainly helps to reduce the length of the focusing zone, as noted by "L_f" in the figure, which becomes closer to what is known as confocal focusing. The former problem provides an important clue in the resolution of that problem. If it is possible to position easily focusable (less energetic) ions close to the centerline and more energetic ions near the inner surface of the tube before their incidence, ions with different KEs are likely to be focused at the nearly the same position. As shown in Fig. 6, conventional ion optics essentially produces a conical ion beam the width of which is proportional to the ion KE. More energetic ions intrinsically fly into the Einzel lens farther from the centerline, which is desirable for the confocal focusing. In order to realize this result for the SPMS, it is necessary to make all incident ion trajectories parallel to the central axis before the entrance to the Einzel lens, which is the only remaining problem.

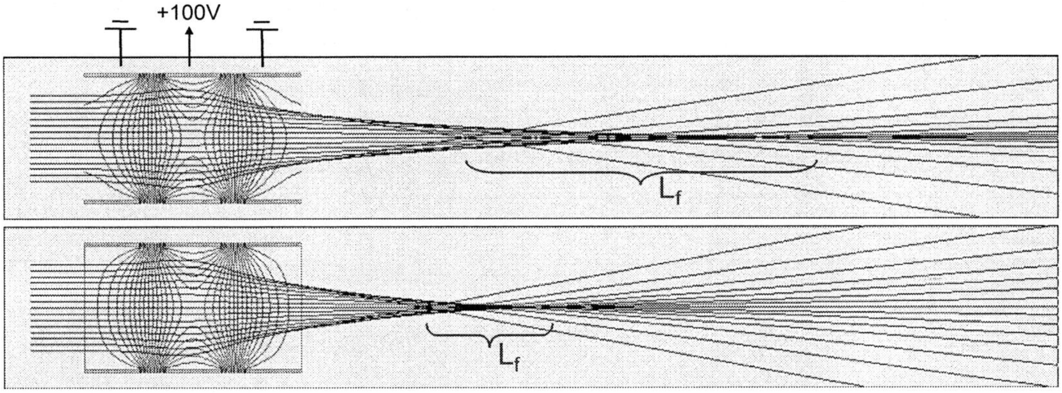

Figure 19. Comparison between (a) a conventional Einzel lens with (b) a lens with a grid for focusing of the incident ion beams at KE = 150eV; L_f represents a focusing zone over which each focal point of ions entering at different radial positions is distributed. Blue and red lines are ion trajectories and isopotential lines, respectively. The diameter and length of each tube are 16mm and 10mm, respectively.

Trial 3: A Tubular Electrode and the Final Design

The third trial would be to locate some type of effective deflector downstream of, but as close as possible to, the extraction grid. One of the best candidates would be the Einzel lens, but it is typically as long as ~ 15 mm. A shorter Einzel lens with a strengthened lens-deflection capability as a strong deflector is needed. Noting the benefit of the Einzel lens with two nets, we propose to insert a tubular electrode between the first and second grid electrodes in Fig. 1. This results in a new minimized assembly including an ion repelling plate and a focusing device. This is termed a "tubular electrode assembly (TEA)", throughout this article. Fig. 20 shows that the TEA deflects ion trajectories progressively with increase in the voltage (V_T) of the tubular electrode, eventually making the trajectories parallel to the TOF axis.

However, the distance between the laser focus and the detector appears to be too short for proper digitization of ion signals. Therefore, an additional Einzel lens is located downstream of the TEA so as to focus ion trajectories to the detector with a diameter of 18mm at the end of the 1-m long TOF tube. A three-dimensional configuration of the complete ion optics is described with dimensions of its components in Fig. 21. The figure confirms that 0-200 eV energetic ions are detected with a repelling voltage as low as 7000V, the typical voltage used for the SPMS.

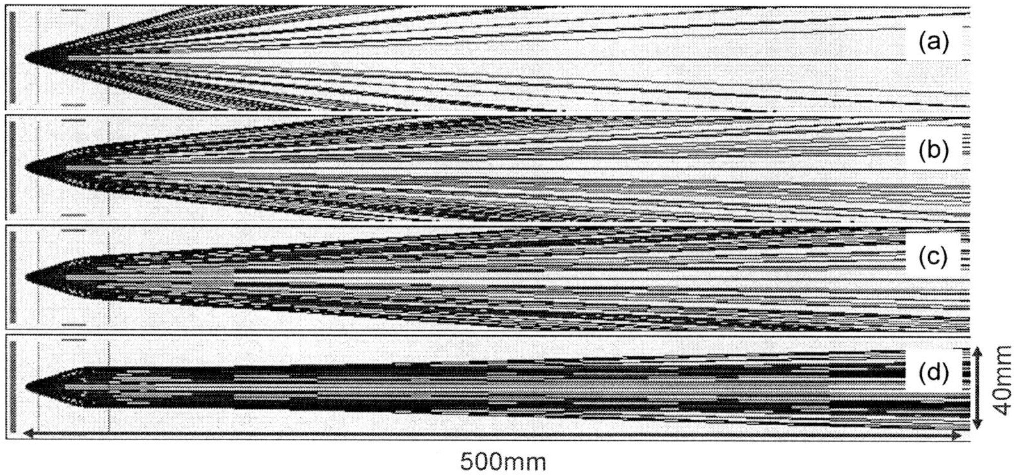

Figure 20. Characteristics of the tubular electrode assembly comprised of blue and red lines corresponding respectively to KE = 200 and 100eV. The voltages to the tubular electrode are altered at (a) 0V, (b) 2000V, (c) 2500V and (d) 3000V, maintaining a repeller plate voltage constant at 4500V together with two grounded grids. The red and blue lines represent ion trajectories at KE = 100 and 200eV, respectively.

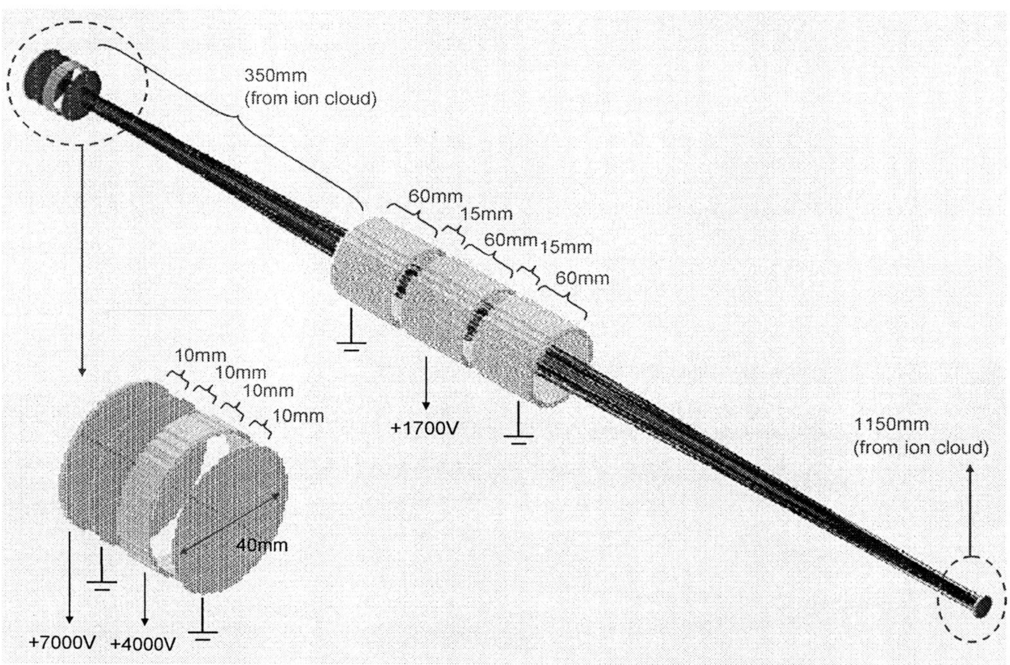

Figure 21. Development of a new ion optics consisting of a tubular electrode assembly and an Einzel lens for 100% detection of 0 – 200eV energetic ions; the MCP detector with an effective size of 18mm is shown as a disc at the right end. The green, red and blue lines correspond to KE = 10, 100 and 200eV, respectively.

4.2. Design of an Aerodynamic Lens

To evaluate and optimize entrainment and focusing of the inlet particle beam, numerical simulation is performed to track particle behavior by utilizing FLUENT software (version 6.2.16). Under typical operating condition of the aerodynamic lens (ADL) inlet for the SPMS, particle concentration is so low that inter-particle interaction is negligible and the existence of particles hardly affect the gas flow. [42-45] Therefore, particle-free gas flow is first calculated under the assumption of axisymmetric, steady, compressible, laminar and viscous flow. Then particles with a specified diameter are launched at various radial positions in the simulated gas flow field. The particles in the present simulation are assumed to be spherical NaCl particles with a density of ~ 2 g/cc. By solving the Langevin equation including inertia and drag forces [42-46], we could simulate particle trajectories through the ADL system. As the particle size of interest in this study ranges from 30 to 300 nm, we don't include in the calculation the Brownian motion of particles that would have a significant effect especially for the size smaller than 30 nm. [45]

Analysis of Single Aerodynamic Lens

The performance of the ADL or the single lens has been often represented by overall contraction ratio (ηc) of particle beam and transmission efficiency (ηt) of particles. The two factors have been expressed as a function of four dimensionless groups [42-46], i.e. df/OD, L/df, flow Reynolds number (Re0) based on OD and particle Stokes number (St) as

$$\eta_c = f(d_f / OD, L / d_f, \mathrm{Re}_0, St)$$
$$\eta_t = g(d_f / OD, L / d_f, \mathrm{Re}_0, St) \qquad (16)$$

Due to the low pressure inside the ADL, the gas mean free path is much larger than 300 nm, leading the particle behavior to the free molecular regime. In that regime, the St can be defined using Epstein's mobility as: [44,46]

$$St = \frac{1}{\left(1+\frac{\pi\alpha}{8}\right)\sqrt{2\pi\gamma^3}} \frac{\dot{m}\rho_p d_p c^3}{p_{up}^2 d_f^3} \propto \frac{\rho_p d_p}{p_{up}^2 d_f^3 M^{3/2}} \qquad (17)$$

where α is the momentum accommodation coefficient (close to the unity), γ is the specific heat ratio of the gas (1.4 for air), ρp is particle density, dp is particle diameter, $c = \sqrt{\gamma R T_{up} / M}$ is the speed of sound in the gas at the upstream temperature (Tup). M is molecular weight of the gas and $\dot{m}$ is mass flow rate of carrier gas, respectively. Notably, Eq 17 implies that the St has a functional dependence on the systematic parameters of dp, df, pup and M. Since the Re0 can not vary even at different dfs where flow characteristics are significantly different, the gas flow should be better characterized by the other Reynolds number based on the df rather than the OD as

$$\mathrm{Re} = \frac{\rho V d_f}{\mu} = \frac{4\dot{m}}{\pi \mu d_f} \propto \frac{1}{d_f} \tag{18}$$

where μ is gas viscosity that does not depend on the pressure. The mass flow rate $\dot{m}$ is not varied much from a typical value of 0.0026 g/s (125sccm) under the current operating condition of the SPMS. Also from mass conservation, $\dot{m}$ is treated as constant throughout this article unless otherwise stated. If the df decreases, the Re and St are both increased (see Eqs 17-18), leading to significant changes in both gas flow characteristics and particle behavior. At the same time, the Pup is expected to increase too because the downstream pressure is fixed constant in the operation of the SPMS. The higher Pup condition reduces the detrimental effect of particle diffusion. More importantly, as the OD is often restricted geometrically for the application to the SPMS, the lens-geometry-related parameter (df/OD) would be reduced into the Re (see Eq 18). The length of spacer L is not a big concern for the design as long as the L is long enough for the gas flow to redevelop after the orifice. Taking these issues into account, Eq 16 is likely reduced to the simpler form as

$$\eta_c = f(\mathrm{Re}, \mathrm{St}),\ \eta_t = g(\mathrm{Re}, \mathrm{St}) \tag{19}$$

At the fixed df where Re is constant (see Eq 18), the ηc must be determined solely by the St. Indeed, Zhang et al.'s simulation [42,43] obviously shows that the curves of the ηc against dp or Pup fall into a single curve of ηc versus St (see Figures 7-8 and Figure 14 in references Zhang et al. [42] and Zhang et al. [43], respectively). Moreover, this implies that there should exist an optimal Stokes number (St0 when ηc = 0) for best focusing aerosols with optimal particle size (dp0). As seen in Fig. 22, particles smaller than the dp0 are not efficiently focused while larger particles (> dp0) are overfocused due to their greater inertia. Rearranging Eq 17 for df, one can estimate the optimal size (df0) of single orifice so as to focus particles with a specified size of dp0 as [44]

$$d_{f0} = \frac{1}{\left(1 + \frac{\pi \alpha}{8}\right)^{1/3} \left(2\pi\gamma^3\right)^{1/6}} \left(\frac{\dot{m} \rho_p d_{p0} c^3}{p_{up}^2 St_0} \right)^{1/3} \tag{20}$$

Analysis of Multi-Lens System

Using Eq 20, we first determine the opening sizes of five successive lenses so that each lens is designed to focus particles with a particular dp0. The present ADL system obtained in this way shows that the df0 decreases gradually in the streamwise direction. This means that the expected optimal sizes (dp0) of particles decrease in that direction, because the dp0 is mainly proportional to df03 as long as the St0 does not change much (see Eq 17 or 20). But, the universal optimal Stokes number St0 (at ηc = 0) exists only when the Re is not varied along all orifices (see Eq 19). This would be true only in the case of using same-size lenses in series which is known to be less effective.

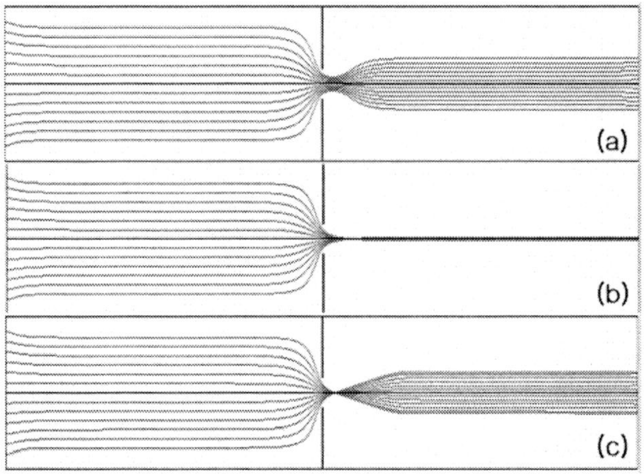

Figure 22. Particle size-dependent focusing of single lens: (a) insufficient focusing smaller particles (d_p=30nm), (b) focusing intermediate sized particles (d_p=150nm), (c) overfocusing larger particles (d_p=500nm) : Q = 125sccm.

In our multi-lens system seen in Fig. 23, Re and St are both drastically increasing along the lenses. Eq 19 indicates that ηc vs St curves at different Reynolds numbers will not collapse into a single line, which is qualitatively consistent with Zhang et al.'s finding [42] from single lens analysis (refer to Figures 9-11 in reference Zhang et al. [42]). As the universality of the St0 is broken in our multi-lens system and the Pup experiences a stepwise decrease to the stable low pressure downstream, the optimal size of particles at each stage of the multi-lens system would deviate from that expected by Eq 20. Hence, one may have to know the St0 at each stage in the whole system, prior to the determination of the orifice diameters. This would be tedious time-consuming iterations of initial guess of df0s and numerical simulation for the whole system. Liu et al. [46] reported that when df/OD < 0.25, the St0 has a weak dependence on the ratio. In Zhang et al.'s design [43] where the ratio varies from 0.4 to 0.5, the St0 is very sensitive to the ratio as confirmed by their simulation. Recalling our inference that the effect of the ratio could be interpreted as another effect of the Re (see the pre-mentioned discussion on Eq 4), it is not a surprise to see a strong dependence of the St0 on the Re. [43] In contrast, Wang et al.'s ADL system [45] with the ratios of 0.17 - 0.27 showed that the St0 had a very weak dependence on the Re (see Figure 2 in reference).

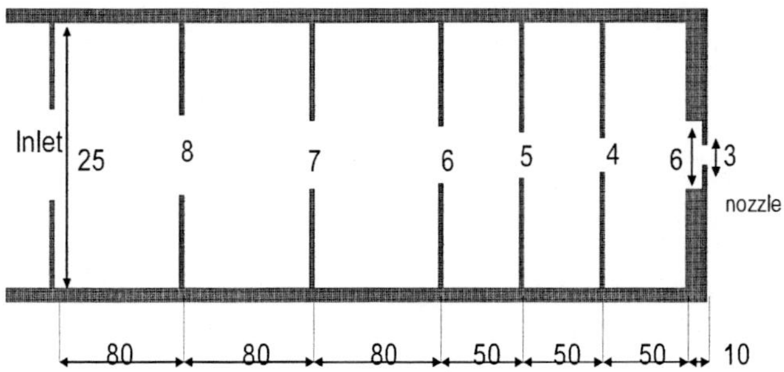

Figure 23. Schematic of an aerodynamic lens designed to focus 30-300nm aerosols. The geometric dimensions are in mm.

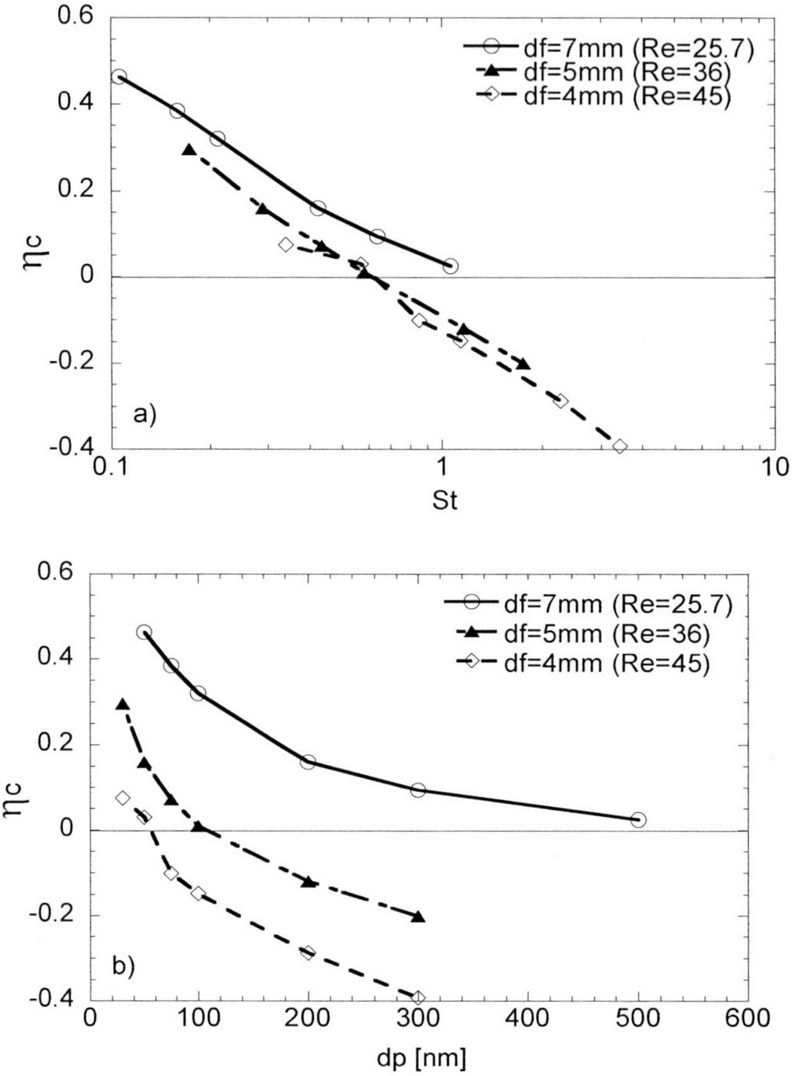

Figure 24. Comparisons of contraction ratios at the 2nd, 4th and 5th lenses as a function of (a) Stokes numbers and (b) particle diameters : Q = 125sccm.

Interestingly, Fig. 24(a) shows that the St0 is not varied around 0.6 from the 3th orifice to nozzle. But, the 1st and 2nd orifice diameters give the ratio of 0.32 and 0.28, respectively, which are slightly larger than the critical value of 0.25 by Liu et al. [46]. Therefore, the universality of the St0 is slightly broken only at these two orifices, moving the St0 to 1.0.

We listed in Table 4 optimal particle sizes (dp0) at each orifice and nozzle for both the optimal Stokes numbers of 0.6 and 1.0. As the Re = 22 at the 1st orifice is pretty close to that at the 2nd orifice, it should be a reasonable approximation to use St0 = 1.0 for the 1st orifice. The optimal particle sizes predicted for the orifices are highlighted with a bold style in the table. The result indicates that the present ADL system is designed for focusing nanoparticles in a wide range of 15 – 700 nm.

Table 4. Comparisons of design parameters between the present design and the previous ones at Q = 125 sccm.

OD[mm] = 25							
lens order	df	df/OD	Ma	Re	Pup[Pa]	dpo[nm]	Sto
1	8	0.32	0.086	22.48	227.2	750	1.036
2	7	0.28	0.112	25.69	225.6	390	0.816
3	6	0.24	0.155	29.97	223.2	210	0.713
4	5	0.2	0.228	35.97	218.6	105	0.642
5	4	0.16	0.386	44.96	209.2	43	0.561

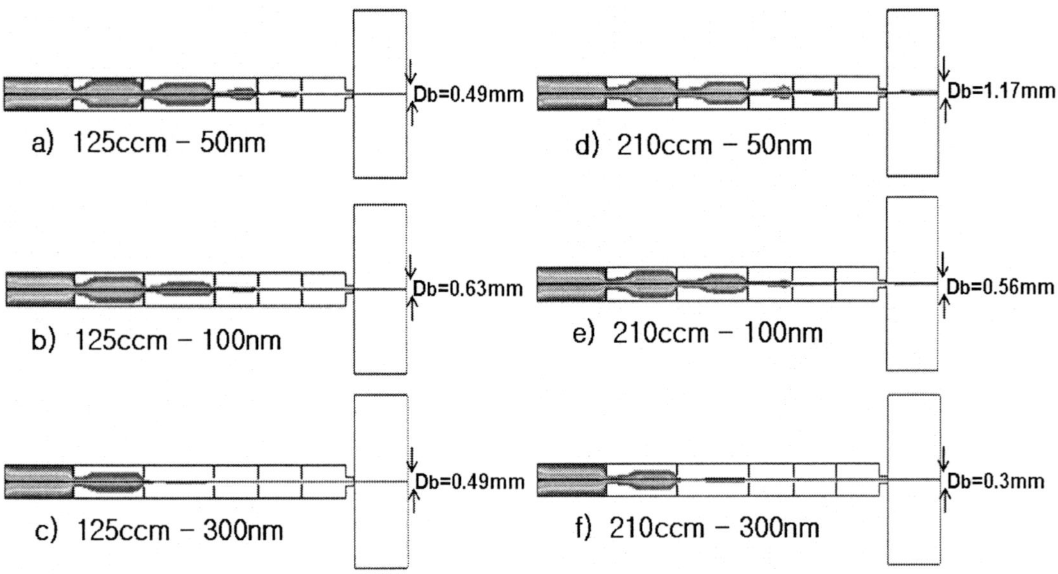

Figure 25. Trajectories of particles with a variety of their sizes at two gas flow rates.

In addition, we consider two aerosol-gas flow rates of 125 and 210 sccm upstream of the flow limiting orifice. The use of a smaller mass flow rate or a smaller flow limiting orifice provides an advantage of focusing smaller particles because it decreases all upstream pressures inside the ADL and thereby the St for the smaller particle increases in the direction of approaching the St0. But, this is achieved at the expense of the hitting rate of aerosols by the SPMS. Fig. 25 presents the trajectories of aerosol particles of 50, 100, and 300 nm together with the final aerosol beam diameters 40mm downstream of the ADL exit. Fig. 25(a) shows 50 nm particles are collimated at the 5th nozzle, which is exactly identical to the predicted location in Table 4. Both the figure and table show that particles of dp0 = 300 nm are completely focused between the 2nd and 3rd orifices. We would like to emphasize that the final beam diameters is still in the sub-millimeter range.

Conclusions

In this article, we offered a broad review for single particle mass spectrometry, starting from the working principle and several concerns for interpretation of the data, through applications to probe nanoscale phenomena, to introduction of some ideas for better design, based on our previous publications on the SPMS. There are a couple of other historical review articles dealing with the application i.e. investigation of accumulate-mode environmental aerosols using aerosol time-of-flight mass spectrometry. However, this article is specifically devoted to explaining the whole process including initial approach, theory, and experimental evaluation for each topic. We hope that this will provide some guideline for readers who are interested in the use and/or the development of the SPMS in the field of nanoparticle technology.

Acknowledgments

This work was supported by Core Environmental Technology Development Project for the Next Generation (Project No. 102-071-058).

References

[1] Friedlander, *S. K. Smoke, Dust*, and Haze; John Wiley & Sons: New York, 1977.
[2] Kodas, T.; Hampden-Smith, *M. Aerosol Processing of Materials*; Wiley: New York, 1999.
[3] Ivanov, G. V.; Tepper, F. *4th International Symposium on Special Topics in Chemical Propulsion*. 1997, 636.
[4] Aumann, C. E.; Skofronick, G. L.; Martin, J. A. J. Vac. *Sci. Technol*. B. 1995, 13 (3), 1178.
[5] Ortega, A. *Int. J. Chem. Kinetics*. 2001, 33, 343.
[6] Mahadevan, R.; Lee, D.; Sakurai H.; Zachariah, *M. R. J. Phys. Chem*. A. 2002, 106, 11083-11092.
[7] Kittelson, D. J. *Aerosol Sci*. 1998, 29, 575-588.
[8] Gross, D. S.; Galli, M. E.; Silva, P. J.; Wood, S. H.; Liu, D. -Y.; Prather, *K. A. Aerosol Sci. Technol.* 2000, 32, 152-163.
[9] Noble, C. A.; Prather, *K. A. Mass Spectrum. Rev*. 2000, 19, 248-274.
[10] Park, K.; Lee, D.; Rai, A.; Mckherjee, D.; Zachariah, *M. R. J. Phys. Chem. B*. 2005, 109, 7290-7299.
[11] Rai, A.; Lee, D.; Park, K.; Zachariah, *M. R. J. Phys. Chem. B.* 2004, 108, 14793-14795.
[12] Lee, D.; Miller, A.; Kittelson, D.; Zachariah, *M. R. J. Aerosol Sci*. 2006, 37, 88-110.
[13] Lee, D.; Miller, A.; Park, K.; Zachariah, *M. R. Int. J. Automotive Technol*. 2006, 7(6), 667-673.
[14] Lee, D.; Park, K.; Zachariah, *M. R. Aerosol Sci. Technol*. 2005, 39, 162-169.
[15] Lee, D.; Cho, S.-W.; Lee, S. G.; *Kim, S.-G.. J. Korean Phys. Soc*. 2006, 49(1), 370-375.
[16] Cho, S.-W.; Lee, *D. Rapid Commun. Mass Spectrom*. 2007, 21(20), 3286-3294.
[17] Lee, K.-S.; Cho, S.-W.; *Lee, D. J. Aerosol Sci*. 2008, in press.

[18] Johnston, *M. V. J. Mass Spectrom*. 2000, 35, 585-595.
[19] Salt, K.; Noble, C. A.; Prather, *K. A. Anal. Chem*. 1996, 68, 230.
[20] Refer to the manual of the ATOFMS available at *http://www.tsi.com*.
[21] Reents, W. D.; *Ge, Z. Aerosol Sci. Technol*. 2000, 33, 122-134.
[22] Reents, W. D.; Schabel, *M. J. Anal. Chem*. 2001, 73, 5403-5414.
[23] Lee, S. G.; Choi, S. M.; Lee, *D. Thermochimica Acta*. 2007, 455(1-2), 138-147.
[24] Smith, R. A.; Tisch, J. W. G.; Ditmire, T.; Springate, E.; Hay, N.; Mason, M. B.; Gumbrell, E. T.; Comley, A. J.; Mountford, L. C.; Marangos, J. P.; Hutchinson, *M. H. R. Phys. Scripta*. 1999, T80, 35-39.
[25] Last, I.; Jortner, *J. Phys. Rev. Lett*. 2001, 87, 033401.
[26] Ditmire, T.; Donnelly, T.; Rubenchik, A. M.; Falcone, R. W.; Perry, *M. D. Phys. Rev*. A. 1996, 53, 3379.
[27] Hutchinson, M. H. R.;, Ditmire, T.; Springate, E.; Tisch, J. W. G.; Shao, Y. L.; Mason, M. B.; Hay, N.; Marangos, *J. P. Phil. Trans. R. Soc. Lond. A*. 1998, 356, 297.
[28] Chen, Z. Y.; Cogley, C. D.; Hendricks, J. H.; May, B. D.; Castleman, *A. W. J. Chem. Phys*. 1990, 93, 3215.
[29] Snodgrass, J. T.; Roehl, C. M.; Bowers, *M. T. Chem. Phys. Lett*. 1989, 159, 10.
[30] McPherson, A.; Luk, T. S.; Thompson, B. D.; Borisov, O. B.; Shiryaev, O. B.; Chen, X.; Boyer, K.; Rhodes, *C. K. Phys. Rev*. Lett. 1994, 72, 1810.
[31] Milchberg, H. M.; McNaught, S. J. et al. *Phys. Rev*. E. 2001, 64(5-2), 056402.
[32] Zhou, L.; Park, K.; Milchberg, H. M.; Zachariah, *M. R. Aerosol Sci. Technol*. 2007, 41, 818.
[33] Vyazovkin, S.; Wight, *C. A. Int. Rev. Phys. Chem*. 1998, 17, 407.
[34] Brown, M. E. *Introduction to Thermal Analysis Techniques and Applications;* Chapman and Hall: London, 1988.
[35] Ettarh, C.; Galwey, *A. K. Thermochimica Acta*. 1996, 288, 203.
[36] Levenspiel, *O. Chemical Reaction Engineering; 3rd ed.*; John Wiley & Sons: 1999.
[37] Hering, S.; Friedlander, S.; Collins, J.; Richards, *W. Environ. Sci. Technol*. 1979, 13(2):184-188.
[38] Miyamoto, N.; Zhixin, H.; Harada, A.; Ogawa, H.; *Murayama, T. SAE Papers* 871612. 1987, 1-7.
[39] Bonczyk, *P. A. Comb. Flame*. 1991, 87, 233-244.
[40] Otto, K; Sieg, M. H.; Zinbo, M.; Bartosiewicz, L. SAE Papers. 800336. 1980, 277-289.
[41] Zhou, L.; Rai, A.; Zachariah, *M. R. Int. J. Mass Spectrom*. 2006, 258, 104.
[42] Zhang, X.; Smith, K. A.; Worsnop, D. R.; Jimenez, J.; Jayne, J. T.; Kolb, *C. E. Aerosol Sci. Technol*. 2002, 36, 617-631.
[43] Zhang, K.; Smith, K. A.; Worsnop, D. R.; Jimenez, J. L.; Jayne, J. T.,; Kolb, C. E.; Morris, J.; Davidovits, *P. Aerosol Sci. Technol*. 2004, 38, 619-638.
[44] Wang, X.; Kruis, F. E.; McMury, *P. H. Aerosol Sci. Technol*. 2005a, 39, 611-623.
[45] Wang, X.; Gidwani, A.; Girshick, S. L.; *McMurry, P. H. Aerosol Sci. Technol*. 2005b, 39, 624-636.
[46] Liu, P.; Ziemann, P. J.; Kittelson, D. B.; McMurry, *P. H. Aerosol Sci. Technol*. 1995a, 22, 293-313.
[47] Wang, X.; McMurry, *P. H. Int. J. Mass Spectrom*. 2006, 258, 30-36.

In: New Trends in Chemical Physics Research
Editor: James N. Lapherty

ISBN: 978-1-61668-535-5

Chapter 2

DIELECTRIC RESPONSE OF POLAR SOLVENT AT AN ELECTRODE- ELECTROLYTE INTERFACE

Om Prakash Thakur and Anjani Kumar Singh*
University of Delhi New Delhi, India.

ABSTRACT

The interfacial region with molecular organization and non-cento symmetry between two phases in contact has in variably different dielectric properties from those of the bulk phases. The response of polar solvent molecules to the changing electric fields is certainly different at different points in the interaction zone as the polar solvent molecules near the electrode surface are rigidly oriented and ordered in comparison to the unoriented or poorly oriented polar molecules towards the bulk in electrolyte. It has been observed that none of the existing model is able to explain the multiple dielectric relaxations exhibited by the polar solvent in the interfacial region. The significant change in dielectric behavior of the polar solvent near the electrode's surface has been mainly attributed to the rearrangement and segregation of polar molecules in the electric double layer and from the continuous rupture and reformation of hydrogen bond due to surface effect. As a result of rearrangement of water molecule dipole, the permittivity of water does not only vary from $\varepsilon=4$ on its surface to $\varepsilon=80$ in the bulk about 10nm away from surface but also shows its oscillatory character due to oscillatory electric field prevailing up to $10A^0$ from the solid surface due to surface effect. The fluctuating nature of permittivity would certainly affect the electromechanical and electrochemical properties of the interface.

Keywords: Dielectric relaxation, Electric double layer, Electrolyte, Interface
PACS: 73.30+Y; 72; 77.22.Gm

---

* Corresponding author: anjaninsit@gmail.com School of Applied Sciences,Netaji Subhas Institute of Technology, University of Delhi Sector 3, Dwarka, New Delhi 110078, India.

1. INTRODUCTION

A solid charged electrode does not only impart its electrostatic forces on the water molecule dipole but also affect the polar molecules due to its surface effect. The arrangement of water molecules at the interface is an important question in electrochemistry, biology and geochemistry. The behavior and the orientation of water molecules are quite different at the surface from those in the bulk. In this paper the dielectric permittivity profile of interfacial water (a dipolar liquid) has been discussed. Among the variety of dipolar liquids, pure water and its mixture with other material are very significant in the current interest of liquid state physics particularly on its applications in biology, medical science, chemical engineering, industry, biotechnology, electrochemistry and environmental science. As the polar solvent constitute more than 80% of total volume of the electrolyte, its contributions for determination of electrochemical properties cannot be ignored. Most often, water molecules are physically impregnated in to other material or chemically bound to the molecule of other material. Depending upon the material structure of other phase, there may be various molecular forces of bound water molecule, differing in binding energy with neighbor atoms and in dielectric properties. For an example the interaction energy of water molecule with other material is different depending on whether the other materials are hydrophilic or hydrophobic as shown in figure 1.

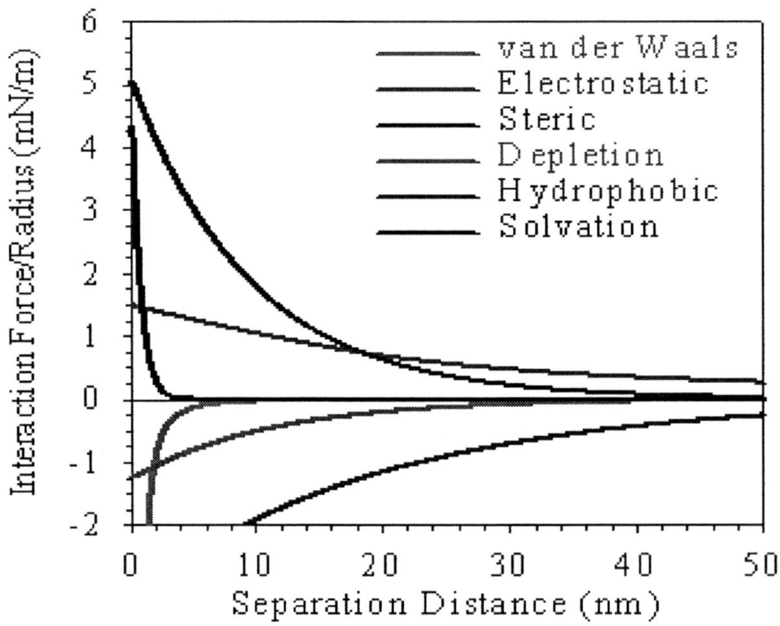

Figure 1. The effect of hydration forces on particle interaction. Attraction is weakened or strengthened depending on whether the surfaces are hydrophilic or hydrophobic.

Dielectric relaxation of water (associating liquids) could reflect changes in its dipole moment resulting from the continuous rupture and reformation of hydrogen bond and the dielectric relaxation time of water around hydrophobic particles increases as a result from reduced density of hydrogen bonding sites at the solvent- solute interface [1]. Using computer simulation studies, a deeper insight in the micro dynamics of water is currently achieved. The dielectric relaxation time for solvent in an aqueous solution is found to be different from its value for the pure water at same temperature as revealed from the dielectric spectra results carried out by the various researchers in the past decade and a great variety of relaxation studies has been performed on the aqueous solutions of high water content. The dielectric spectra exhibited by the solution (electrolyte) reflect two relaxation regions [1], in which one is attributed to the solvent and the other one to the solute.

2. Segregation of the Interfacial Polar Solvent Molecules

The existence of ordering, layering and segregation of water molecules (polar solvent molecules) at an electrode- electrolyte interface has been experimentally verified using X-ray scattering [2]. The orientational distribution of water dipole on the solvation layer is governed by the polarity of the electrode as indicated in figure 2. The diffuse double layers and the solvation layers formed at the electrode-electrolyte interface determine the dielectric behavior along with other electromechanical and electromechanical properties. Large changes in permittivity have been resulted [3] from two mechanisms:

1. The polarization due to Maxwell-Wagner effect i.e. due to contact of two phases of contrasting permittivities.
2. Polarization of charged double layers formed due to segregation of charged particles (ions) in response to the electric field gradient existing at the interface.

In the study [4] of the dielectric relaxations exhibited by the clay-suspensions and electrolyte, three models namely, equivalent circuit models (ECMs)[5]; empirical Debye-relaxation models (DMs) [6,7] and mixing models (MMs) [8] are not able to explain all the relaxation mechanisms completely but each one of them is able to explain at least one relaxation process.

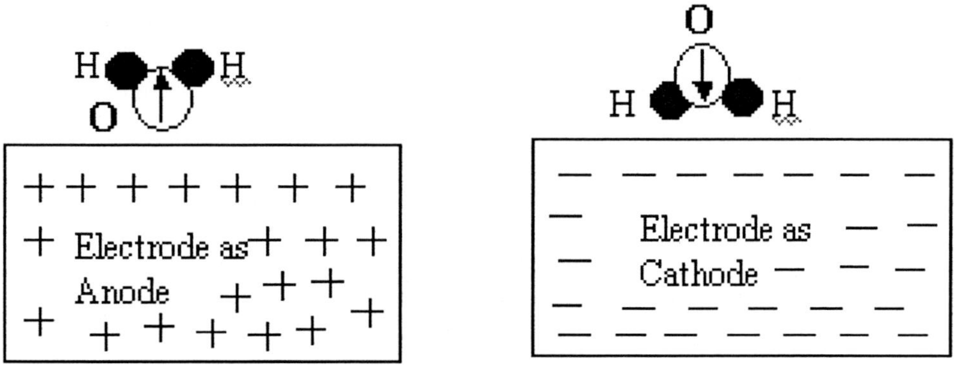

Figure 2. Schematic diagrams illustrating the orientation of water molecules on the surface of electrode depends on its polarity.

When the frequency of an electromagnetic field applied to a system containing phases of contrasting permittivities is changed, a large variation in permittivity has been observed. Also the variation in orientation of water molecules varying from strong orientation at the electrode's surface to the weak or poor orientations in the bulk towards electrolyte (figure 3) has resulted in an anisotropic behavior of the polar solvent in the space charge regions of the interaction zone.

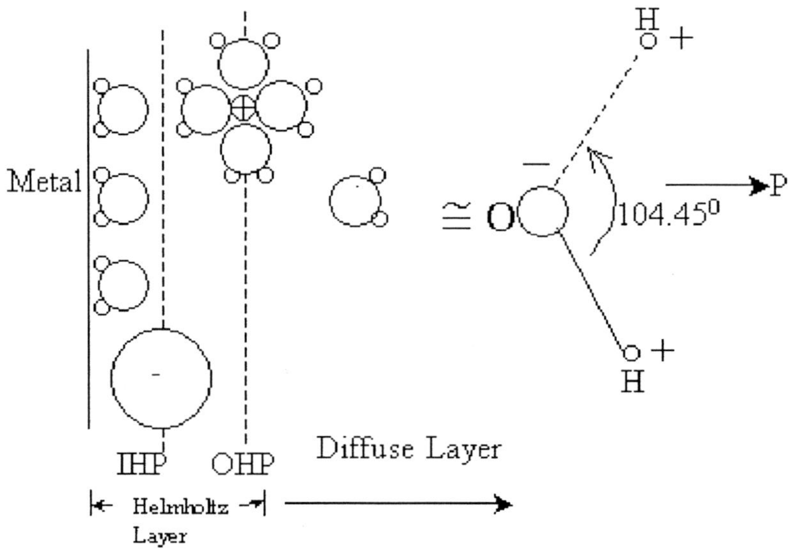

Figure 3. Schematic diagram illustrating orientation of water molecules such that hydrogen atoms orient towards negatively charged metallic surface and formed first solvation shell for the metal electrode.

3. Dielectric Relaxation

The interfacial region containing materials of different dielectric constant such as soils, rocks, solutions or electrolytes containing micro molecules or colloidal particles exhibits a large change in permittivity and impedance as the frequency of the applied electromagnetic field changes [9,3]. This change in dielectric properties may be used to study the electrochemical & electromechanical properties of the interfacial material. Polarization due to the charge imbalance created at the interface between the two phases of different permittivity (Maxwell- Wagner effect) and the polarization of the counterians in diffuse double layers (figure3) are responsible for the frequency dependent dielectric behavior.

The orientation of polar molecules under the influence of time varying electric fields at microwave frequency (MHz-GHz) is the primary mechanism for its dielectric behavior [10]. At low frequency (Hz-MHz), the segregation of charges at the solution- solid interface is the main mechanism for its dielectric behavior. In the presence of an electric field of strong amplitude, the electric polarization is no longer proportional to electric field due to unknown molecular forces and the dielectric permittivity depends [11] on the field strength as follows:

$$\varepsilon(E) = \varepsilon_1 + \Delta\varepsilon \tag{1}$$

Where ε_1 is the linear value of ε, and $\Delta\varepsilon$ is the nonlinear dielectric increment varying as the square of the strong field and producing a negative contribution due to decrease in the potential energy of the polar molecule in electric field (Langevin saturation).

4. Field Dependent Dielectric Behavior at the Interface

Several researchers [12,13,14] have studied the field dependent and distance dependent dielectric constant of water (considering hydration of ions) near the charged surface, and biological system such as water in membrane spanning pores. The non-linear relation between the applied electric field and the dielectric constant has experimentally confirmed [15] for the first time. A simple expression [16] for the field dependent dielectric constant using Onsagar and Kirkwood theories has been given as follows:

$$\varepsilon = n^2 + \frac{\alpha\pi N_0(n^2+2)\mu_v}{E} L\left(\frac{\beta\mu_v(n^2+2)E}{kT}\right), \tag{2}$$

where n is the optical refractive index, N0 is the number of molecules per unit volume, μ_v is the dipole moment of the water molecule, E is the field strength, T is the absolute temperature, L(x) is the usual Langevin function and α & β are numerical factors.

Bockris and Reddly [17] observed the variation in dielectric constant varying from 6 in the first layer of water to 78 in the bulk at the liquid-solid interface in the electric double layer. Such type of variation in water permittivity has been experimentally measured using atomic force microscopy [18], which suggests the reorientation of water molecule dipoles at the interface. An equation for the variation of dielectric constant has been given by Teschke [19], as follows:

$$\varepsilon_{DL} = \varepsilon_{bulk} \Big/ 1 + B\phi^2(x) \tag{3}$$

where $\phi(x)$ is an exponential function of distance characterized by the Stern's length λ, ε_{bulk} is the dielectric constant in the bulk and B is an adjustable parameter. The variation of dielectric constant with distance from the electrode is indicated in figure 4.

In the study [20] of the field dependent dielectric properties of pure water using molecular dynamics computer simulations, the influence of the solid electrode's surface on the orientational distributions of the water molecules is up to a distance of 100nm from the surface of the electrode and an oscillating electric field with decreasing amplitude exists up to a distance of 100 nm from the electrode's wall

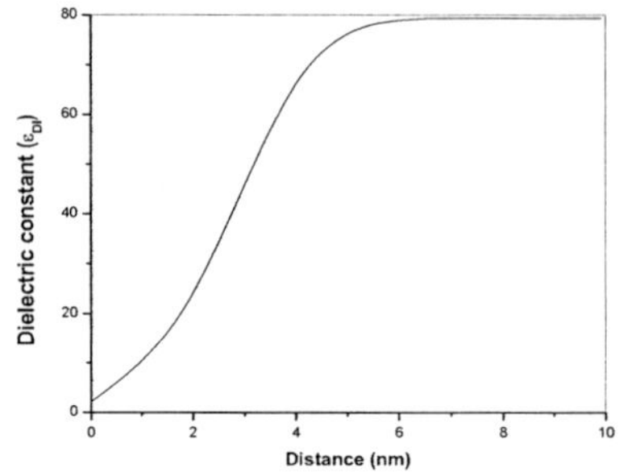

Figure 4. Schematic plot indicating the variation in dielectric constant from the electrode's surface

. The profile for density, charge and electric fields distribution indicates the influence on water molecules up to 10 A^0 from the wall. The effect of the surface is erased after 10 A^0 away from the surface as shown in figure 5. A schematic diagram of an electrode-electrolyte interface along with a schematic plot between the electric fields and distance showing the oscillatory electric fields as shown in figure 6.

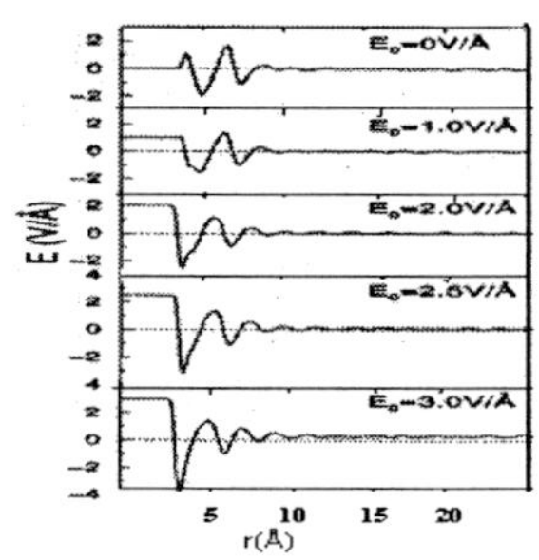

Figure 5. The variation in electric fields with distance from the electrode's surface, showing its oscillatory nature with decreasing amplitude (similar plot for density verse distance has been observed) [20].

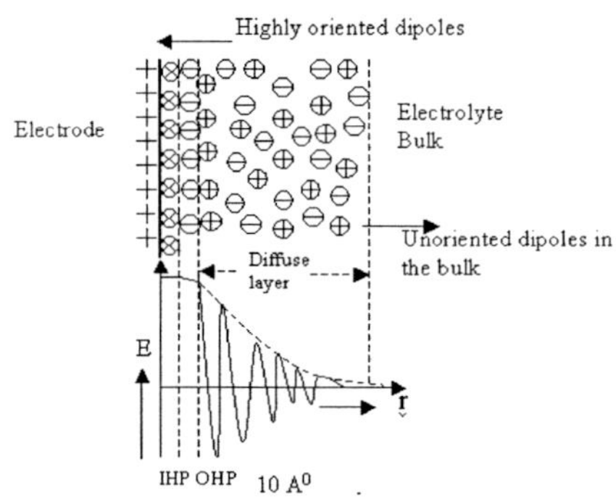

Figure. 6. (a) Electrode-Electrolyte interface indicating the location of inner Helmholtz plane (IHP), Outer Helmholtz plane (OHP) and diffuse double layer with the dipole orientations varying from strong orientation on the surface to the poor orientation towards the bulk. (b) The schematic plot between the electric fields and the distance shows the oscillatory nature of the field with decreasing amplitude.

Result from different simulations (figure 7) the theoretical prediction suggests that the dielectric constant of water decreases almost exponentially on increasing electric field. But the electric fields is oscillatory up to 10 A0 from the surface of electrode, thus the dielectric constant must be oscillatory up to 10A0 from the surface of the electrode. The oscillatory nature of permittivity can't be ignored particularly in the case where the Debye-Huckel length (which is a measure of thickness of electric double layer) is very small as in the case of 10-3M NaCl and KCl solutions [19] in which debye length is around 10nm and the oscillatory nature of permittivity does certainly affect the electromechanical & electromechanical properties of the interface.

Fluctuating density [20] of water must fluctuate the body force [21] and hence the dielectric constant of an aqueous solution of high water content at the interface. Thus the dielectric permittivity of the aqueous solution will not only has a tensor form (electrostriction) [21], but also has fluctuating nature up to 100 nm from the surface of the electrode due to its surface effect. The variable nature of dielectric permittivity is the area of main concern for proper estimation of electrostriction and for exploiting the interface as an electromechanical transducer in Nanometric dielectrics [22]. Proper precautions must be taken in the interpretation of the empirical results of the electromechanical behavior in nanometric dielectrics particularly for an aqueous solution of higher water content at an electrode-electrolyte interface.

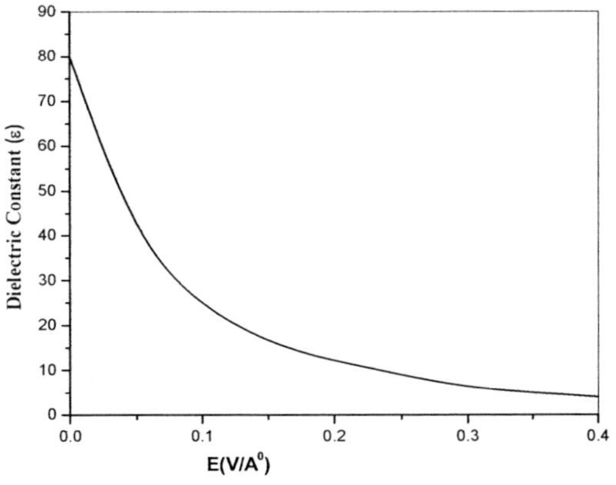

Figure7. Schematic plot of field dependent dielectric constant of water

5. Conclusion

The various short ranges and long ranges inter molecular forces, known and unknown influence the dielectric behavior of the solvent due to unidentical surrounding environment of solute and due to sudden change in material at the interface. The response of the dipoles to the applied field is different if one move from electrode surface (with maximum orientation of polar molecules) to the bulk electrolyte (poor orientation). Due to differential response exhibited by the dipoles, the dielectric constant is not uniform in the interaction zone. The critical analysis on the structures and orientations of water molecules and its changes with temperature, pressure, frequency, and solute and electrode surface has produced a variety of models and as yet, no single theory or model appears to be both widely accepted capable of explaining a satisfactory range of dielectric behaviors. The solid electrode does not only affect the dielectric response due to its charged nature but also due to its surface effect. The variable & oscillatory dielectric permittivity of water at the interface affect the electromechanical, electrochemical and the various other properties which are very important in the field of electrochemistry, material science, biotechnology, industry and environmental sciences.

REFERENCES

[1] Kaatze, U. (1995). Microwave Dielectric Properties of Liquids, *Radiat. Phys. Chem.*; 45; 549-566.

[2] Toney, M.F., Howard, J.N., Richer, J., Borges, G.L., Gordon, J.G., Melroy, O.R., Wiesler, D.G., Yee, D. and Sorensen, L.B. (1994). Voltage –dependent Ordering of water Molecule at an Electrode-electrolyte Interface, *Nature, Vol. 68*, pp. 444-446.

[3] Hasted, J.B., *Aqueous dielectric*, Chapman and Hall, London, (1973).

[4] Dudley, L. M., Bialkowski, S., Or, D., and Junkermeier, C. (2003). Frequency impedance behavior of montmorillonite suspensions: Polarization mechanisms in the low frequency domain. *Soil Sci. Soc. Am. J.* 67, pp. 518-526.

[5] Bonanos, N., Steele, B.C.H., Butler, E.P., Johnson, W.B., Worrel, W.L., Macdonald. D.D., and BcKubre, M. C. H., In. J. R. Macdonald (ed.) *Impedance spectroscopy*. John Wiley and Sons, New York (1987) pp.191-237.

[6] Kuang, W. and Nelson, S. O. (1997). *J. Colloid Interface Sci.*193, pp-242-249.

[7] Kuang, W. and Nelson, S. O. (1998). *ASAE 41*, pp- 173-184.

[8] Nettelblad, B. and Niklasson, G. A. (1996) *J. Colloid Interface Sci.* 181, pp-165-168.

[9] Mudgett, R E., Goldblith, S. A., Wang, D. I. C. and Westphal, W. B. (1977). Dielectric properties in. solid foods of high moisture content at ultra high microwave frequencies J. *Food Proc. Preserv.* 1: pp-119-151.

[10] Hippel, A. R. V. *Dielectric Materials and Applicutions*. MIT Press, Cambridge (1993).

[11] Dejardin, J. L., (2002) *J. Chem. Phy. Vol. 117*, 1750.

[12] Conway, *B.E. Ionic Hydration In Chemistry And Biophysics* (Elsevier Amsterdam 1981).

[13] Sansom, M.S., Smith, G.R., Adcock, C. and Biggin, P.C. (1997) *Biophysical Journal* 73: 2404.

[14] Hyun, J.K., Babu, C.S. and Ichiye, T. (1995). Apparent local dielectric response around ions in water: a method for its determination and its applications, *J. Phys. Chem.* 99, 5187.

[15] Kolodziej, H. A. and Jones, B. P. (1975). *J. Chem. Soc.*, Faraday Trans. 2 71, 269.

[16] Booth, F. (1951). *Journal of Chemical Physics, Vol. 19*, Issue 4, 391.

[17] Bockris, J. O' and Reddy, A. K. N. *Modern Electrochemistry* (Plenum, New York), (1970).

[18] Teschke, O., Ceotto, G. and de Souza, E. F. (2001). Interfacial water dielectric-permittivity-profile measurements using atomic force microscopy *Physical Review E,* Vol. 64, 011605.

[19] Teschke, O. and de Souza, E. F. (1999). Dielectric exchange: The key repulsive or attractive transient forces between atomic forces microscope tips and charged surfaces, *Applied Physics letter* Vol. 74, No. 12, pp. 1755-1758.

[20] Yeh, I. C., and Berkowitz, M. L. (1999). *Journal of Chemical Physics*, Vol. 110, Issue 16, 7935.

[21] Stratton, J. A. *Electromagnetic Theory*, McGraw-Hill, New York, (1941).

[22] Thakur, O. P. and Singh, Electromechanical Phenomena at the Interface in Nanometric Dielectrics, *Proceedings of 2nd International Conference on Sensing Technology*, New Zealand, page 188-192, ICST (2007).

In: New Trends in Chemical Physics Research
Editor: James N. Lapherty
ISBN: 978-1-61668-535-5

Chapter 3

AN OPEN FIELD: QUANTUM DYNAMICS IN BASE PAIR DNA SYSTEMS

Giovanni Villani[*]
Istituto per i Processi Chimico-Fisici,
Pisa, Italy

In the DNA, the base pairs are involved in two qualitatively different mutual interaction types, hydrogen bonding and base stacking, and it is well known that these two major factors contribute to the stability of DNA double helix. The role of hydrogen bonds is well known in providing structure and directionality in the DNA base pairs. Base stacking interaction is one of the driving forces responsible for the stabilisation of the three-dimensional structure of DNA and RNA molecules. A very few studies are present in literature to study with a quantum approach these two processes and only our papers have considered the time-dependent hydrogen transfer between the base pairs in a quantum way.

In the A-T case there are two parallel hydrogen bonds and in the guanine-cytosine system three. The most stable structure of both systems is named Watson-Crick tautomer. From this structure, while keeping each monomer in its neutral form, a double hydrogen atom transfer, as proposed by Löwdin gives the imino-enol tautomer A*-T* and G*-C* and the imino-enol-imino-enol G#-C#, in adenine-thymine and guanine-cytosine base pairs respectively (Fig. 1).

The idea that imino-enol tautomeric forms of DNA bases play a role in mutagenesis was formulated at same time that the Watson-Crick DNA structure. If it exists an equilibrium in the two tautomeric forms of the base pair and if the rare tautomer remain stable during the time period needed for replication process, the double hydrogen transfer would play an important role for the occurrence of spontaneous mutations. Up to now, the question of whether point mutation could be formed spontaneously via the tautomerisation process remains unanswered. We believe that this is a typical problem where a theoretical approach with reliable calculations can be very useful.

[*] Istituto per i Processi Chimico-Fisici, IPCF-CNR Via G. Moruzzi, 1 I-56124 Pisa (Italy) villani@ipcf.cnr.it

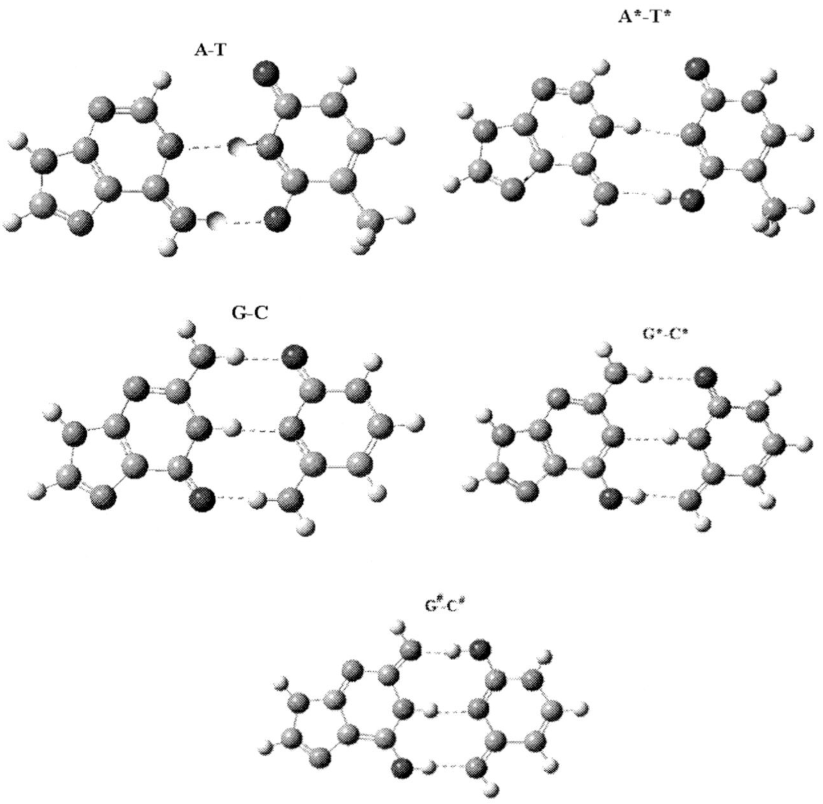

Figure 1.

In a hydrogen-bonded system a quantitative evaluation of the hydrogen transfer probability needs the knowledge of the potential energy surface (PES) in the coordinates involved in this process and, as demonstrated by Florián and al., the use of an adiabatic potential that allows the geometry relaxation of the system during the hydrogen transfer process is necessary to estimate both the amount of the two tautomers and the dynamics of the process. Our theoretical electronic calculation of the adenine-thymine and guanine-cytosine base pairs has been shown in[1,2]. In these studies, we have found two static stable structures for both systems: the A-T and A*-T* tautomers with a difference of energy of 12.9 kcal/mol and the G-C and G*-C* tautomers with a difference of energy of 8.83 Kcal/mol. There are not minima for the zwitterionic structures and for the $G^{\#}$-$C^{\#}$ tautomer, but (see later) this does not mean the irrelevance of these tautomers in the dynamics. The dynamical calculations on these systems have been performed propagating an initial state by a Lanczos scheme (with Householder reorthogonalisation) described in previous papers. The basis set has been formed by tensorial product of the eigenstates of the three anharmonic oscillators for G-C and the two for A-T base pairs. These eigenstates have been computed with an accurate numerical procedure based on a linear variational approach with a Gaussian basis set. The initial state is the tensorial product of the two (or three) ground states of the monodimensional oscillators. These initial states have the same hydrogen bond situation of the ground states of the Watson-Crick tautomers (A-T and G-C).

For discussing a possible mechanism from the Watson-Crick to imino-enol tautomers, we must note that the not all the hydrogen bonds have the same behaviour. In fact in the case of adenine-thymine system, the PES as a function of the hydrogen position in N---O bridge coordinate is steeper than that in the hydrogen position in N---N bridge coordinate. We have propose to call *hard hydrogen bond* the N-H---O, where the position of the hydrogen is practically frozen, and *soft hydrogen bond* the N-H---H, where the position of the hydrogen can be change more easily. The same different behaviour in the hydrogen bridges can be found in the guanine-cytosine system: the N-H---N and the O---H-N bonds are *soft* and the N-H---O bond is *hard*. This different behaviour of the hydrogen bridges is not due to intrinsic characteristic of these bonds since it disappears when the other coordinates not directly involved in the hydrogen bridges are frozen.

We have found that the suitable mechanism for the conversion of A-T to A*-T* can not be the concerted double hydrogen transfer. Only a two steps mechanism is possible since the movement of the hydrogen in N---O bridge is frozen when the hydrogen in the other hydrogen bridge is around the equilibrium position of the A-T structure. Instead, the change of energy due to the movement of the hydrogen in N---N bridge is softer. The two step mechanism of A*-T* formation is not efficient and only a small amount of this tautomer can be found at any time, but in any case this process generate several hundreds of mutation points in a DNA with one million of A-T (or T-A) bases. Beyond these two tautomers, and with larger amount than the A*-T* structure, there is the dynamical $A^{\delta+}$-$T^{\delta-}$ tautomer related to the transfer of the hydrogen atom in the N---N bridge from the thymine to the adenine[1]. Instead, the suitable mechanism for the conversion of G-C to G*-C* can be both the concerted double hydrogen transfer and the two steps process since these two hydrogen bonds are *soft*, in our nomenclature. Contrary, the conversion of G-C to $G^{\#}$-$C^{\#}$ are principally a two step process since a bridge is *hard* an the other is *soft*. Some amount of these tautomers can be found at any time, but in any case this generate several thousands of mutation points of the G-*-C* structure and hundreds of the $G^{\#}$-$C^{\#}$ one in a DNA with one million of G-C (or C-G) bases. Beyond these three tautomers, there are the dynamical $G^{\delta+}$-$C^{\delta-}$ and $G^{\delta-}$-$C^{\delta+}$ tautomers[2], related to the transfer of the hydrogen atom in the N---O, N---N and O---N bridges, i.e. from the guanine to the cytosine and vice versa. There is a very large difference between the probabilities of the imino-enol tautomers of both adenine-thymine and guanine-cytosine computed in these papers and the same literature dates since we have computed the quantum and the others the thermal probabilities.

The consequence of the large amount and time evolution of rare tautomer populations on biology properties can be summarised as follow. First of all, there is a large difference between these quantum probabilities and the frequency of spontaneous mutations since the value of this frequency in DNA is from 10^{-8} to 10^{-10}. Of course we must also remember that the rate of spontaneous mutation is related to biological condition, for example the occurrence of mutations in the nonreplicated genome is significantly slower than the one taking place during DNA replication. In any case, we believe that this large difference can be an indication of a minor role of the base pair tautomerisation in the mutation process. Second, the difference of stability of A-T and G-C has been related to the deficiency of G-C content of the DNA of higher organisms, which is generally as low as 0.5 times (0.4 in humans) the A-T content. We believe that also this consideration is not reasonable, since our calculation supports the same stability for the A-T and G-C systems. Finely, a very periodic behaviour of the imino-enol structures must be underlined and an aperiodic behaviour for the imino-enol-imino-enol $G^{\#}$-$C^{\#}$ tautomer and with a large oscillations of the amount of these structures. This is a first time where a coherent quantum process with a well defined periodic time has been evidenced in this essential biological system, but also the aperiodic behaviour of the $G^{\#}$-$C^{\#}$ tautomer is important. One of the most important conclusion of the scientific area of the Complexity Science is the presence of both these opposite behaviours in the same system. Here, we find exactly this situation.

During these processes of hydrogen transfer a general consideration on the atomic charge variation can be done. First of all, in these hydrogen bonds there is not a proton transfer. This is the motif because I always say "hydrogen atom transfer" and never "proton transfer". This means that the hydrogen atomic charge changes through the transfer from the donor to the acceptor, but never one has a proton passage and an electron that remains in the original fragment. Hence, the process of hydrogen transfer in the base pairs systems is a case of proton-coupled electron transfer. Moreover, in both cases the heavy atoms supporting the bridge have complementary charge variation with the acceptor that becomes less negative and the donor more negative.

After we have studied the relation between the hydrogen atoms transfer and the base stacking interaction. Two phenomena are present in the polynucleotide system: the stacking coupling among a base pair and their neighbors and the dynamical process of hydrogen atoms transfer in the base pair. These two phenomena interact since the base-base stacking coupling is modified by the hydrogen atoms passage from complementary bases in a pair. Of course, this modification is maximum for the nearest bases and decrease for farther bases hence we assume that only the nearest bases are coupled to a specific base in the chain. These modifications are time-dependent since the process of hydrogen transfer is time-dependent and a simple sinusoidal behaviour is assumed.

The base stacking energy consists of two parts. The van der Waals component of stacking is overlap-dependent and includes the dominating dispersion attraction as well as steric effects. The electrostatic component of stacking reflects the mutual interaction of molecular electrostatic potentials of bases. In contrast to the variable electrostatic stacking terms, the van der Waals contributions are essentially independent of the sequence. Their total values are highly conserved, ranging from −16.6 to −16.9 kcal/mol. There are moderate variations in the Pyr-Pyr and Pur-Pyr(3') terms, correlated with the presence or absence of the N2-amino group. In addition to the base-base interactions it would be possible to analyse other contributions to the potential energy as well, e.g., sugar-base interactions. However, the interbase interactions most probably contain the majority of stacking coupling effects.

There are five different situations for a base pair in a polynucleotide in the assumption of base-stacking coupling only with the nearest bases[3]. If X is the base pair considered and x and y its nearest ones, one has:

1. A terminal base followed by an equal base: X-x-
2. A terminal base followed by a different base: X-y-
3. A central base with equal neighbors: -x-X-x-
4. A central base with different neighbors: -y-X-y-
5. A central base with an equal and a different neighbors: -x-X-y-

The main result of this study can be considered the evidence of a long-time coherent process of complete transfer of hydrogen atoms in a base pair in the condition of consecutive identical base pair. In this case the amount of the so called "rare tautomers", the imino-enol tautomeric forms of DNA base pairs, can become much more larger, until the complete passage, in the oligomer systems.

If we applied the quantum dynamics study to the two hydrogen bridges of the adenine-thymine base pair in the case where the possibility of nonlinear hydrogen atom movement has been explicitly considered[4], five are the principal differences.

1. The amount of tautomers, both zwitterionic and neutral (imino-enol), are largely increased;
2. The periodicity behaviour of the tautomer probabilities has been lost;
3. The time-dependent average probabilities reach the constant values;
4. In the simpler model the imino-enol tautomer is obtained with a concerted two hydrogen atoms transfer from the bases, while in this model a two steps process is the most probable.
5. Two differences in the time-dependent interaction of this system in these two models must be evidenced. First of all, if one considers an interaction of this system with a rapid process in the femtosecond scale a coherent hydrogen transfer process can be found in the simpler model, but not in this last model. Second, a slower process in the hundred of femtosecond scale finds a periodic change of the asymptotic average probabilities in the simpler model and a constant amount of these tautomer species in the other model. These suggestions can be proved experimentally and can be relevant in the biological roles of the DNA.

Otherwise, if we applied the quantum dynamics study to the three hydrogen bridges of the guanine-cytosine base pair with the possibility of nonlinear hydrogen atom movements[5], four are the principal differences.

1. The amount of tautomers, both zwitterionic and neutral (imino-enol and imino-enol-imino-enol), is largely increased. In particular, the insignificant $G^{\#-}C^{\#}$ tautomer of the simpler model becomes more important in this model.
2. When present in the simpler model, the periodic behaviour of a tautomer probability has been conserved in this model.
3. The time-dependent average probability reaches a constant value at different time for the different hydrogen atom transfer. In particular, the constant value is reached first for the *hard* than for the *soft* hydrogen bridge.
4. The imino-enol tautomer is obtained with a two-step mechanism of hydrogen atom transfer in both the simpler and more complex model.

REFERENCES

[1] G. Villani, Chem Phys 316 (2005) 1-8.
[2] G. Villani, Chem Phys 324 (2006) 438-446.

[3] G. Villani, Chem Phys 325 (2006) 389-396.
[4] G. Villani, Chem Phys 336 (2007) 143-149.
[5] G. Villani, J Chem Phys 128 (2008) 114306-114313.
[6] P. O. Löwdin, Rev. Mod. Phys. 35 (1963) 724.
[7] P. O. Löwdin, Adv. Quantum Chem. 2 (1965) 213.
[8] J. D. Watson and F. H. C. Crick, Nature 171 (1953) 737 and 964.
[9] J. Florián, V. Hrounda and P. Hobza, J. Am. Chem. Soc. 116 (1994) 1457.
[10] Lami and G. Villani, Chem. Phys. 172 (1993) 285.
[11] Ferretti, G. Granucci, A. Lami, M. Persico and G. Villani, J. Chem. Phys. 104 (1996) 5517.
[12] P. Hamilton and J. C. Light, J. Chem. Phys. 84 (1986) 306; Z. Bacic and J. C. Light, *ibid*, 85 (1986) 4594.
[13] E. Nir, C. Plützer, K. Kleinermanns and M. S. de Vries, Eur. Phys. J. D 20 (2002) 317.
[14] Auerbach, C., *Mutation Reaserch, Problems, Results and Perspectives* (John Wiley and Sons, New York, 1976).
[15] J. Florián and J. Leszczyński, J. Am. Chem. Soc. 118 (1996) 3010.
[16] Komberg, A. and Baker, T. *DNA Replication* (W. H. Freeman, New York, 1992).
[17] G. Villani, Chem. Phys. 302 (2004) 309.

In: New Trends in Chemical Physics Research
Editor: James N. Lapherty
ISBN: 978-1-61668-535-5

Chapter 4

ADVANCED RAMAN SPECTROSCOPY FOR COUNTERACTING PREMATURE WEAR FAILURE OF BIOMEDICAL POLYETHYLENE

***Leonardo Puppulin**[*a]* ***and Giuseppe Pezzotti**[†a,b]*

[a]Kyoto Institute of Technology, Kyoto, Japan
[b]Loma Linda University, Loma Linda, CA 92354, USA

ABSTRACT

During its historical evolution, hip arthroplasty have encountered and still occasionally encounters enormous obstacles, which mainly deal with the quality of the used prosthetic materials. Since the first attempts, which used prostheses made of primitive materials at the end of the nineteenth century, many different materials have been experimented and improved surgical approaches undertaken. Hip prostheses have been conceived according to different design criteria, but they all have been following the wide underlying principle of low-friction arthroplasty. Such a basic concept yet represents the standard in total hip arthroplasty. In this paper, an analytical overview is given from a materials science perspective of the major problems encountered in the recent history of polyethylene, the most used material in hip arthroplasty. An outline is also given of an emerging Raman spectroscopic technology aimed at non-destructively monitoring the structural and environmental reliability of advanced biomaterials.

---

* Ceramic Physics Laboratory & Research Institute for Nanoscience Kyoto Institute of Technology, Sakyo-ku, Matsugasaki 606-8585 Kyoto, Japan

† Ceramic Physics Laboratory & Research Institute for Nanoscience Kyoto Institute of Technology, Sakyo-ku, Matsugasaki 606-8585 Kyoto, Japan and Department of Orthopaedics, Orthopaedic Research Center, Loma Linda University, Loma Linda, CA 92354, USA

INTRODUCTION

Nowadays, total hip arthroplasty (THA) represents a very spread and effective surgical procedure. Surgeons and technologists make daily efforts in improving the outcomes of THA, with the ultimate goal of creating a prosthesis that reliably lasts at least as long as a human lifetime. While the results of primary hip arthroplasty are generally very good, revision surgeries might score variable success with regards to their clinical outcomes [1]. In addition, they invariably represent an expensive procedure and a severe burden to the patients [2]. Thus, a reduction of the failure rates of only a few percents can, due to the large number of patients involved, have a vast influence on the accumulated costs and patient suffering. In other words, the key issue in hip arthroplasty resides in the improvement of the prostheses with regard to their long-term in vivo reliability. These circumstances amply justify the search for new hip prostheses with improved structural/functional characteristics and elongated lifetimes. Most recent innovative trends in THA have focused on the improvement of the tribological behavior of hip joints and challenged the achievement of a longer durability, with the potential for a service-life spanning several decades [3]. Such trends have naturally led to an increase in the use of ceramic materials, either as ceramic femoral heads coupled with advanced acetabular cups made of polyethylene with improved molecular structure and quality. The greater driving force in using ceramic bearings is a systematic reduction in periprosthetic osteolysis (i.e., mainly arising from polyethylene wear debris [4, 5]), which in turn can greatly reduce the number of surgical revisions.

Comprehensive reports on the long-term clinical performance of advanced polyethylene materials with improved molecular structures are available in the recent medical literature [6-8]; however, such polyethylene structures may intrinsically differ from each other and, more remarkably, even from piece to piece in the same production batch. This is due to a somewhat sluggish technological development of quantitative methods for the control of the polyethylene surface structure after irradiation by gamma rays or after being subjected to other types of sterilization process. In addition, a concern remains that, despite a clear reduction in the amount of worn volume upon cross-linking, the amount of polyethylene debris particles with fine sizes comprised in the osteolytic potential range may remain high [9]. Thus, the potential for a reduction in clinical cases of osteolysis could be possibly lower than theoretically expected from the lowered volumetric wear rate. Furthermore, poor control of the sterilization process may also trigger the presence of free residual radicals on the irradiated surface of the polyethylene structure after cross-linking, thus causing local oxidation gradients with subsequent material embrittlement [10, 11]. In this latter context, retrieval studies of explanted highly cross-linked acetabular cups after short-term in vivo service have eventually shown trends that were not predicted by long-term simulator studies [12].

In this paper, we shall first review some statistics on hip-joint surgery and have a brief survey on the main cases of failure of hip prostheses. Then, a detailed description of an advanced technological approach in reliability control and non-destructive analysis of the polyethylene component of hip prostheses will also be given. The main aim of this paper is to drive the attention of the international orthopaedic community on the need for a highly interdisciplinary approach to the study of hip joint arthroplasty. In this context, we provide here a vivid example of how newly developed Raman technology may provide final solutions to historical problems related to the chemical and the structural reliability of polyethylene materials employed in total hip arthroplasty.

HISTORICAL BACKGROUND ON POLYETHYLENE IN THA

The polyethylene part of hip prostheses, referred to as an acetabular cup or a liner, is responsible for absorbing the impacts experienced by the implant during its lifetime and may become subjected to extensive wear (i.e., resulting in the release of foreign debris into the patient). Oxidation of polyethylene liners unavoidably takes place in vivo and greatly affects the structural integrity of the liner. Extensive wear of the liner will result in the need for a revision surgery. Modern liners are manufactured from ultra-high molecular weight polyethylene (UHMWPE), a material with extreme durability in its non-oxidized state [13]. Orthopaedic implant industry usually employs a sterilization method involving the exposure of the liner to gamma radiation. However, when this method is applied in air or in a non-completely controlled inert atmosphere, it may cause severe oxidation of the polyethylene surface, with the liner becoming highly brittle, defective and with a surface highly prone to wear [14]. These negative effects may be strongly amplified when irradiation in presence of air is coupled with an extended shelf life.

In 1987, a modified polyethylene developed to reduce wear in total joint replacement components, including hip, knee, shoulder, and elbow joint replacements, was introduced in the market. Initially, a gamma-ray process in air was used to sterilize all components, while subsequently the components were sterilized by different processing methods including gamma-sterilization in nitrogen environment and gas plasma. In the late 1990s, all defective products were withdrawn from the market. In 2001, the British Government's Medical Devices Agency issued a voluntary recall of liners sterilized by gamma irradiation in air. Following a review of hip patients at sites with exceptionally high failure rate (up to more than thirteen times higher than the expected rate), a warning was also issued and patients instructed to proceed to a peer check of their implanted liners. It is now well known that gamma-ray sterilization of UHMWPE can induce oxidative degradation and that this effect increases during shelf storage and during implantation, significantly shortening the clinical life span of the prosthesis [13, 14]. Analytical studies of polyethylene wear debris have been recently published for a statistically meaningful number of patients who had received defective polyethylene implants sterilized by gamma rays in air (i.e., of the same type of those recalled by the British Government) and had suffered prosthetic loosening [15]. A comparison was carried out with implants made of traditional polyethylene and sterilized by the same method. Interestingly, the frequency distribution of globular and fibrillar particles was similar in both groups, but the globular particles of the defective samples had a mean surface area significantly lesser than that of the implants used for comparison. In other words, the two materials, despite undergoing the same type of sterilization, produced wear debris with significantly different morphologies. Such a difference can be regarded as the origin of a more intensive biological response, namely of an earlier and more massive osteolysis. In other words, experience teaches us that oxidative degradation of UHMWPE joint components is a complex phenomenon, which is strongly affected by the intrinsic molecular structure of the material. In addition, along with increased wear resistance, there can be deleterious changes in ductility and fracture resistance in UHMWPE.

Nowadays, while some manufacturers prefer to sterilize their implants using ethylene oxide, gas plasma or electron beam irradiation, others continue to use gamma irradiation, but with the implant sealed in a low-oxygen package [16]. The reason for yet selecting sterilization by gamma rays resides in the understanding that crosslinking, induced as a by-product of radiation sterilization, can improve the wear resistance of UHMWPE. Clearly, enormous progress has been made in material processing technologies since the disclosure of the failures of liners sterilized by gamma irradiation in air. However, clinical reports continue to show that the extent of oxidation scatters markedly, depending on type of resin, atmosphere in which the component was irradiated and subsequently stored, duration of shelf storage, and duration of use in vivo [17]. Some of the radiation-sterilized UHMWPE implants show little or no oxidation, even after long-term shelf storage or implantation. In other words, an open issue yet remains on how can we be assured on a deterministic basis, namely with respect to each very piece of polyethylene implanted in the human body, about safety and lifetime concerns. These concerns are particularly true if one thinks that the numbers of total hip replacements in the world are increasing. Therefore, even assuming that the revision rate has not significantly increased during the last decades, there is an overall increase in the total number of revisions. In other words, we are not improving in avoiding our mistakes. Clearly, an even minimum reduction of the failure rates by only a few percents can have a vast influence on the reduction of the number of patient suffering. This observation should represent the main driving force for improvements to be continuously pursued by surgeons and technologists.

Need for Raman Analysis of Polyethylene Liners

From a materials science point of view, the general classification of UHMWPE refers to a polymeric substance whose repeating unit is the hydrocarbon ethylene group ($-C_2H_4-$) and whose molecular weight exceeds 1.75 million g/mol. The UHMWPE materials used for joint implants are generally in the order of 3–6 million g/mol. The increase in molecular weight is generally accompanied by an improvement of mechanical and functional properties, including a low coefficient of friction and a high resistance to wear. In particular, these latter two characteristics make UHMWPE suitable for use in bearing surfaces of joint implants [18, 19]. Since sterilization is needed in medical devices, the material needs to be further treated after manufacturing and Co^{60} γ-sterilization is nowadays the most widely used industrial method for sterilization. The main advantages of this method reside in its reliability, flexibility, cost-effectiveness, and in the total absence of chemical residues after sterilization. The effects of sterilization on the structure of polyethylene components for joint implants has been deeply investigated [20-23]; unfortunately, the results have not so far produced a fully conclusive verdict on how these structural changes affect the success or failure rates of the joint implant. However, all studies agree on the mechanism that produces the change in mechanical properties, which is the rupture of the molecular chain bonds through the mechanism of chain scission. Broken chains allow free radicals to form within the polymeric structure, thus creating potential sites of bonding to neighboring molecules. This phenomenon is usually referred to as cross-linking, namely the structural process through which adjacent linear chains are joined to each other by covalent bonds.

Cross-linked (namely, γ- or electron-irradiated or, more in general, surface treated) polyethylene materials, would be expected to be stronger but also to possess decreased ductility and increased susceptibility to fatigue damage and wear [24]. As a matter of fact, concurrent to cross-linking, irradiation-driven chain scission can also lead to oxidative degradation, in which the free ends of the structural lamellas react with oxygen present in the surroundings. The degree to which oxidation occurs depends on the amount of oxygen diffusion and thus on the material structure on a mesoscopic scale. It should be noted that medical grade UHMWPE must be a pure polyethylene because of the highly required level of biocompatibility; accordingly, it must not contain any stabilizers and, as a result, it may be more prone to oxidative degradation than other commercially available grades. In addition, as already mentioned in the previous section, oxidation may take place during irradiation (i.e., immediate oxidative degradation), shelf life and *in vivo* operative life (i.e., long-term oxidative degradation) [13,14,17]. Immediate oxidative degradation can be minimized by elevating the radiation dose rate and/or by irradiating in inert atmosphere (e.g., nitrogen). On the other hand, long-term oxidative degradation can be delayed by carefully storing the samples in the dark at low temperature and in nitrogen atmosphere. In spite of all these efforts, even sophisticated methods of irradiation, storage, remelting, etc., have not been proved so far capable of prevent oxidation completely. Accordingly, it becomes clear how the method of sterilization becomes a primary concern with regard to total joint replacements. Some manufacturers favor the method of γ-sterilization in the presence of an inert gas since oxidation is minimal and the additional cross-linking improves abrasive wear resistance [28-32]. Conversely, some other companies prefer EtO [32] or gas plasma [33] sterilization claiming that fatigue wear, causing pitting and delamination, is more likely to occur in γ-radiated materials (i.e., owing to its increased oxidation state). The extent to which structural changes contribute to failure rates of joint implants has been a source of conflict in recent years, many of these conflicts mainly dealing with a dualism between wear and deformation properties [34].

Two types of mechanism significantly contribute to hip joint degradation: creep deformation and abrasive wear [35, 36]. Creep deformation refers to a permanent deformation that occurs under loading and does not completely recover after load release. In this type of mechanism, no mass is lost from the sample, but the polyethylene molecules pack and adjust their reciprocal positions under pressure. Conversely, abrasive wear is accompanied by an irreversible loss of mass from the material. Molecules from the surface of the polyethylene body become partly delaminated and can be peeled away, involving volumetric changes, which in turn result in polyethylene debris. The degree to which the dimensional change is caused by these two mechanisms is difficult to determine quantitatively. Although somewhat competitive with respect to the design of an ideally efficient polymeric structure, creep and wear resistances may lead to the same disadvantages, namely increased femoral head penetration and, ultimately, joint loosening. In addition, as previously discussed, debris from abrasive wear is a direct cause of osteolysis.

It is beyond the scope of this paper to provide conclusive statements on how sterilization-related structural changes in UHMWPE can affect the *in vivo* lifetime of hip joint implants. While we consider a full rationalization of the effect of different sterilization techniques as a final goal to be thoroughly sought until an unequivocal clarification is reached, we shall drive here the reader's attention on the new possibilities in non-destructive analysis offered by recent advances in confocal vibrational spectroscopy [37-39]. Such new spectroscopic techniques allow visualizing in a fully non-destructive manner and with microscopic spatial resolution (both *in-depth* and *in-plane*) the salient structural and micromechanical characteristics of biomedical UHMWPE grades.

In past years, Fourier-transform infrared (FT-IR) spectroscopy has been extensively used to characterize the hydroperoxide concentration and distribution in UHMWPE biomedical samples [40, 41]. According to such a spectroscopic method, different oxidation behaviors and mechanisms have been identified and discussed in correlation with processing, sterilization, and packaging conditions. Raman spectroscopy can enable one to make the same quantitative assessments of UHMWPE but in a non-destructive manner. [37]. As showed in Figure 1, FT-IR requires the hip cup to be cut at different depths in slices of some tens of microns thickness while confocal Raman analytical technique enables to non-destructively perform *in-depth* scanning analysis by simply moving the laser focus position throughout the thickness of the hip cup. Raman spectra can identify crystalline and amorphous phase fractions, oxidation state and residual strain in the material. In other words, through Raman spectroscopic experiments, it is possible to verify the actual structural changes that occur during sterilization by γ-radiation in UHMWPE; but, more importantly, by taking advantage of its fully contactless character, its high scanning flexibility, and non-destructiveness, it is possible to scrutinize each single UHMWPE component according to a fast procedure before implantation. According to this practice, a new possibility is opened to introduce an advanced quality control procedure for biomedical implants of deterministic nature, thus fully eliminating the inability of randomized controlled trials to detect infrequent events and unexpected side effects.

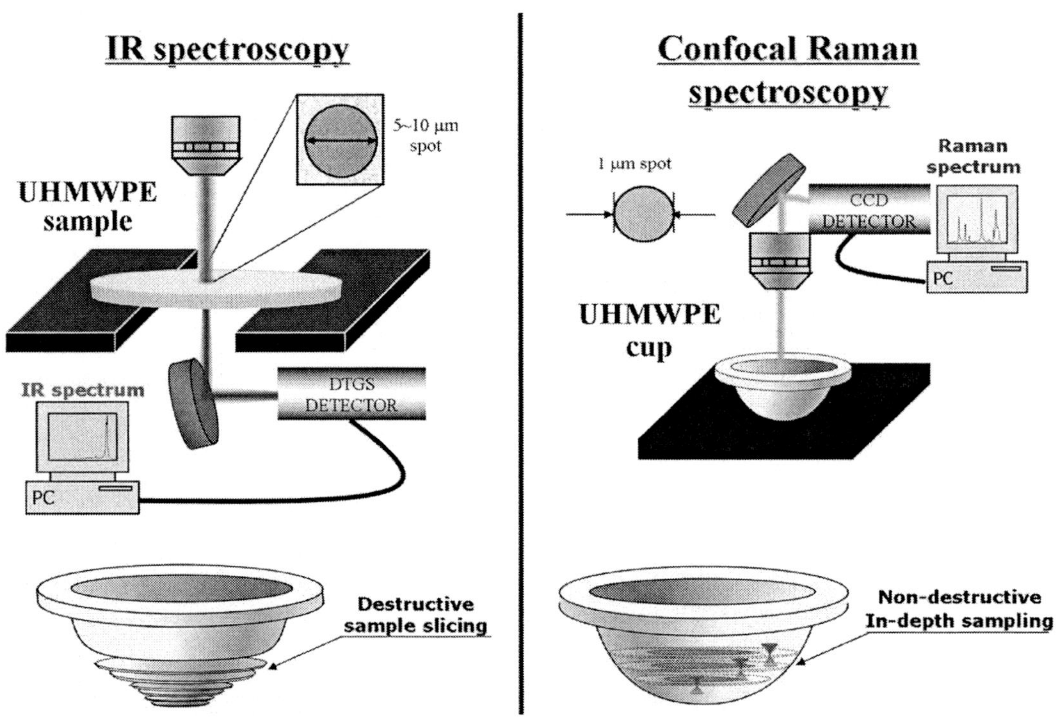

Figure 1. Schematics showing a comparison between IR and confocal Raman analytical techniques for quantitatively assessing in-depth oxidation profiles in polyethylene acetabular cups.

EXPERIMENTAL PROCEDURES

In Raman microprobe spectroscopy, a research grade optical microscope is coupled to an excitation laser and to a spectrometer producing a platform capable of obtaining conventional visible images and, in addition, to generate hyperspectral Raman images from selected sample areas [42]. The combination of a confocal Raman microscope, a motorized three-axis stage, and an autofocusing device enables the generation of 2D and 3D maps that can yield straightforward information on the distribution of different crystallographic phases, stress/strain fields, chemical species, and oxidation gradients in heterogeneous samples over selected areas. In recently published papers [37-39], we have described the analytical basis and made a review of the main applications of Raman spectroscopy/imaging in the biomaterials field. In this paper, we emphasize the importance of the Raman spectroscopic technique for the non-destructive analysis of oxidation and strain in UHMWPE. In order to characterize oxidation patterns not only on the surface but also along the material sub-surface, *in-depth* Raman spectroscopic analyses can be performed, according to the confocal method.

Raman spectra were collected at room temperature by a triple-monochromator (T-64000, Jovin-Ivon/Horiba Group, Kyoto, Japan) equipped with a charge-coupled device (CCD) detector, and analyzed by using commercially available software (Labspec, Horiba/Jobin-Yvon, Kyoto, Japan). In the experimental analysis of acetabular cups, a confocal pinhole with an aperture-diameter of 100 μm was placed in the optical path to exclude photons scattered from out-of-focus regions of the probe. Accordingly, only signals from a limited in-depth region close to the selected abscissa of the focal plane were brought to the detector. The confocal configuration of the probe adopted throughout the present experiments corresponded to a $\times 100$ objective lens; numerical objective aperture and confocal pinhole diameter were fixed as: NA=0.9 and $\Phi = 100\,\mu$m respectively. An automated sample stage with sub-micrometric step precision, was employed, making it possible to record spectra at each depth with focusing below the sample surface, or to map spectra with lateral line scanning on the sample surface. Square maps of Raman spectra were collected from different *in-depth* planes (from 0 to 50 μm inside). The map size was 50 μm x 50 μm and the square mesh was characterized by a 5 μm step. Considering the micrometric size of the probed area, the position *in-depth* of each point of a map is not affected by the surface curvature of the hip cup.

In-depth oxidation and strain profiles were also collected with translating the focal plane along the in-depth (sub-surface) direction up to 3000 μm inside.

With concurrently performing Raman and infrared spectroscopy characterizations on a series of UHMWPE oxidized samples, we have previously shown the validity of a relationship between the oxidation index, *OI*, of UHMWPE and the relative intensity of selected Raman bands, as follows [37]:

$$OI = \exp\left\{1.19 \times \tan\left[14.26\left(\frac{I_{1414}}{I_{1293} + I_{1305}} - 0.26\right)\right] - 0.13\right\} \qquad (1)$$

where I_{1414}, I_{1293}, and I_{1305} represent the intensities of the Raman bands of UHMWPE associated with -CH_2- bond wagging and -CH_2- twisting vibration. Explanations about the experimental procedure followed for validating Eq. (1) and details on the reliability of the Raman measurements of oxidation index have been given in the Appendix of Ref. [37]. It should be noted that Raman measurements of oxidized synthetic polymers might be very difficult due to the presence of a broad fluorescence background, which results from oxidative degradation. However, we have not only shown that Eq. (1) can be used to a degree of precision for *OI* assessments but also that this practice enables the determination of *OI* non-destructively from the same spectral records used for analyzing microstructural features in polyethylene [37].

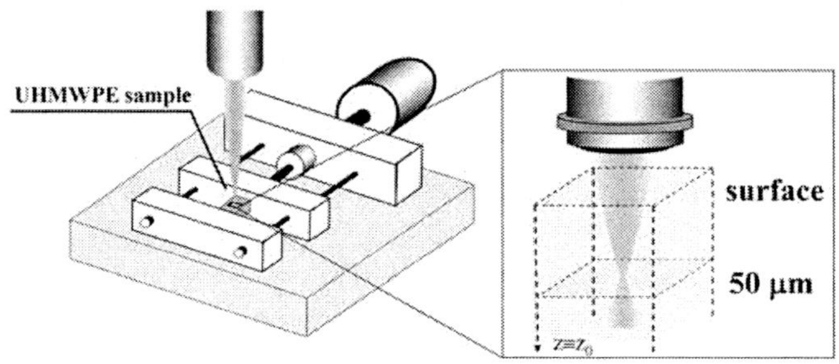

(A)

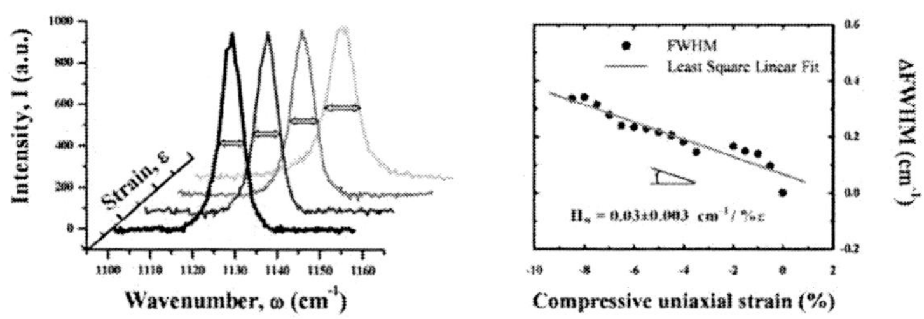

(B)

Figure 2. The scheme in (A) shows the uniaxial compressive jig by means of which the strain calibration of UHMWPE was carried out. The laser was focus at 50 □m in-depth to collect spectra not affected by surface roughness and shear effect. In (B), the broadening of the Raman band located at 1127 cm-1 and its linear dependency on the applied compressive strain are shown.

In a recent study conducted in our laboratories we have developed a new technique to assess residual compressive strain in polyethylene. Strain calibrations were carried out in order to assess the relationship between compressive strain and the *full width at half maximum* (FWHM) of the Raman band located at 1127 cm^{-1} which is due to the all-symmetric C-C stretching vibration of crystalline and amorphous phase. As shown in figure 2(A), UHMWPE hip cups have been cut into rectangular bars whose ratio between the height and the side was a little less than 2, in order to avoid buckling, shearing and barreling during compression. Utilizing a uniaxial compressive *jig* a linear correlation between band width and residual compressive strain of polyethylene have been found. Broadening of the Raman band under compressive strain is directly visible on the spectrum (figure 2(B)). The linear regression line of FWHM data calculated at different percentage of compressive strain is described by the following equation:

$$\Delta w = \Pi_w |\varepsilon|$$

where Δw represents the variation in band width (ΔFWHM) of the symmetric C-C stretching band of UHMWPE; $|\varepsilon|$ is the absolute magnitude of the applied strain; and Π_w is a proportionality coefficient explicitly given in the inset of Figure 2(B). Plastic strain profiles in confocal configuration as a function of depth along the subsurface can be easily calculated by the same Raman spectra collected to assess the oxidation profiles.

RESULTS

In this investigation, we have scrutinized a large number of virgin UHMWPE cups as received from the makers, before exposure to atmosphere. Figure 3 shows a comparison between two acetabular cups of the same type fabricated by the same maker (both treated nominally in an inert atmosphere with the same amount of γ-radiation and under exactly the same nominal irradiation conditions: 33 kGy) and belonging to the same batch. Through confocal experiments on a large number of acetabular cups obtained from various makers, the existence of a weak oxidation profile along the sub-surface has been typically observed (as shown in the series of sub-surface profiles in Figure 3(A)); however, we could also identify a small fraction of new (i.e., unused) cups (i.e., typically comprised between 1 and 3 % (from 3 to 9 cups), depending on the investigated batch), in which the immediate sub-surface region experienced a relatively high oxidation gradient (as in the in-depth profile series for the acetabular cup shown in Figure 3(B)). It can be thought that the structural degradation occurring *in vivo* in the near-surface region may be directly affected by a pre-existing oxidation state in the acetabular cup, thus shortening the lifetime of the acetabular component, leading to the early formation of polyethylene debris upon wear contact *in vivo*, and triggering osteolysis. The cycle of oxidation products and radical species formation has been shown to occur over several years as oxidation levels continuously increase in components implanted *in vivo* [43, 44]. The resulting formation of chain scission products creates shorter molecular chains, degrading the mechanical properties and the wear performance of the surface slab of material, which thus represents the most critical part for the *in vivo* lifetime of the joint [44, 45]. Figures 4(A) and (B) show typical maps of sub-surface oxidation profiles as measured in the main wear zone of UHMWPE acetabular cups produced by the same maker and both retrieved from female patients after 2.8 and 12.2 y, respectively. Maps of *in-depth* profiles for non-wear areas belonging to the same cups are given in Figures 4(C) and (D), respectively. It should be noted that both these implants were used in successful surgeries from the viewpoint of the relatively small amounts of volumetric wear detected. The reasons for the revision surgery were dislocation and aseptic loosening for the short-term and the long-term *in vivo* exposed acetabular cups, respectively. As from a comparison among the four profiles shown in Figure 4, it is clear that, in the same implant, the wear zone undergoes a greater near-surface oxidation than the non-wear zone. In addition, a time elongation of the exposure *in vivo* clearly increases the amount of oxidation along the material subsurface. Unfortunately, data showing the oxidation state of the very same cup before implantation (i.e., in its unexposed status) and after its retrieval from the patient are not available at the present time. One may attempt to grasp the oxidation state of an explanted acetabular cup from its oxidation profile in the non-wear zone, although this practice might lead to quite approximate assessments since other factors may also influence the overall oxidation rate of the cup. Acetabular cups with a stronger near-surface oxidation state in the wear zone were found to usually experience higher oxidation profiles also in their non-wear zones. A puzzling observation for environmentally degraded UHMWPE components is the occurrence of the most severe oxidation typically below the exposed surface of the component [46-50]. Spectroscopic analyses for ketones, which are the primary oxidation products in polyethylene materials, revealed maximum concentrations of this species several millimeters below the material surface [49]. This circumstance has been explained by considering that alkyl radicals, which are uniformly generated in polyethylene during irradiation, react with peroxyl radicals to form ketones. This chemical reaction encounters the most favorable kinetic conditions in the material sub-surface because at the very surface high oxygen concentration depletes alkyl radicals (creating an excess of peroxyl radicals) while in the bulk the low oxygen concentration inhibits peroxyl radicals formation. As a result, a maximum ketones concentration, and thus oxidation rate, is noticed in the material sub-surface, where the concentrations of both alkyl and peroxyl radicals are balanced. However, the depths extensively reported so far for such a maximum are of the order of several mm [49, 50]. Recent advances in confocal Raman spectroscopy, including the development of precise mathematical deconvolution procedures [37], have enabled us to locate an additional oxidation maximum at depths of the order of the tens of microns. Typical oxidation profiles, as revealed by the Raman confocal probe in unused, short-term, and long-term *in vivo* oxidized acetabular cups (produced by the same maker and subjected to the same amount of γ-radiation dose), are shown in Figure 5. The existence of two distinct oxidation maxima of similar intensity in the sub-surface is clearly seen. At the present time it is not clear whether or not the chemical mechanism leading to the oxidation maximum closer to the material surface is of the same nature of that observed farther below the surface, which was already reported in previous literature papers. However, the results of compressive residual strain calculated by the same spectra collected to assess oxidation profiles in the 3 hip cups suggest the following explanation. As shown is Fig. 6, a peak of compressive strain can be observed in the sub-surface profiles of retrieved hip cups, at locations roughly coincident with the depth at which maxima of oxidation index were noticed. It is likely that oxidative species continuously generate during *in-vivo* lifetime at several tens of microns *in-depth* where the damages of the mechanical action of the femoral head and the oxygen diffusion kinetic encounter the most favorable condition to trigger oxidation of UHMWPE.

Discussion

Nowadays, national registers have been developed in many countries for follow-up and surveillance of joint replacement surgery. Registers are primarily intended to function as quality control systems, their main objective being that of detecting inferior prosthetic implants, bone cements, routines and procedures as early as possible after their introduction to the market. The intrinsic quality of joint devices and replacement surgeries can be promptly assessed owing to hospital-specific results that are in turn reported yearly to all hospitals involved with joint replacement surgery. Statistical methods are commonly used in the evaluation of the data, which typically include survival analyses and sample regression. The end-point in these analyses usually is the revision of prosthetic components. However, it should be noted that all those kinds of "quality control" methods only work *a posteriori* and strongly rely on a statistical sampling rather than on a control of deterministic nature. It is evident that inferior as well as defected implants have been used, still are used, and will continue to be used in the lack of an obligatory system not only for the introduction but also for the *a priori* quality control of each individual prostheses implanted in the clinical practice. Severe lessons from the past are teaching us that regulations should not only allow future revisions in the same hip of the same patient to be linked to the corresponding primary procedure, but also the initial quality of the implanted components to be systematically measured, followed up and evaluated in terms of reliability parameters and residual durability. Therefore, there is a clear hint that health authorities should engage in the development of fully deterministic quality control procedures, including advanced spectroscopic screening especially for those components subjected to high risk of degradation (e.g., polyethylene). Concurrently, the research community together with the orthopaedic associations should strive to widely spread understanding and to influence this process with clear research outputs and outline suggestions for the selection of more advanced evaluation methods.

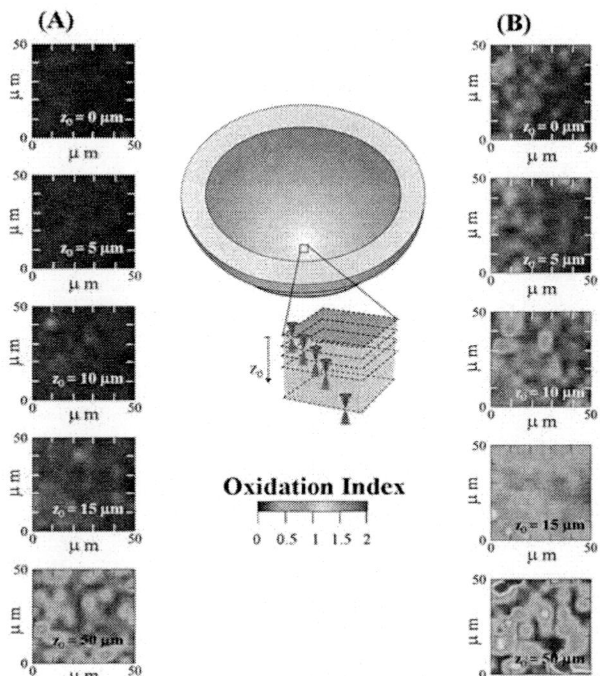

Figure 3. (A) and (B) show series of near-surface maps of the in-depth state of oxidation as detected in two new acetabular cups fabricated by the same maker, sterilized by γ-ray (33 kGy) and belonging to the same batch. Despite the common processing and surface irradiation treatment, a clear difference in the in-depth oxidation gradients can be noticed.

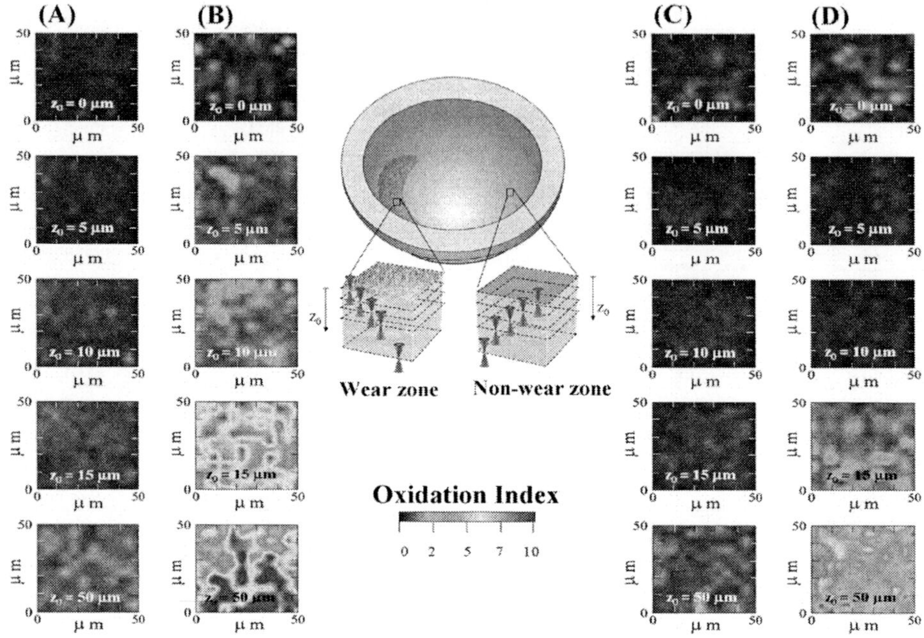

Figure 4. In-depth oxidation profiles from two different retrieved hip cups. (A) and (B) show series of typical maps of sub-surface oxidation profiles as measured in the main wear zone of two UHMWPE acetabular cups produced by the same maker and both retrieved from female patients after 2.8 and 12.2 y, respectively. Series of in-depth profiles for non-wear zones belonging to the same cups are shown in (C) and (D), respectively.

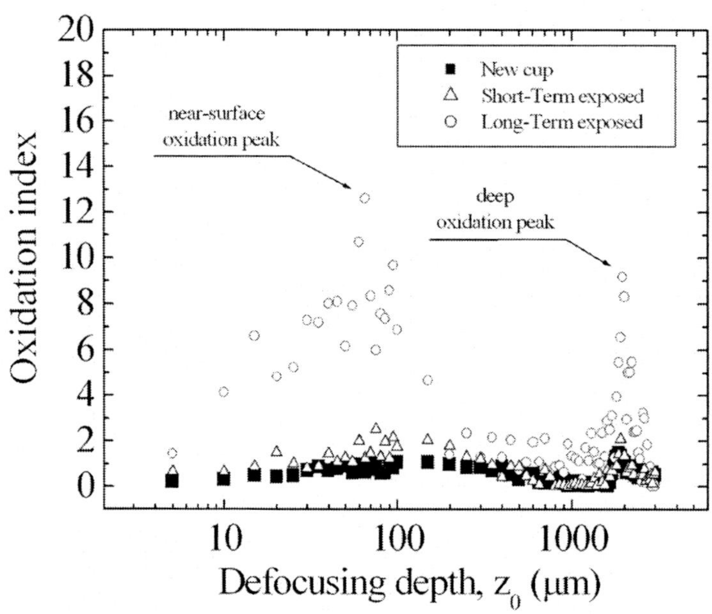

Figure 5. Linear assessment of in-depth oxidation profiles, as non-destructively revealed by a confocal Raman probe in unused, short-term, and long-term in vivo oxidized acetabular cups of the same type.

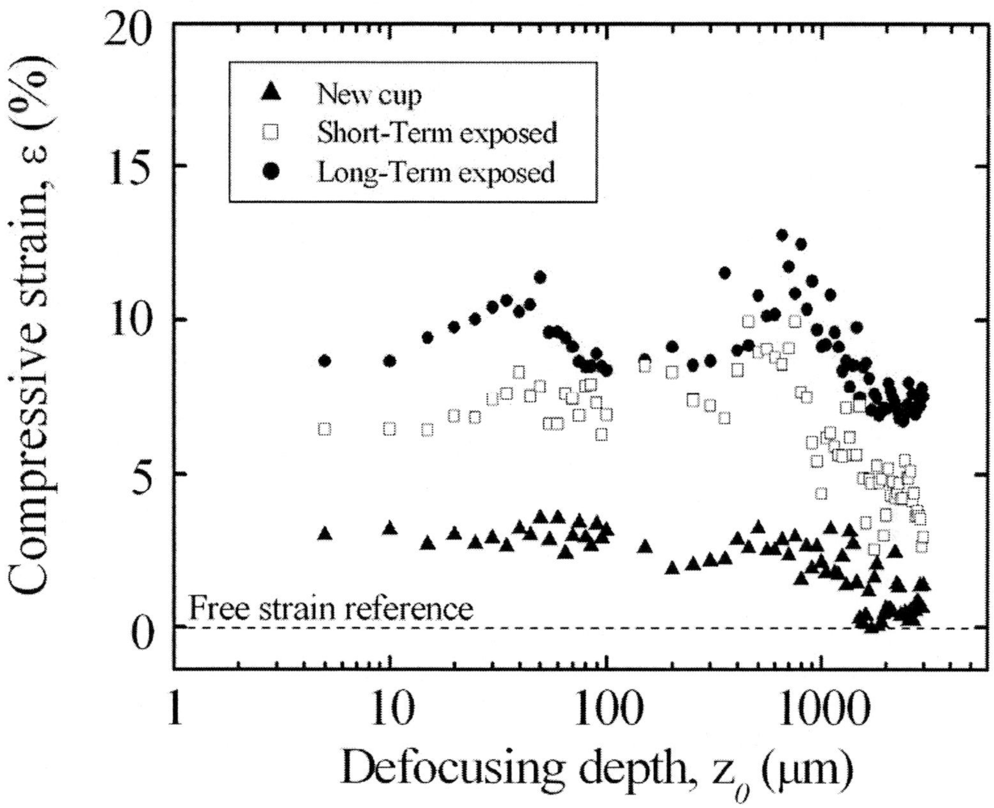

Figure 6. In-depth residual compressive strain profiles calculated by the same spectra collected for oxidation index assessments. The two maxima of strain are roughly coincident with the two maxima observed in the oxidation analysis.

CONCLUSION

In conclusion, the present study, has provided a direct method to quantitatively and non-destructively assess the in-depth oxidation state of UHMWPE acetabular cups; in addition, this study has clarified some peculiar aspects of the *in vivo* development of such profiles, as well as the existence of a fraction of already oxidized acetabular cups in the as-received state from the maker. Although a final answer about the micromechanical origin of two maxima in oxidation index, as observed along the in-depth profile in the main-wear zone, could not be given, the possibility to non-destructively and quantitatively measure the degree of oxidation from confocal Raman spectra in polyethylene acetabular cups was clearly validated. The confocal Raman approach may enable one screening the as-received liners in order to avoid implanting those already oxidized and defected. In addition, Raman spectroscopy can also help materials technologists to put forward new guidelines for the development of polyethylene grades with improved resistance to oxidation and wear, with particular relevance to the estimation of the amount of cross-linking and crystallinity needed to potentially eliminate the clinical problem of osteolysis.

ACKNOWLEDGMENT

Reviewed by Prof. Dr. Nobuhiko Sugano, MD, Department of Orthopaedic Surgery, Osaka Graduate School of Medicine, 2-2 Yamadaoka, Suita 565-0871, Japan, JP

REFERENCES

[1] Kurtz S, Mowat F, Ong K, Chan N, Lau E, Halpern M. Prevalence of primary and revision total hip and knee arthroplasty in the United States from 1990 through 2002. *J. Bone Joint Surg. Am.* 2005; 87(7): 1487-1497.

[2] Furnes A, Lie SA, Havelin LI, Engesaeter LB, Vollset SE. The economic impact of failures in total hip replacement surgery: 28,997 cases from the Norwegian Arthroplasty Register, 1987-1993. *Acta Orthop. Scand.* 1996; 67(2): 115-121.

[3] Jones CP III, Kelley S. Cementless fixation of the femoral stem. *Hip reconstruction. Current Opinion in Orthopedics* 2001; 12(1): 52-56.

[4] Oparaugo PC, Clarke IC, Malchau H, and Herberts P. Correlation of wear debris-induced osteolysis and revision with volumetric wear-rates of polyethylene: A survey of 8 reports in the literature. *Acta Orthopaedica* 2001; 72(1): 22-28.

[5] Zhu YH, Chiu KY, Tang WM, Review article: Polyethylene wear and osteolysis in total hip arthroplasty. *J. Orthopaedic Surg.* 2001; 9(1): 91-99.

[6] Gordon AC, D'Lima D Colwell Jr. CV. Highly cross-linked polyethylene in total hip arthroplasty. *J. Am. Acad. Orthop. Surg* 2006; 14(9): 511-523.

[7] Jacobs CA, Christensen CP, Greenwald AS, McKellop H. Clinical Performance of Highly Cross-Linked Polyethylenes in Total Hip Arthroplasty. *J. Bone Joint Surg. Am.* 2007; 89(12): 2779-2786.

[8] Gordon AC, D'Lima DD, and Colwell Jr. CW. Highly Cross-linked Polyethylene in Total Hip Arthroplasty. *J. Am. Acad. Orthop. Surg.* 2006; 14(9): 511-523.

[9] Endo MM, Barbour PS, Barton DC, Fisher J, Tipper JL, Ingham E, Stone MH. Comparative wear and wear debris under three different counterface conditions of crosslinked and non-crosslinked ultra high molecular weight polyethylene. *Biomed. Mater. Eng.* 2001;11(1): 23-35.

[10] Premnath V, Harris WH, Jasty M and Merrill EW. Gamma sterilization of UHMWPE articular implants: an analysis of the oxidation problem. *Biomaterials* 1996; 17(18): 1741-1753.

[11] Heisel C, Silva M, and Schmalzried TP. Bearing Surface Options for Total Hip Replacement in Young Patients. *J. Bone Joint Surg. Am.* 2003; 85(7): 1366-1379.

[12] Bradford L, Baker DA, Graham J, Chawa A, Ries MD, Pruitt LA. Wear and surface cracking in early retrieved highly cross-linked polyethylene acetabular liners. *J. Bone Joint Surg. Am.* 2004; 86-A(6): 1271-1282.

[13] Sutula LC, Collier JP, Saum KA, Currier BH, Currier JH, Sanford WM, Mayor MB, Wooding RE, Sperling DK, Williams IR. Impact of gamma sterilization on clinical performance of polyethylene in the hip. *Clin. Orthop.* 1995; 319: 28-40.

[14] Oonishi H, Kadoya Y, Masuda S. Gamma-irradiated cross-linked polyethylene in total hip replacements—analysis of retrieved sockets after long-term implantation. *J. Biomed. Mater. Res. Part A* 2001; 58(2): 167-171.

[15] Visentin M, Stea S, Squarzoni S, Reggiani M, Fagnano C, Antonietti B, Toni A. Isolation and characterization of wear debris generated in patients wearing polyethylene Hylamer inserts, gamma irradiated in air. *J. Biomater. Applications* 2005; 20(2): 103-121.

[16] Bargmann LS, Bargmann BC, Collier JP, Currier BH, Mayor MB. Current sterilization and packaging methods for polyethylene. *Clin. Orthop*. 1999; 369: 49-58 .

[17] Fisher J, Hailey JL, Chan KL, Shaw D, Stone M. The effect of ageing following irradiation on the wear of UHMWPE. *Trans. Orthop. Res. Soc*. 1995; 20:120-128.

[18] James SP, Blazka S, Merrill EW, Jasty M, Lee KR, Bragdon CR, and Harris WH. Challenge to the concept that UHMWPE acetabular components oxidize in vivo. *Biomaterials* 1993; 14(9): 643-647.

[19] Affatato S, Zavalloni S, Taddei P, Di Foggia M, Fagnano C, Viceconti M. Comparative study on the wear behaviour of different conventional and cross-linked polyethylene's for total hip replacement. *Tribology International 2008*; 41(8): 813-822.

[20] McKellop HA, Shen FW, Yu YJ, Lu B, Salovey R. Effect of sterilization method on the wear rate of UHMWPE acetabular cups in a hip simulator. In: Transactions of the 43rd Annual Meeting, *Orthopedic Research Society* 1997. San Francisco: 94-99.

[21] V. Premnath, W. H. Harris, M. Jasty, E. W. Merrill, Gamma sterilization of UHMWPE articular implants: an analysis of the oxidation problem. *Biomaterials* 1996; 17(18): 1741–1753.

[22] Tabb DL, Sevcik JJ, Koenig JL. FTIR study of the effects of irradiation on polyethylene. *J. Polym. Sci. Polym. Phys. Ed*. 1976; 13: 815-824.

[23] del Prever EB, Crova M, Costa L, Dallera A, Camino G, Gallinaro P. Unacceptable biodegradation of polyethylene in vivo. *Biomaterials* 1996; 17(9): 873– 878.

[24] Fanconi BM. Chain scission and mechanical failure of polyethylene. *J. Appl. Phys*. 1983; 54(10): 5577–5582.

[25] Sutula LC, Collier JP, Saum KA, Currier BH, Currier JH, Sanford WM, Mayor MB, Wooding RE, Sperling DK, Williams IR. Impact of gamma sterilization on clinical performance of polyethylene in the hip. *Clin. Orthop*. 1995; 319: 28-40.

[26] Oonishi H, Kadoya Y, Masuda S. Gamma-irradiated cross-linked polyethylene in total hip replacements—analysis of retrieved sockets after long-term implantation. *J. Biomed. Mater. Res. Part A* 2001; 58(2): 167-171.

[27] Fisher J, Hailey JL, Chan KL, Shaw D, Stone M. The effect of ageing following irradiation on the wear of UHMWPE. *Trans. Orthop. Res. Soc*. 1995; 20:120-128.

[28] Bapst JM, Valentine RH, Vasquez R. Wear simulation testing of direct compression molded UHMWPE irradiated in oxygenless packaging. *In: Transactions of the 23rd annual meeting of the Society for Biomaterials* 1997; New Orleans: pp. 72-78;

[29] St John KR, Takeuchi MJ, Poggie RA. Eff.ct of radiation dose and packaging conditions on the oxidation of UHMWPE. *In: Transactions of the 23rd annual meeting of the Society for Biomaterials* 1997; New Orleans: pp. 46-51.

[30] Streicher RM. Influence of ionizing irradiation in air and nitrogen for sterilization of surgical grade PE for implants. *Radiat. Phys. Chem*. 1988; 31(4-6): 693– 698.

[31] Sun DC, Wang A, Stark C, Dumbleton JH. The concept of stabilization in UHMWPE. *In: Transactions of the Society for Biomaterials 5th World Congress* 1996; Toronto: pp. 195-201.

[32] Trieu HH, Avent R. Effect of sterilization on shelf-aged UHMWPE tibial inserts. *In: Transactions of the 21st annual meeting of the Society for Biomaterials* 1995; San Francisco: pp. 109-114.

[33] Fisher J, Reeves EA, Isaac GH, Saum KA, Sanford WM. Comparison of the wear of aged and non-aged UHMWPE sterilized by gamma irradiation and by gas plasma. *In: Transactions of the Society for Biomaterials 5th World Congress* 1996; Toronto: pp. 971-976.

[34] Li S, Burstein AH. Current concepts review: UHMW polyethylene. *J. Bone Joint Surg. Am.* 1994; 76(7): 1080-1090.

[35] Schmalzreid TP, Callaghan. Current concepts review: Wear in total hip and knee replacements. *J. Bone Joint Surg. Am.* 1999; 81(1): 115-136.

[36] Edidin AA, Pruitt L, Jewett CW, Crane D, Roberts D, Kurtz SM. Plasticity-induced damage layer is a precursor to wear in radiation-crosslinked UHMWPE acetabular components for total hip replacement. *J. Arthroplasty* 1999; 14(5): 616-627.

[37] Pezzotti G, Kumakura T, Yamada K, Tateiwa T, Puppulin L, Zhu W, Yamamoto K. Confocal Raman spectroscopic analysis of cross-linked ultra-high molecular weight polyethylene for application in artificial hip joints. *J. Biomed. Opt.* 2007; 12(1): 014011-1-14.

[38] Pezzotti G. Stress microscopy and confocal Raman imaging of load-bearing surfaces in artificial hip joints. *Expert Rev. Med. Devices* 2007; 4(2): 165-189.

[39] Pezzotti G. Raman piezo-spectroscopic analysis of natural and synthetic biomaterials. *Anal. Bioanal. Chem.* 2005; 381: 577-590.

[40] Tabb DL, Sevcik JJ, Koenig JL. FTIR study of the effects of irradiation on polyethylene. *J. Polym. Sci. Polym. Phys. Ed.* 1975; 13: 815– 824.

[41] Shen F-W, Yu Y-J, McKellop H. Potential Errors in FTIR Measurement of Oxidation in Ultrahigh Molecular Weight Polyethylene Implants.J. *Biomed. Mater. Res.* 1999; 48(3): 203-210.

[42] Lewis IR, Edwards HGM. *Handbook of Raman Spectroscopy: From the Research Laboratory to the Process Line.* New York(USA) and Basel (Switzerland); Marcel Dekker Ltd: 2001.

[43] Taddei P, Affatato S, Fagnano C, Bordini B, Tinti A, and Toni A. Vibrational spectroscopy of ultra-high molecular weight polyethylene hip prostheses: influence of the sterilization method on crystallinity and surface oxidation. *J. Mol. Struct.* 2001; 613: 121-129.

[44] Eyerer P and Ke YC. Property changes of UHMW polyethylene hip cup endoprostheses during implantation. *J. Biomed. Mater.* Res. 1984; 18: 1137-1151.

[45] Costa L, Luda MP, Trossarelli L, Brach del Prever EM, Crova M, and Gallinaro P. *Oxidation in orthopaedic UHMWPE sterilized by gamma-radiation and ethylene oxide. Biomaterials* 1998; 19(7-9): 659-668.

[46] Yeom B, Yu Y-J, McKellop HA, Salovey R. Profile of oxidation in irradiated polyethylene. *J. Polym. Sci.: Part A: Polym. Chem.*1998; 36: 329-339.

[47] Blanchet TA, Burroughs BR. Numerical oxidation model for gamma radiation-sterilized UHMWPE: Consideration of dose-depth profile. *J. Biomed. Mater. Res.* (Appl. Biomater.) 2001; 58: 684-693.

[48] Buchanan FJ, Sim B, Downes S. Influence of packaging conditions on the properties of gamma-irradiated UHMWPE following accelerated ageing and shelf ageing. *Biomaterials* 1999; 20(9): 823-837.

[49] Daly BM, Yin J. Subsurface oxidation of polyethylene. *J. Biomed. Mater. Res.* 1998; 42(4): 523-529.

[50] O'Neill P, Birkinshaw C, Leahy JJ, Buggy M, and Ashida T. The distribution of oxidation products in irradiated ultra-high molecular weight polyethylene. *Polymer Degradation and Stability* 1995; 49: 239-244.

In: New Trends in Chemical Physics Research
Editor: James N. Lapherty

ISBN: 978-1-61668-535-5

Chapter 5

GROUND STATE ISOMERISM AND DUAL FLUORESCENCE OF 2-METHYLHARMINE ANHYDRONIUM BASE (7-METHOXY-1, 2-DIMETHYL-2H-PYRIDO[3,4-B]INDOLE) IN 2-METHYLBUTANE AND OTHER APROTIC SOLVENTS

***Emilio García-Fernández*[*], *María A. Muñoz*[†],**
***Carmen Carmona*[‡] *and Manuel Balón*[§]**
Universidad de Sevilla, Sevilla, Spain.

ABSTRACT

Ground state isomerism and dual fluorescence of 2-methylharmine anhydronium base (7-methoxy-1, 2-dimethyl-2H-pyrido[3,4-b]indole), HIA, have been studied in 2-methylbutane and other aprotic solvents. Absorption, steady state and time resolved fluorescence measurements reveal the existence of a ground state equilibrium between two HIA isomeric species, namely Q and Z, of presumed quinoid and zwitterionic structures, respectively. Upon light absorption, the Q isomer emits fluorescence at short wavelengths from the locally excited state, LE, resulting from a vertical Franck-Condon transition, with a lifetime around 4 ns. The Z species, once in its LE excited state, quickly relaxes to an internal charge transfer state, ICT, emitting long wavelength fluorescence with a lifetime around 11 ns.

[*] Departamento de Química Física, Facultad de Farmacia, Universidad de Sevilla, 41012 Sevilla, Spain.
[†] Departamento de Química Física, Facultad de Farmacia, Universidad de Sevilla, 41012 Sevilla, Spain.
[‡] Departamento de Química Física, Facultad de Farmacia, Universidad de Sevilla, 41012 Sevilla, Spain.
[§] Corresponding author. Tel.: +34954 55 67 33, Fax: +34954 55 71 74E-mail address: balon@us.es *Departamento de Química Física, Facultad de Farmacia, Universidad de Sevilla, 41012 Sevilla, Spain.*

INTRODUCTION

Carbolines, or pyridoindoles, constitute a class of diaza-aromatic drug-binding alkaloids resulting from the condensation of a pyridine and an indole ring [1,2]. Because the presence in their conjugated rings of pyrrolic and pyridinic nitrogen atoms, carbolines can hydrogen bond with a variety of substrates and solvents experiencing different ground state prototropic equilibria [3,4]. In the lowest singlet excited state, the charge density of the carboline ring nitrogen atoms changes profoundly, becoming the pyridinic nitrogen much more basic and the pyrrolic nitrogen much more acid [3,5]. Thus, upon light excitation, carbolines can experience photo induced double proton transfers to give zwitterionic phototautomers. Over the past decades, the excited state proton transfer reactions, ESPT, and the phototautomerism of the carbolines have been profusely studied [4-13].

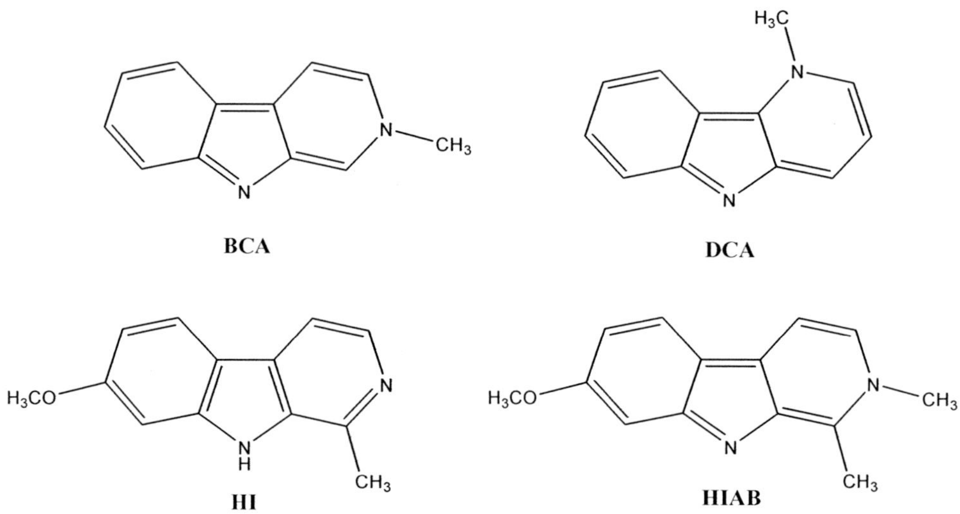

Scheme 1.

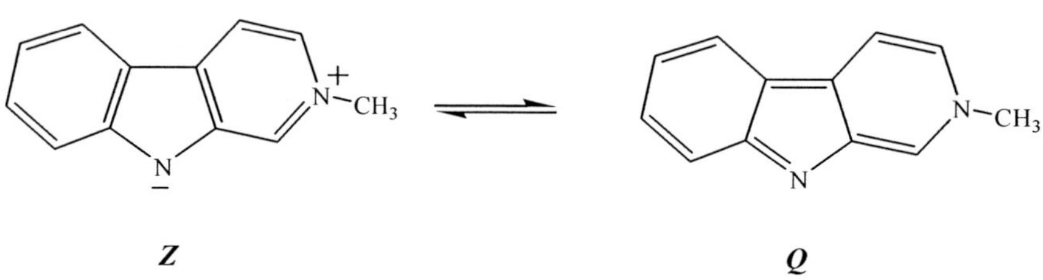

Scheme 2.

Recent studies of our research group have demonstrated that the anhydronium bases, or anhydrobases, of the N-pyrido methyl derivatives of the delta- and betacarboline, DCA and BCA, Scheme 1, possess interesting photophysical properties, such as ground state isomerism and dual fluorescence, not previously observed in other carboline derivatives [14-16]. Thus, our experimental results provided conclusive proofs on the existence of two ground state BCA isomers whose equilibrium concentrations change with the medium polarity. We have proposed that these isomers possess the quinoid, Q, and dipolar zwitterionic, Z, structures depicted in Scheme 2. Upon excitation, the isomers of the carboline anhydrobases emit dual fluorescence, at short and long wavelengths, from locally, LE, and intra molecular charge transfer, ICT, excited states.

To learn more about the interesting properties of the carboline anhydrobases, we thought interesting to study the spectral and photophysical properties of the 2-methylharmine anhydrobase, 7-methoxy-1,2-dimethyl-2H-pyrido[3,4-b]indole, HIA, Scheme 1, in 2-methylbutane, 2MB, and other low polar aprotic solvents. It is well known that the presence of a 7-methoxy group in the betacarboline ring of the parent harmine, 7-methoxy-1-methyl-9H-pyrido[3,4-b]indole, HI, Scheme 1, a representative member of the betacarboline family, gives distinctive spectral and reactivity patterns to the betacarboline ring [17,18]. Therefore, we thought that the presence of this group in the HIA skeleton could also significantly affect the photophysics of the carboline anhydrobases providing further insights on their intriguing properties.

Experimental

HI and HIA of the best available quality (>99%) were obtained from Aldrich Chemical and were used without further purification. Spectral grade aprotic solvents 2-methylbutane (0.479), 2MB, hexane (0.519), HX, cyclohexane (0.557), CHX, methylcyclohexane (0.563), MCH, toluene (0.655), TL, and dimethylsulfoxide (1.000), DMSO, were commercial products of the best available quality and were stored on molecular sieves to prevent water contamination. Between parentheses, the corresponding SPP solvent polarity parameters are indicated [19]. As other related solvent polarity scales, the SPP scale is based on the solvatochromism of a probe, 2-dimethylamino-7-nitrofluorene, and its homomorph 2-fluoro-7-nitrofluorene. It has been chosen because, differently as it typically occurs in the other solvent polarity scales, the wider span of the SPP scale allows better differentiating the solvent polarity of the very low polar alkane solvents 2MB, HX, CHX and MCH. Absorption and fluorescence spectra of the solvents were periodically checked for impurities.

The room temperature UV-VIS absorption and steady state fluorescence measurements were recorded, using Suprasil quartz cells of 1 cm light path, in a Cary 100 Conc spectrophotometer and a Hitachi F2500 spectrofluorometer, respectively. The excitation spectra were corrected as in elsewhere [16]. Solutions for spectroscopic measurements were freshly prepared and kept in the dark to avoid photodecomposition. Because the quenching by molecular oxygen was checked to be negligible, the measurements were carried out with non-degassed solutions at room temperature. Due to the scarce solubility of HIA in the studied solvents, the concentrations of the solutions were unknown. Therefore, absorbance measurements are expressed in arbitrary units. However, to avoid inner filter effects and re-absorption phenomena, the absorbances of the HIA solutions at the excitation wavelengths used for the fluorescence measurements were maintained below 0.1.

Time resolved fluorescence studies were investigated by using the time-correlated single-photon counting, SPC, technique in a FL-900CD Edinburgh Analytical Instrument. The excitation source, a nanosecond, nF900, flash lamp filled with 0.4 bar H2, operated at 40 kHz with 6kV applied across 1 mm electrode gap. Fluorescence decays from the samples were acquired to (5-10) 103 counts in the peak and were fitted by deconvolution analysis to a sum of exponential functions with amplitudes, αi, and lifetimes, τi.

$$I(t) = \Sigma \alpha_i \exp(-t/\tau_i) \quad (1)$$

The magnitude of the reduced $\chi 2r$ and the shape of the autocorrelation function of the weighted residuals were judged the goodness of the individual fits.

RESULTS AND DISCUSSION

For comparison purpose, the room temperature absorption spectra of HIA, HI, BCA and DCA in 2MB are depicted in Figure 1. As this Figure shows, the absorption spectrum of HIA notably differs from that of HI, but is similar to those of BCA and DCA. However, it must be noted that the presence of the 7-methoxy group produces an important batochromic shift in the absorption spectrum of the BCA. It is also noteworthy the close correspondence between the HIA and DCA absorption spectra.

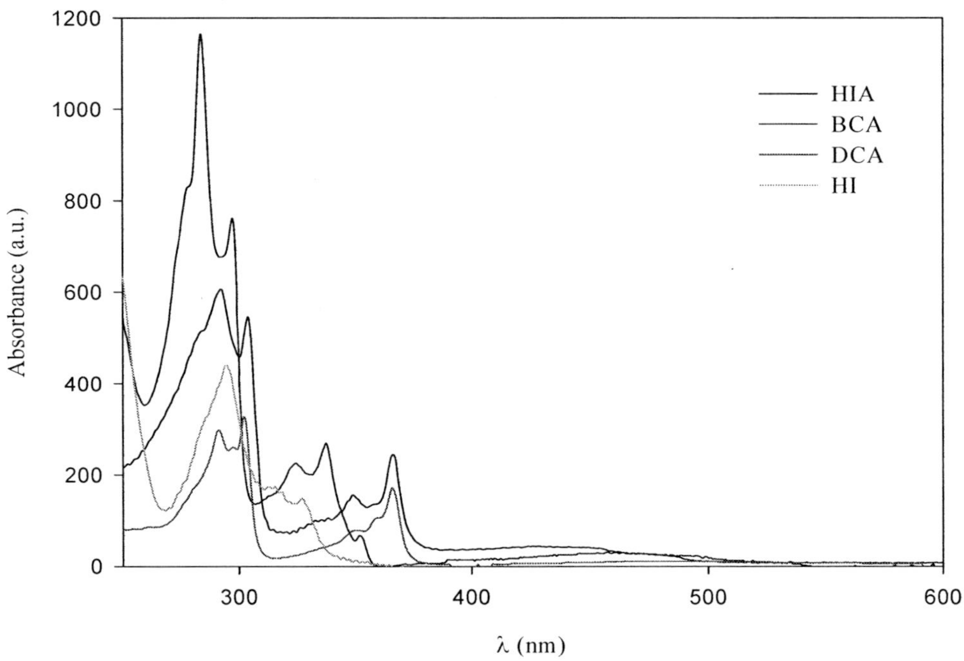

Figure 1. Room temperature UV-Vis absorption spectra of HI, HIA, BCA and DCA in 2MB.

The room temperature absorption spectrum of HIA consists of three bands; a first medium intensity band centred at 290 nm, A1, a second structured weak band at 330-380 nm, A2, and a third very weak and wide absorption band at 450 nm, A3, whose position, shape and extremely weak intensity suggest a forbidden charge transfer transition. In the slightly more polar solvents HX, CHX or MCH, neither the position nor the shape of the HIA absorption bands change, but there is an enhancement of the absorbance in the 320-350 nm spectral region, Figure 2. However, as this Figure shows, in the more polar solvents TL and DMSO the A1 and A2 bands of HIA shift to the blue, whereas the A3 band does not significantly change.

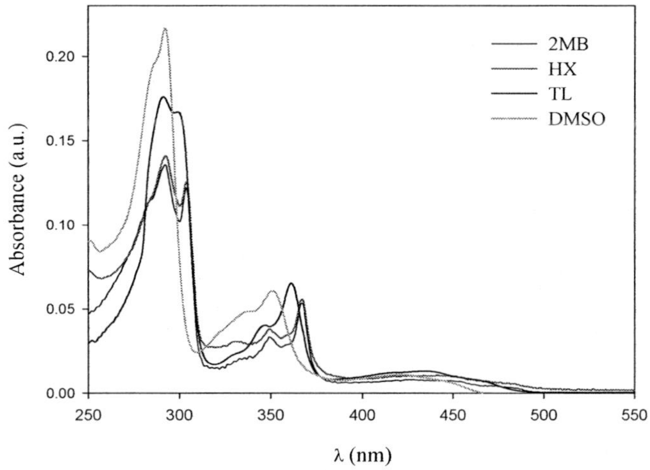

Figure 2. Room temperature absorption spectra of HIA in 2MB, HX, TL and DMSO.

The fluorescence emission spectra of HIA in 2MB strongly depend on the monitored excitation wavelength. Thus, as depicted in Figure 3, the HIA fluorescence spectrum obtained exciting the A1 absorption band shows dual emission with a structured band, F1, at short wavelengths and other broad band, F2, at long wavelengths. Differently, the fluorescence spectra obtained exciting the A2 and A3 absorption bands only show the F2 band. Interestingly, these F1 and F2 bands in the HIA emission spectra render rather different excitation spectra. Thus, as Figure 4 shows, the excitation spectra obtained by monitoring the F1 band is entirely similar, but around 5 nm red shifted, to the absorption spectra of HI. Conversely, the excitation spectra obtained monitoring the F2 emission band is much more close to the absorption spectrum of the proper HIA. This result clearly shows the heterogeneity of the HIA ground state.

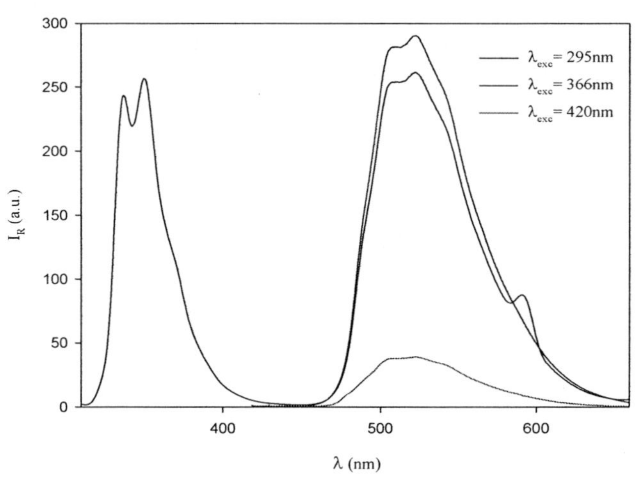

Figure 3. Room temperature fluorescence emission spectra of HIA in 2MB obtained by monitoring the A1 (λexc=295 nm), A2 (λexc= 325 nm) and A3 (λexc=366 nm) absorption bands.

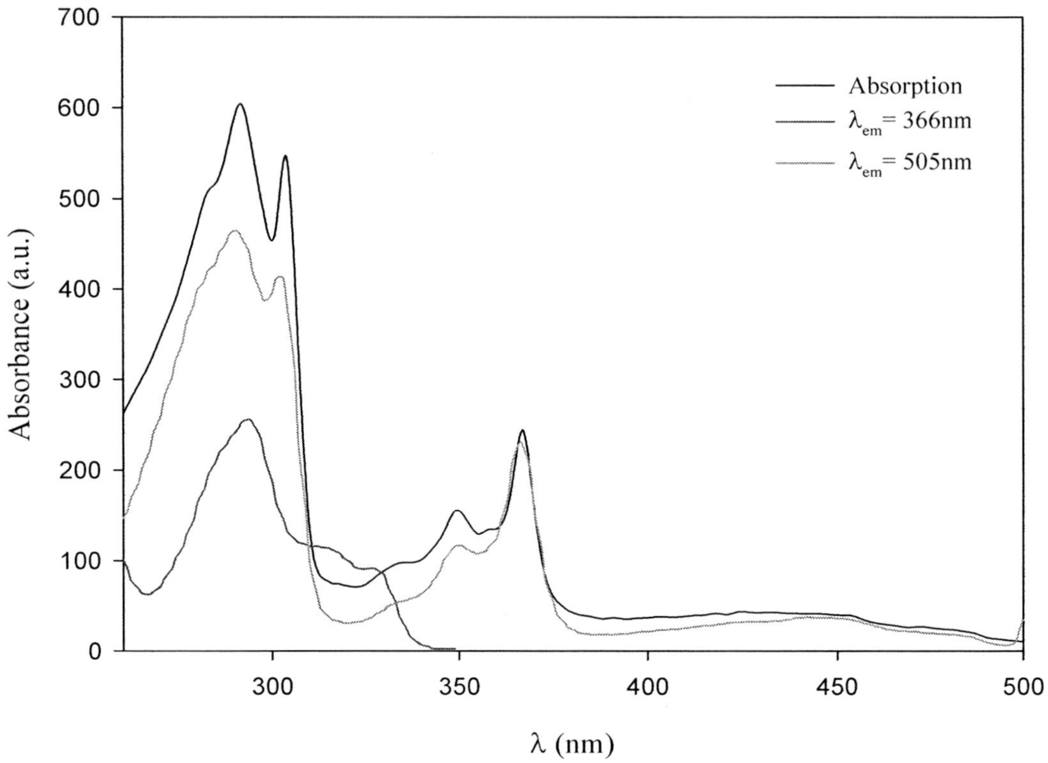

Figure 4. Room temperature absorption and fluorescence excitation spectra of HIA in 2MB obtained by monitoring the F1 (λexc= 366 nm) and F2 (λexc= 505 nm) fluorescence emission bands.

Time resolved measurements in the F1 and F2 emission bands plentifully confirm the heterogeneity of the HIA ground state in 2MB. Thus, the F1 and F2 emission bands decay mono-exponentially with different decay times. The decay time of the F1 band, 4±1 ns, is very close to that reported for HI in benzene, 4.6 ns [5]. Conversely, the decay time of the F2 band, 11±1 ns, is close to that of 14.3 ns reported for the HI zwitterion in aqueous basic solutions [5].

Regarding to the observed heterogeneity of the HIA ground state in 2MB, it is worthwhile to mention that the successive dilution of the HIA solutions only produces the expected intensity decrease of the HIA emission and excitation spectra, but it does not modify neither the shape nor the position of their bands. Therefore, it seems rather unlikely that the ground state heterogeneity of HIA could be due to self-aggregation phenomena. Conversely, these results point out to the existence of two different ground state HIA species. Hereafter, we will name Q to the HIA isomer emitting the F1 fluorescence band at short wavelengths with a decay time of 4±1 ns and Z to the other isomer emitting the F2 fluorescence band at long wavelengths with a decay time of 11±1 ns.

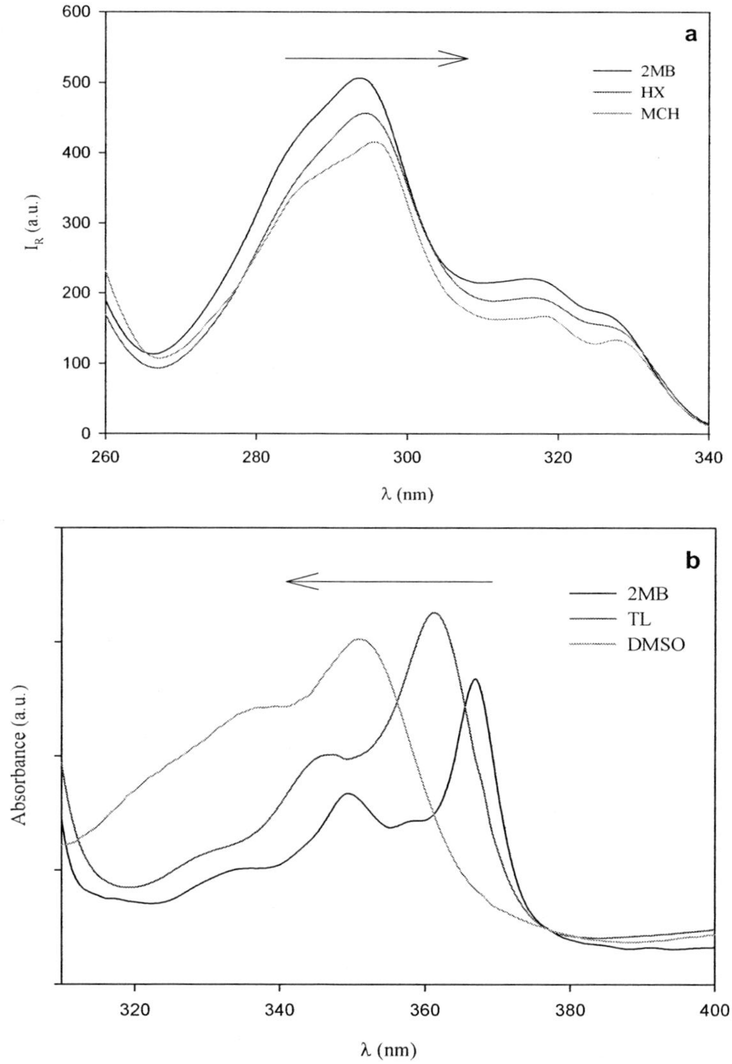

Figure 5. Solvatochromic shifts of the Q (a) and Z (b) excitation spectra.

The close correspondence between the HIA absorption spectrum and the Z excitation spectrum indicates that the proportion of the Q isomer in the 2MB solutions of HIA is very small. Furthermore, the proportion of this isomer apparently diminishes as the solvent polarity increases. Thus, on going from 2MB to HX, CHX and MCH, the intensity of the F1 band of the HIA fluorescence spectra attributed to the Q isomer emission progressively diminishes, results not shown, and practically disappears in TL and DMSO. Regarding the influence of the solvent polarity on the HIA spectra, it is also interesting to highlight the different solvatochromism of the excitation spectra of the Q and Z isomers of HIA. Thus, as Figure 5a shows, upon increasing the solvent polarity, the excitation spectrum of the Q isomer shifts to the red and that of Z to the blue, Figure 5b.

To sum up, the above results reveal that, as BCA and DCA, HIA also shows ground state isomerism and dual fluorescence emission in low polar aprotic solvents. However, as we discussed in previous papers [14-16], the ground state isomerism of the carboline anhydrobases cannot be accommodated into the classical models of isomerism. Thus, it is difficult to imagine isomeric structures resulting from simple structural modifications of the carboline anhydrobase rings without disrupting the ring aromaticity. Moreover, these rather unlikely partially aromatic structures could hardly explain the spectral and photophysical properties of the carboline anhydrobase isomers.

Therefore, in an attempt to rationalise the intriguing spectral and photophysical properties of the carboline anhydrobases we have postulated the existence of ground state isomers with quinoid and zwitterionic structures [14-16]. This hypothesis implicitly assumes that the electronic changes accompanying the zwitterionic-quinoid transformation render different nuclear configurations and energies for the carboline anhydrobase rings. Accordingly, the proposed quinoid and zwitterionic structures must be considered different chemical species and not, as long time ago Armitt and Robinson [20] suggested, resonance hybrids without physical significance. In this sense, the isomerism of the carboline anhydrobases could be considered as a particular case of valence isomerism in which the pyrrolic and pyridinic nitrogen atoms of the carboline ring inter change their electronic configurations. Thus, in the zwitterionic structure, the pyridinic and pyrrolic nitrogen atoms contribute to the π-aromatic system of the carboline ring with one and two electrons, respectively. Conversely, in the quinoid structure the reverse occurs.

On this basis, we think reasonable to assign the quinoid and zwitterionic structures depicted in Scheme 3 to the Q and Z ground state isomers of HIA, respectively. Accordingly, the different spectral and photophysical properties of the Q and Z isomers would be related with the distinct electronic transitions originating their different electronic configurations. Thus, the red shift of the Q excitation spectra upon increasing the solvent polarity suggests a ππ* transition, whereas the blue shift of the Z isomer points out to a nπ* transition. A similar behaviour was previously observed for the isomers of BCA [16]. Therefore, it can be deduced that the nπ* transition of the Z isomer of HIA involves the no bonding electron pair of the pyrrolic nitrogen atom. This result clearly indicates that the electronic density increase experienced by the pyrrolic nitrogen atom on going from the Q to the Z structures has a destabilising effect on the energy of its no bonding electron pair. Therefore, this result supports our presumption that the quinoid-zwitterionic isomerisation process produces profound changes on the electronic configuration and energy of the carboline ground state.

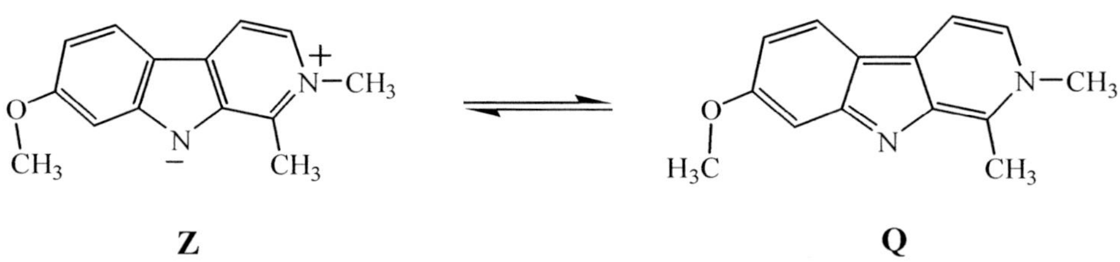

Scheme 3.

On the other hand, the marked differences between the fluorescence spectra of the Q and Z isomers suggest that their singlet excited emitting states are also of very different nature. Thus, as Figure 6 shows, the emission band of the Q isomer in 2MB is mirror image and 800 cm-1 red-shifted with respect to its low energy excitation band. This behaviour is typical of molecules emitting from locally excited states reached from vertical Franck-Condon transitions. It indicates that neither the geometry nor the electronic structure of the molecule importantly change upon excitation. Conversely, the features of the emission band of the Z isomer and, particularly, its long Stokes shift, point out to this isomer experiences much more profound geometrical and electronic changes upon excitation.

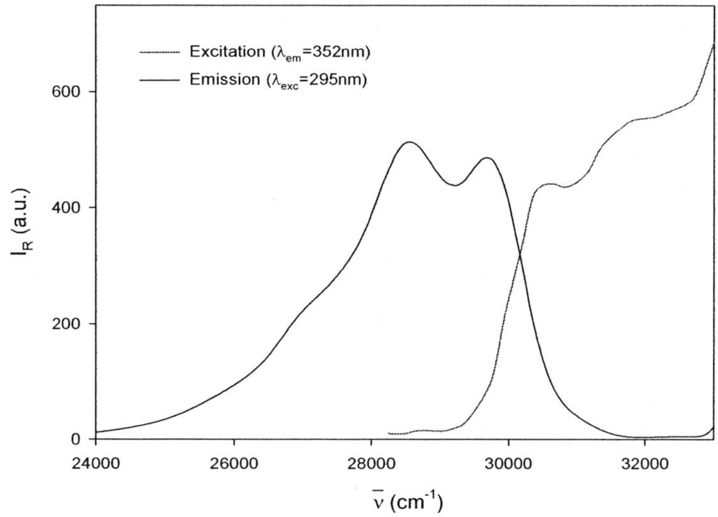

Figure 6. Stokes shift of the F1 emission band of the Q isomer of HIA in 2MB.

The different nature of the Q and Z emitting excited states, could be explained assuming, as in previous papers [14,16], that the coupling of the La and Lb states of the HIA isomer gives rise to two low-lying singlet excited states: a locally excited state, LE, and an intra molecular charge-transfer state, ICT. This charge-transfer process would logically involve the pyrrolic and pyridinic nitrogen atoms in the carboline skeleton. Thus, as depicted in Scheme 4, the absorption of a light pulse promotes the ground state of the HIA isomers to their corresponding LE excited states via a Franck-Condon vertical transition. The rapid charge redistribution after excitation results in an electronic potential surface possessing two energy minima, one of them shifted toward the nuclear configuration of the ICT state. As a result, the excited molecule can emit F1 fluorescence at short wavelengths from its LE excited state, or it moves along the excited state energy surface toward the ICT equilibrium position and emits F2 fluorescence at long emission wavelengths. The relative contributions of these emissions to the fluorescence spectrum will depend on the rate constant and, consequently, on the energy barrier of the LE-ICT process.

Accordingly, because the Q isomer only emits from its LE state, it can be inferred that the great height of the LE-ICT energy barrier impedes the LE→ICT conversion. Conversely, in the case of the Z isomer, which only emits from its ICT state, it would be assumed that the extremely fast LE→ICT process takes place practically without energy barrier. The different behaviours of the HIA isomers can be easily rationalized considering that the zwitterionic structure is much more prone to the ICT processes than the quinoid one. Moreover, it is evident that the 7-methoxy group of the HIA skeleton also play a key role in the relative stabilization of the LE-ICT states of the HIA isomers. Thus, differently as for the Q and Z isomers of HIA, the isomers of the parent BCA, lacking of this group, simultaneously emit from both LE and ICT states.

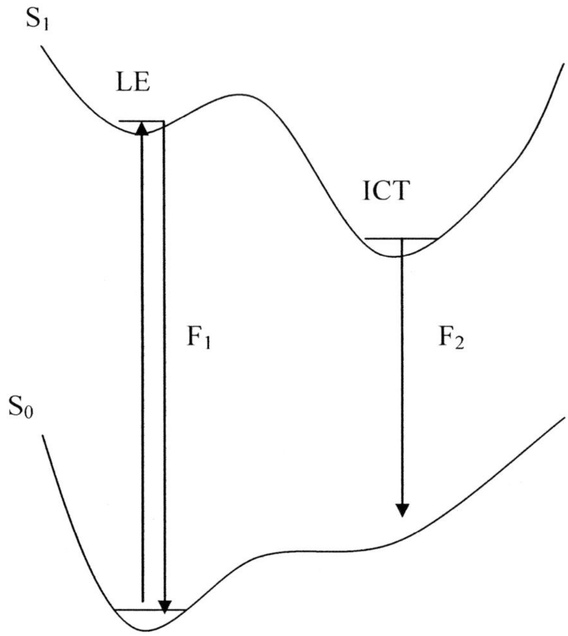

Scheme 4.

To sum up, the intriguing spectral and photophysical properties of HIA in low polar aprotic solvents can be tentatively rationalised through the mechanism depicted in Scheme 5. Accordingly, the HIA ground state in these solvents is heterogeneous, coexisting in equilibrium two different isomeric species, namely Q and Z, whose presumed quinoid and zwitterionic structures are depicted in Scheme 3. Upon light absorption, the Q isomer emits short wavelength fluorescence from the LE excited state resulting from a vertical Franck-Condon transition, with a lifetime around 4 ns. The Z species, once in its LE excited state, quickly relaxes to an internal charge transfer state, ICT, emitting long wavelength fluorescence with a lifetime around 11 ns.

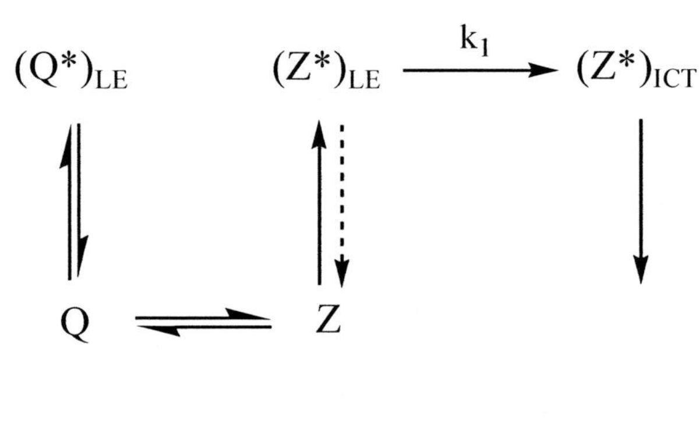

Scheme 5.

ACKNOWLEDGMENTS

We gratefully acknowledge financial support from the Dirección General Científica y Técnica MEC, CTQ2006-13539, and Junta de Andalucía, 20005/FQM-386.

REFERENCES

[1] Abramovitch R.A.; Spencer I.D. *Adv Heterocycl Chem* 1964, 3, 79-207.

[2] Balón M.; Muñoz M.A.; Guardado P.; Hidalgo J.; *Carmona C. Trends Photochem Photobiol* 1994, 3, 117-138.

[3] Balón M.; Hidalgo J.; Guardado P.; Muñoz M.A.; *Carmona C. J Chem Soc Perkin Trans* 2, 1993, 99-104.

[4] Carmona C.; Galán M.; Angulo G.; Muñoz M.A.; Guardado P.; *Balón M. Phys Chem Chem Phys* 2000, 2, 5076-5083.

[5] Dias A.;, Varela A.P.; Miguel M.G.; Maçanita A.L.; Becker R.S. *J Phys Chem* 1992, 96, 10290-10296.

[6] Sakuros R.; Ghiggino K.P. *J Photochem* 1982, 18, 1-8.

[7] Dias A.; Varela A.P.; Miguel M.G.; Becker R.S.; Burrows H.D.; Maçanita A.L. *J Phys Chem* 1996, 100, 17970-77.

[8] Draxler S.; Lippitsch M.E. *J Phys Chem* 1993, 99, 11493-11496.

[9] Reyman D.; Pardo A.; Poyato J.M.L. *J Phys Chem* 1994, 98, 10408-10411.

[10] Reyman D.; Viñas M.H.; Poyato J.M.L.; Pardo A. *J Phys Chem A* 1997, 101, 768-775.

[11] Carmona C.; Balón M.; Galán M.; Angulo G.; Guardado P.; Muñoz M.A. *J Phys Chem A* 2001, 105, 10334-10338.

[12] Carmona C.; Balón M.; Galán M.; Guardado P.; *Muñoz M.A. Photochem Photobiol* 2002, 76, 239-246.

[13] Chou P.T.; Liu Y.I.; Wu G.R.; Shiabo M.Y.; Yu W.S. *J Phys Chem B* 2001, 105, 10674-10683.

[14] Balón M.; Carmona C.; *Muñoz M.A. Chem Phys* 2004, 302, 13-20.

[15] Balón M.; Muñoz M.A.; *Carmona C. Chem Phys* 2004, 302, 67-77.

[16] Sánchez-Coronilla A.; Carmona C.; Muñoz M.A.; *Balón M. Chem Phys* 2006, 327, 70-76.

[17] Wolfbeis O.; Fürlinger E. *Z Phys Chem Neue Folge* 1992, 129, 171-183.

[18] Khalil G.E.; Mayouf A. Can J Spectrosc 1984, 29, 104-108.

[19] Catalán J. *Handbook of Solvents:* Solvent effects based in pure scales, G. Wypich Ed., Inc. and Chem. Tec. Publishing, Toronto-New York, 2001

[20] Armitt J.W.; Robinson R. *J Chem Soc* 1925, 127, 1604-1608.

In: New Trends in Chemical Physics Research
Editor: James N. Lapherty

ISBN: 978-1-61668-535-5

Chapter 6

CHEMICAL-PHYSICAL PROPERTIES OF BIOPROTECTANT-BIOMOLECULE SYSTEMS

***S. Magazù*[1a], *F. Migliardo*[2a,b], *D. Barreca*[3a], *E. Bellocco*[4a] *G. Galli*[5a] *and G. Laganà*[6a]**

[a] University of Messina, Messina, Italy

[b] University of Lille , CEDEX, France

ABSTRACT

The present work is addressed to investigate the effects of some bioprotectant systems on an enzyme, the ornithine carbamoyltransferase (OCTase), and a protein, lysozyme.

The stabilizing effects of a class of bioprotectant systems, such as trehalose, PEG, NaCl, ornithine, sodium phosphate and ammonium sulphate, on OCTase and lysozyme structure and thermal inactivation are analysed by using small angle neutron scattering (SANS), circular dichroism and UV spectroscopy.

In particular, the behaviour of residual activity of OCTase has been determined as a function of time, while changes in secondary and tertiary enzyme structure are monitored by circular dichroism and UV spectroscopy. Finally SANS allow to study the

---

[1] Corresponding author: Salvatore Magazù, Dipartimento di Fisica, Università di Messina, P.O. Box 55, S.ta Sperone C.da Papardo, 98166 Messina, Italy, smagazu@unime.it, phone: +390906765025; fax: +39090395004

[2] Department of Physics, University of Messina, P.O. Box 55, Sperone 31, 98166 Messina, Italy Laboratoire de Dynamique et Structure des Matériaux Moléculaires, University of Lille I, UMR CNRS 8024 - 59655 Villeneuve d'Ascq CEDEX, France

[3] Department of Organic and Biological Chemistry, University of Messina, P.O. Box 55, Sperone 31, 98166 Messina, Italy

[4] Department of Organic and Biological Chemistry, University of Messina, P.O. Box 55, Sperone 31, 98166 Messina, Italy

[5] Department of Physics, University of Messina, P.O. Box 55, Sperone 31, 98166 Messina, Italy

[6] Department of Organic and Biological Chemistry, University of Messina, P.O. Box 55, Sperone 31, 98166 Messina, Italy

conformational properties of lysozyme as a function of temperature in absence and in presence of trehalose and PEG.

Keywords: bioprotection; activity; denaturation; proteins; enzymes; spectroscopy.

INTRODUCTION

It is well known that extremophiles are organisms living and loving extreme environments [1-9]. They had to adapt to one or several physical-chemical extreme parameters: high temperatures for thermophiles and hyperthermophiles that live near geysers and hydrothermal vents; psychrophiles that grow at temperatures below 15°C, in glacier water and polar seas; halophiles that thrive in hypersaline environments like the Lac Rose in Senegal. Other physical-chemical extreme parameters are, for example, high pressure, high radiation activity, high and low pH, etc…

Extremophiles have evolved biomolecules (proteins, membranes) and molecular mechanisms that ensure metabolic flux is maintained at conditions that would ordinarily be considered abiological.

Life under extreme conditions is limited primarily by the stability of macromolecules and their ability to exert their biochemical functions under nonconventional conditions that normally lead to the collapse of structures or to functional inactivation [6,10-13].

The present work is collocated in this framework with the goal to better understand the mechanisms making the bioprotectant molecules effective in inhibiting thermal denaturation or inactivation.

We have first investigated the ornithine carbamoyltransferase (OCTase) which catalyzes the carbamoylation of L-ornithine to form citrulline in the sixth step of the arginine biosynthesis pathway [14-16]. OCTase has a trimeric structure composed of identical subunits and catalyzes the synthesis of citrulline from ornithine and carbamoyl phosphate with the release of orthophosphate during the first step in the de novo biosynthesis of arginine. Most anabolic OCTases form trimers of about 105 kDa, whereas the anabolic OCTase of *Pyrococcus furiosus*, an aquatic anaerobic hyperthermophiles archaeon first isolated in a hydrothermal vent near Vulcano Island (Messina, Italy), is a dodecamer composed of four catalytic trimers (420 kDa) [17,18].

Biological molecules by hyperthermophiles, living at very high temperature, have different properties: it was found that some proteins are generally denser, having a shorter surface loop length and solvent exposed surface area, whereas some enzymes are generally more rigid in structure to prevent from environmental adverse effect and are stabilized by the pressure, raising its resistance against thermal inactivation [6,10-13]. *Pyrococcus furiosus* has other remarkable mechanisms to protect and proliferate itself: some of its enzymes are tungsten dependent, a very rare element to be found in biological system and it can use a wide range of compounds as a carbon source, such as peptides and carbohydrates [17,18].

The second biomolecule which we have studied is lysozyme, a well known protein which has been extensively studied by theoretical, experimental and simulation methods [19-23]. It has been also pointed out that, in lysozme water solutions, ordered and disordered hydration sites extended over the protein surface alternate, suggesting the presence of a dynamic hydration layer with ionic "flip-flop" occurring between bound waters [21-23]. Small Angle

Neutron Scattering (SANS) measurements revealed that the average interparticle distance increases in lysozyme unsaturated solutions due to the increased interaction between molecules progressed as the salt concentration, while in supersaturated solutions crystallisation processes are activated [23].

As it is known, the analysis of thermodynamic parameters offers some useful information on protein stabilization process. In this frame cosolutes have a remarkable effect on the interplay between solvent entropy and protein stability, influencing chemical potential of both protein and solvent [24-27].

In the present work the investigated cosolutes are bioprotectant systems, such as a glass forming disaccharide, trehalose, a polymer, PEG, and some salts, such as NaCl, ornithine, sodium phosphate and ammonium sulphate.

It has been demonstrated that the bioprotective effectiveness of trehalose in protein stabilization is higher than that achieved using other sugars such as maltose, sucrose, glucose or fructose [28-31]. This extraordinary effectiveness could be due to a large hydrated volume in comparison to other sugars and consequently its major exclusion from the protein hydration shell, i. e. less trehalose is necessary to decrease the solvation layer of proteins and thus to stabilize and modulate enzyme activity [28-30].

Some experimental findings obtained by several spectroscopic techniques indicate that the structural and the dynamical properties of water result perturbed by disaccharides, and in particular by trehalose [32-43]. The obtained findings are consistent with the picture of a disaccharide perturbing effect on the H_2O tetrabonded network of water molecules [32-37]. Furthermore Elastic Neutron Scattering and viscosity measurements allowed for the "strongest" character in the Angell's classification scheme of trehalose in comparison with the other disaccharides to be pointed out [38-43].

In this paper the effects of the presence of bioprotectant systems on the stabilization mechanisms of OCTase and lysozyme have been studied by using complementary spectroscopic techniques, such as SANS, circular dichroism and UV spectroscopy.

EXPERIMENTAL SECTION

Thresher shark (*Alopias vulpinus)* liver was obtained from animals died for natural causes along the Italian coast. δ-N-(phosphonacetyl)-L-ornithine (δ-PALO) was synthesized and further purified essentially as described by De Martines et al. [44].

Chicken-egg-white lysozyme, trehalose, bovine albumin, citrulline, L-ornithine hydrochloride, carbamoyl phosphate dilithium salt, N-ethylmorpholine, MES, diethanolamine, diacetylmonoxime, thiosemicarbazide, tris (hydroxymethyl) aminomethane, ammonium sulphate, sodium phosphate, guanidine hydrochloride, PEG, sucrose, maltose and glycerol were purchased from SIGMA (St. Louis, MO), epoxy-activated Sepharose 6B was obtained from Pharmacia LKB.

The enzyme was purified as described previously [16] with minor modification. 4 g of thresher shark (*Alopias vulpinus*) liver was homogenate with 1:4 (w/v) of Tris-HCl 50 mM pH 7.8 with1 mM β-mercaptoethanol. Homogenate was heated with stirring to 60°C for 1 minute in a water bath and immediately cooled in an ice bath. Denatured protein and membrane fragment were removed by centrifugation (10000 rpm x 10 minutes at 4°C). The supernatant was mixed with the affinity adsorbent δ-PALO immobilized on epoxy-activated Sepharose 6B. The column was equilibrated with 50 mM Tris-HCl (pH 7.8) and incubated for 30 minutes at 27°C. The mixer was washed with 80 ml of Tris-HCl (pH 7.8) and 80 ml of the same buffer plus 50 mM KCl. The enzyme was eluted with 50 ml of 10 mM carbamoyl phosphate in 50 mM Tris-HCl (pH 7.8). Eluted was dialysed in tubings (MWCO 12-14 kDa) overnight at 4° C against Tris-HCl 50 mM pH 7.8 with 1 mM β-mercaptoethanol and concentrated with Amicon Ultrafiltration Cell using Amicon YM-10 membrane. The protein purified was stored at −20°C.

Enzyme purity vas monitored by SDS/PAGE electrophoresis according to Laemli [45]. Protein determination was performed according to Bradford [46] using bovine serum albumin (BSA) as standard.

OCT activity was measured as the rate of citrulline formation from ornithine and carbamoyl phosphate [47]. The reaction mixtures, in a final volume of 1.0 ml, contained the tri-buffer (diethanolamine/MES/N-ethylmorpholine 0.051 M/0.1 M/0.051 M) pH 7.8, 10 mM ornithine, 10 mM carbamoyl phosphate a 20 μl of enzyme solution (6 μM). After incubation for 5 minutes at 37°C, the reaction was terminated by the addition of 2.0 ml acid reagent (3.7 g of antipyrine,2.5 g of ferric ammonium sulphate in 500 ml H_2O, 250 ml of concentrated H_2SO_4 and 250 ml 85% H_3PO_4). One millilitre of 0.4% diacetylmonoxime was added and the samples were incubated for 15 minutes in a boiling water bath. Formed citrulline was detected through spectroscopic analysis at 464 nm. One unit of OCT activity was defined as the amount of the enzyme that catalyzes the formation of 1 μmole of citrulline per minute at 37 °C.

The absorption spectra were obtained by using a DU/70 spectrophotometer with an integrated temperature programmer at 60°C, both in the absence and presence of selected additives. Enzyme solution (~0.35 mg/ml of protein in 50 mM Tris-HCl buffer pH 7.8) with 2 mM β-mercaptoethanol was heating in absence and in presence of 1.0 M NaCl, 0.5 M trehalose, 0.025 M sodium phosphate, 0.2 M ammonium sulphate, 0.05 M ornithine and the absorbance values were recorded in the range from 240-310 nm after 20 min of incubation. Enzyme solution (~0.35 mg/ml of protein in 50 mM Tris-HCl buffer pH 7.8) without β-mercaptoethanol was also analyzed in the same condition. Absorption spectrum shows a number of absorption bands corresponding to structural groups within the molecule. Proteins absorb ultraviolet or visible light, yielding Lambert-Beer's Law. The energies noted, at these wavelengths, are sufficient to promote or excite a molecular electron to a higher energy orbital. The main protein absorbing groups, referred to as chromophores, are tryptophan (λ_{max}=280 nm), tyrosine (λ_{max}=274 nmm), and phenylalanine (λ_{max}=250 nmm), involving $\pi \rightarrow \pi^*$ transitions. Perturbations at these wavelengths are proportional to an alteration of hydrophobic environments that consequently bring to the loss of tertiary protein structure.

The thermal stability of OCT in the presence and absence of selected additives was determined by the inactivation rate constant (kr) as a function of temperature at 60°C, 64°C, 66°C, 68°C and 70°C. 200μl (0.5 mg/ml) of enzyme solution were added to 0.6 ml of 50 mM Tris-HCl buffer (pH 7.8) with 2 mM β-mercaptoethanol and kept in a water bath at the desired temperature values. 10μl of enzyme solution were withdrawn at appropriate intervals of time and cooled in an ice bath, diluted 50-fold and assay for determination of OCT activity. The residual activity was measured and expressed as a percentage of initial activity. The temperature dependence of kr was analyzed from Arrhenius plot; the activation energy (E_a)

was obtained from the slope of the plot. Activation enthalpy (ΔH^*) was calculated according to the equation:

$$\Delta H^* = E_a - RT \quad (1)$$

where R is the universal gas constant and T is the absolute temperature.

The values for free energy of inactivation (ΔG^*) at different temperatures were obtained from the equation:

$$\Delta G^* = -RT \ln\left(k_T h/kT\right) \quad (2)$$

where h is the Planck constant and k is the Boltzmann constant.

Activation entropy (ΔS^*) was calculated from:

$$\Delta S^* = \left(\Delta H^* - \Delta G^*\right)/T \quad (3)$$

Circular dichroism measurements were performed by using a Jasco J-810 automatic recording spectropolarimeter. The measurements were performed at different temperature values. The scans were recorded twice. Each spectrum was corrected for baseline by subtracting the spectral contribution of the buffer solution and buffer plus additives.

The far-UV spectra were recorded between 190-260 nm using a quartz cell of 1 mm. The protein concentration was 0.16 mg/ml. The samples were prepared in 15 mM Tris-HCl buffer (pH 7.8). For the thermal unfolding measurements, data were collected at 222 nm every second at a heating rate of 1°C/min.

SANS experiment has been performed by using the LOQ spectrometer at the ISIS Facility (Rutherford Appleton Laboratory, Chilton, UK) on lysozyme/D_2O mixtures, lysozyme/D_2O/trehalose mixtures and lysozyme/D_2O/PEG mixtures for temperature values of T=37°C and T=60°C. Lysozyme was dissolved in 40 mM Na acetate buffer at pH 4.25 in D_2O.

Measurements have been performed for different contrast values (20÷80%D_2O/80÷20%H_2O and 100%D_2O). As an example in this paper we show results relative to 100%D_2O solutions. The contrast variation technique collecting data at different D_2O/H_2O molar ratio has been employed in order to determine the protein scattering density length. Taking into account that the amount of labile hydrogen atoms in lysozyme in the native state is 193 and that the scattering length b in D_2O is $5.42\cdot10^{-3}$ nm, the obtained mean coherent scattering length density of the protein ρ values are ρ=$0.39\cdot10^{10}$cm^{-2} at T=310K and ρ=$0.29\cdot10^{10}$cm^{-2} at T=333K.

The Q-range covered by the LOQ spectrometer in this experiment is from 0.007Å^{-1} to 0.287Å^{-1}. Incoming neutrons are monochromatized by a mechanical velocity selector with variable wavelength from 2.2 to 10.0 Å, the wavelength resolution (FWHM) being $8\% < \Delta\lambda/\lambda < 18\%$.

Corrections for inelastic effects, multiple scattering, absorption, sample container and background have been taken into account. For the analysis, the COLETTE software package has been used [48].

RESULTS

We first focus our attention on the study of OCTase activity and structure in presence of cosolutes. The thermal inactivation kinetics of OCT was analyzed in presence of trehalose, ammonium sulphate, sodium phosphate, NaCl and ornithine in the temperature range of 60-70°C. Figure 1 shows the thermal inactivation of OCT. Enzymatic solutions were incubated at the above temperature in absence and in presence of 0.2 M ammonium sulphate, 0.5 M of trehalose, 1.0 M of NaCl, 0.05 M of ornithine and 0.025 M phosphate. By observing Figure 1, it is evident that the enzyme activity follows an exponential decay at all the investigated temperature values.

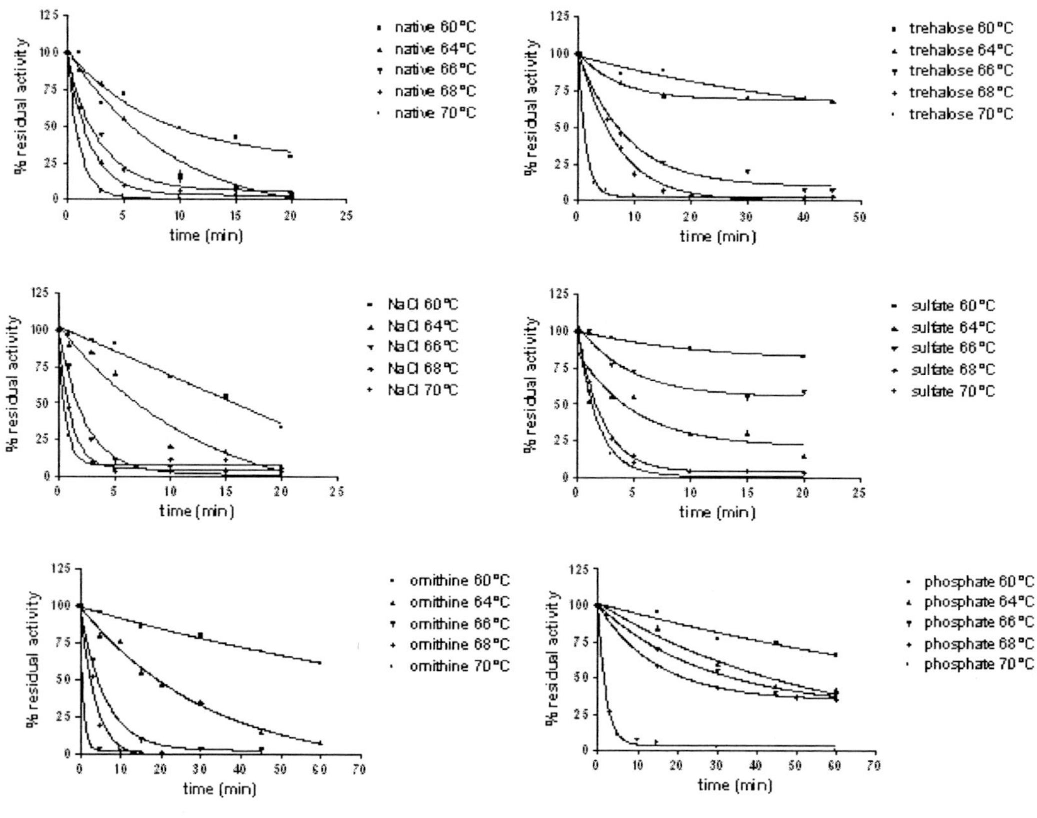

Figure 1. Thermal inactivation of OCT in absence and in presence of 0.2 M ammonium sulphate, 0.5 M of trehalose, 1.0 M of NaCl, 0.05 M of ornithine and 0.025 M phosphate in the temperature range of 60-70°C.

The Arrhenius plots, shown in Figure 2, are linear in the investigated temperature range and no break out point is observed. From these plots the activation parameters, such as free energy (ΔG^*), enthalpy (ΔH^*), and entropy (ΔS^*) of activation, can be evaluated.

The obtained E_a values are: 64.92 ± 2.54, 106 ± 5.34 , 93.13 ± 3.82, 78.30 ± 3.13, 117.09 ± 4.97 and 93.07 ± 3.75 Kcal mol^{-1} for enzyme in absence and in presence of 0.2 M ammonium sulphate, 0.5 M of trehalose, 1.0 M of NaCl, 0.05 M of ornithine and 0.025 M phosphate. The magnitude of free energy activation reflects the effectiveness of relative stabilization due to the various additives.

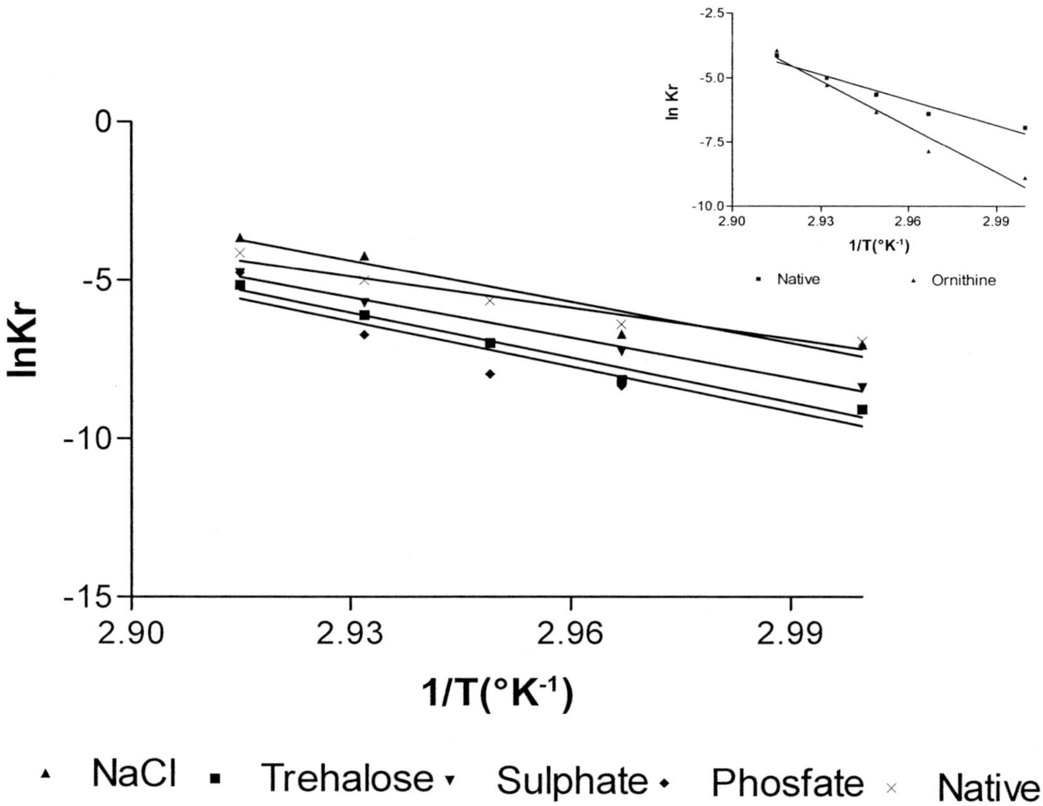

Figure 2. Arrhenius plots of OCT in absence and in presence of 0.2 M ammonium sulphate, 0.5 M of trehalose, 1.0 M of NaCl, 0.05 M of ornithine and 0.025 M phosphate.

Table 1 reports the thermodynamic parameters obtained at different temperature values. At 64°C ΔS* value increases from 119.4 without additives to 199.6, 159.7, 239.6, 199.1, 271.3 cal deg $^{-1}$mol in presence of trehalose, NaCl, ammonium sulphate, sodium phosphate and ornithine, respectively.

Also the ΔG* value (24 Kcal mol^{-1}) increases of about 1.1, 0.58, 1.29, 098 Kcal mol^{-1} with trehalose, ammonium sulphate, sodium phosphate, ornithine, but it remains unchanged in the presence of NaCl.

The ΔH* value (64.25 cal mol^{-1}) for the native enzyme is highly increased in the presence of NaCl, trehalose, ammonium sulphate, sodium phosphate and ornithine with the following results: 77.63, 92.46, 105.33, 92.4 and 116.42 cal mol^{-1}, respectively.

The different variation of these parameters points out the different role played by these cosolutes in stabilization process depending on their chemical–physical properties.

Table 1 Variation of ΔH* (Kcal mol^{-1}), ΔG* (Kcal mol^{-1}), ΔS* (Kcal mol^{-1}) and Kr (x 10^{-4} s^{-1}) values as a function of temperature and additives in solution.

	T (°C)	Enz. Nat.	Trehalose 0.5 M	$(NH_4)^+{}_2SO_4$ 0.2 M	NaCl 1.0 M	Phosphate 0.025 M	Ornithine 0.05 M
ΔH*	60	64.26	92.47	105.34	77.64	92.41	116.43
	64	64.25	92.46	105.33	77.63	92.40	116.42
	66	64.25	92.45	105.32	77.62	92.39	116.41
	68	64.25	92.45	105.32	77.62	92.39	116.41
	70	64.25	92.45	105.32	77.62	92.39	116.41
ΔG*	60	24.10	25.47	25.61	24.12	25.47	25.36
	64	24.00	25.17	24.58	23.78	25.29	24.98
	66	23.65	24.54	23.62	23.75	25.37	24.10
	68	23.36	24.10	23.45	22.83	24.51	23.55
	70	22.91	23.60	23.35	22.57	23.33	22.78
ΔS*	60	120.7	201.2	239.4	160.6	201.0	273.4
	64	119.4	199.6	239.4	159.7	199.1	271.3
	66	119.7	200.3	241.0	158.9	197.7	271.3
	68	119.8	200.4	240.0	160.6	199.0	272.3
	70	120.4	200.7	238.9	160.4	201.3	272.9
Kr	60	9.80	1.10	0.93	8.87	1.15	1.37
	64	16.70	2.89	7.04	23.20	2.42	3.85
	66	35.01	9.34	36.68	29.90	2.69	17.86
	68	66.78	22.20	57.70	146.20	12.03	50.23
	70	158.74	57.70	83.54	260.80	84.95	192.56

Figure 3 shows the UV-spectra of enzyme and enzyme-additives. The absorbance values were recorded in the range from 240-310 nm after 20 min by using a DU/70 spectrophotometer with an integrated temperature programmer.

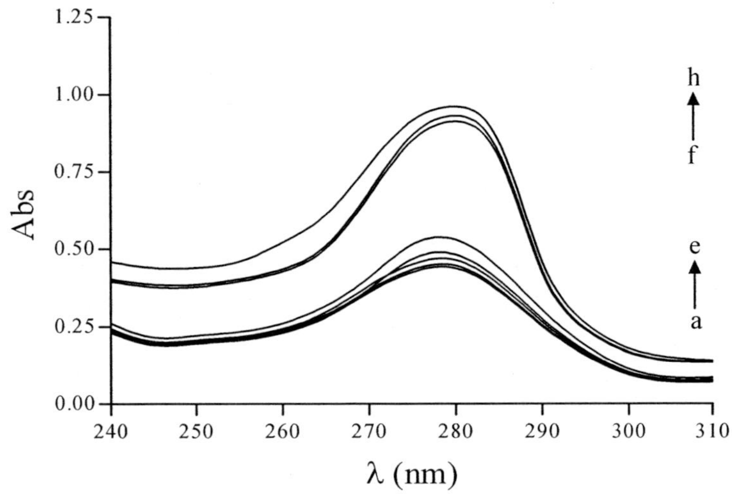

Figure 3. UV-spectra of enzyme and enzyme-additives: a) enzyme at 25°C; b,c,d,e,f) enzyme incubated at 60°C plus: ammonium sulphate; sodium phosphate; ornithine; trehalose; sodium chloride. g,h) enzyme incubated at 60°C with or without 2-mercaptoethanol.

The following features can be observed. Enzyme at 25°C shows a well defined spectrum with a sharp band between 260 and 280 nm, mainly due to tryptophan and tyrosine residues. UV spectroscopy at 60°C, after 20 min of incubation, shows a significant absorbance increase in the spectrum of enzyme alone between 260-280 nm, due to an increased exposure of hydrophobic groups. True to form, the unfolding without a reducing agent is faster, demonstrating the importance of SH groups for the enzyme integrity and activity. Changes in exposure of these buried sites bring also to the loss of enzyme activity and of all tertiary protein structure. At the same wavelength range, no variation is revealed in absorbance value in the presence of 0.2 M ammonium sulphate, 0.025 M sodium phosphate and 0.05 M ornithine. In presence of 0.5 M trehalose an intermediate trend is highlighted, with a little increase between 260-280 nm. Finally the absorbance variation is remarkable in presence 1.0 M NaCl.

Far-UV CD spectra, shown in Figure 4, have been collected during thermal inactivation of OCT. The data analysis of native enzyme reveals the presence of typical α–helical, β-sheet, turn and random structure in accordance with the data obtained by structure analysis of other OCTase [49,50].

OCT CD spectrum shows a well-defined ellipticity value at 222 nm and a positive band around 195 nm. In particular, the mean residue ellipticity at 222 nm is used as an indicator of protein α-helix content [51], since an increase in the ellipticity indicates that the α-helix content decreases. The CD spectrum of native enzyme at room temperature and at 60°C does not show significant changes in the first minutes, but after 1 hour the original secondary structure was almost completely lost.

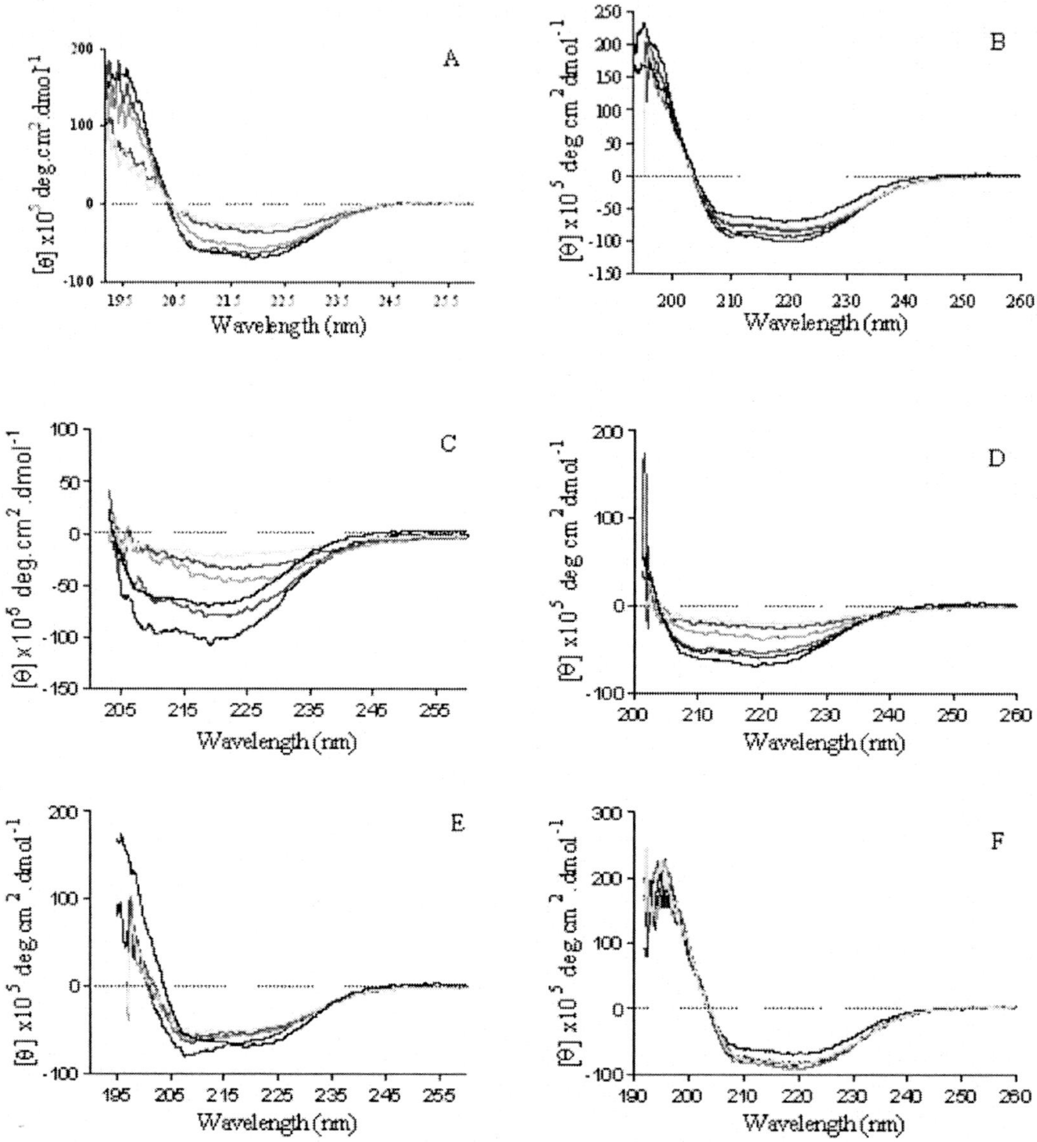

Figure 4. Far-UV CD spectra during thermal inactivation of OCT. Enzyme (0.16 mg/ml) a) in absence and b) in presence of 0.2 M ammonium sulphate, c) 0.5 M of trehalose, d) 1.0 M of NaCl, e) 0.05 M of ornithine and f) 0.025 M phosphate at the temperature of 60°C. Each spectrum was recorded at the following time: 0, 10, 20, 40 and 60 minutes.

Also at 195 nm an intensity decrease as a function of the incubation time was observed, showing the tendency to acquire the spectrum of a non-structured polyptide. Enzyme spectra show also the presence of isodichroich point around 203 nm, revealing a two state transition reaction process from folding to unfolding state.

The enzyme incubation with trehalose, NaCl or ornithine solution shows both changes in the shape of CD spectrum and in molar ellipticity, while in the presence of sodium phosphate and ammonium sulphate very little changes are detected, also after 1 hours of incubation at 60°C, due to the formation of a more compact structure.

Now we analyse the lysozyme behaviour as a function of temperature in presence of cosolutes.

In a SANS experiment, the position of each nucleus can be written taking into account the centre of mass of the molecule $\underline{R}_n$ and results $\underline{r}_{i(n)}=\underline{R}_n+\underline{R}_{i(n)}$. The coherent scattering cross-section can be written as [52]:

$$\left(\frac{\mathrm{d}\sigma}{\mathrm{d}\Omega}\right)_{coh} = \left\langle \sum_{n,m} \exp i\underline{Q}\cdot(\underline{R}_n - \underline{R}_m) \sum_{i(n)} \sum_{j(m)} b_i b_j \exp\left[i\underline{Q}\cdot(\underline{R}_{i(n)} - \underline{R}_{j(m)})\right]\right\rangle \tag{4}$$

Often one can apply the so called "decoupling approximation" which, assuming no correlation between position and orientation of the molecules, gives:

$$\left(\frac{\mathrm{d}\sigma}{\mathrm{d}\Omega}\right)_{coh} = N\Delta(\underline{Q}) + N\left\langle\left|P_i(\underline{Q})\right|\right\rangle^2 S(\underline{Q}) \tag{5}$$

where $\Delta(\underline{Q}) = \left\langle P(\underline{Q})^2\right\rangle - \left\langle\left|P_i(\underline{Q})\right|\right\rangle^2$ refers to the deviation from sphericity and monodispersity of the samples, N is the number concentration of scattering centres, $S(\underline{Q})$ is the static structure factor and $\left\langle\left[P(\underline{Q})\right]^2\right\rangle$ is the form factor which contains information about the size, the shape and the internal structure of the molecule [52].

For rigorous monodispersed spheres $\left\langle P(\underline{Q})^2\right\rangle = \left\langle\left|P_i(\underline{Q})\right|\right\rangle^2 = \Phi(\underline{Q})$ and the scattering cross section per unit volume becomes:

$$\left(\frac{\mathrm{d}\sigma}{\mathrm{d}\Omega}\right)_{coh} = N\Phi(\underline{Q})S(\underline{Q}) \tag{6}$$

For Q values smaller than the inverse of the characteristic dimensions of the particle, the Guinier relation allows to get information about the size:

$$\Phi(Q) = \Delta\rho^2 (V')^2 \exp\left(-\frac{Q^2 R_g^2}{3}\right) \tag{7}$$

where V' is the volume of the particle and R_g is the gyration radius [52].

Figure 5 shows the intensity profiles *I(Q)* as a function of the exchanged wave-vector *Q* of lysozyme/D_2O solution for T=37°C and T=60°C, together with the Guinier and Zimm plots.

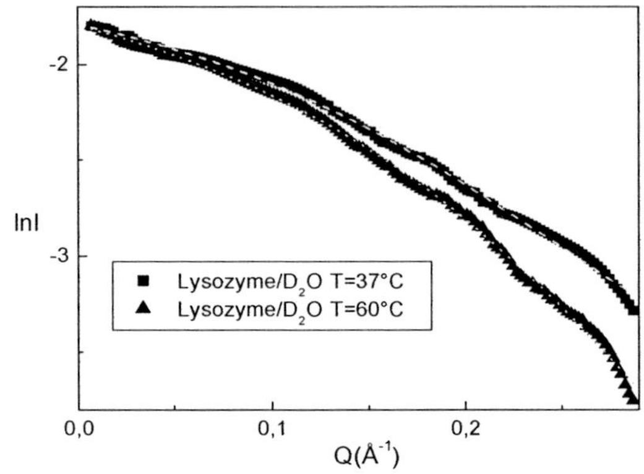

a)

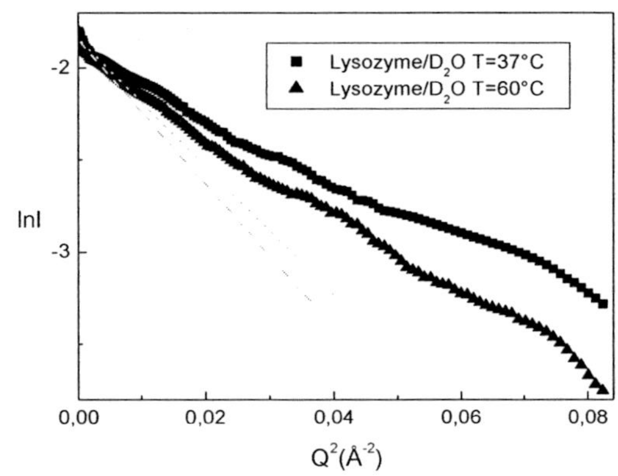

b)

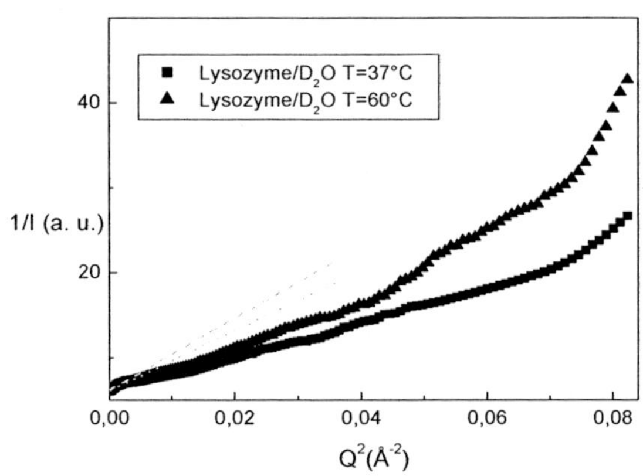

c)

Figure 5. SANS intensity profiles as a function of the exchanged wave-vector Q for lysozyme/D2O solution for T=37°C and T=60°C. In a) the dots represent the experimental values while the grey lines are the result of the fitting procedure using a prolate ellipsoidal model; b) and c) are the Guinier and Zimm plots.

The fitting procedure has been firstly performed using a prolate ellipsoidal model, where the form factor, normalized by the particle volume and averaged over all possible orientations of the ellipsoid, is expressed by:

$$P(Q) = \frac{const.}{V_{ell}} (\rho_{ell} - \rho_{solv})^2 \int_0^1 f^2 \left[Qr_0 \left(1 + x^2 (v^2 - 1)\right)^{1/2} \right] dx + BKG \tag{8}$$

where $f(z) = 3V_{ell} \frac{(\sin z - z\cos z)}{z^3}$, $V_{ell} = (4\pi/3) r_a r_b^2$ and $v = \frac{r_a}{r_b}$.

According to the model, the formula $R_g^2 = 1/5(2r_a^2 + r_b^2)$ allows to estimate the gyration radius values, obtaining the values of R_g=11.8Å and R_g=12.7Å for T=37°C and T=60°C, respectively. Very similar values are obtained by using the Guinier and Zimm procedures, i. e. the values of R_g=11.8Å at T=37°C and of R_g=12.6Å at 60°C by the Guinier plot, and the values of R_g=11.9Å at T=37°C and of R_g=13.0Å at 60°C by the Zimm plot. The values obtained by different procedures are in good agreement with literature data [22] and furnish the confirmation that by increasing temperature lysozyme has lost its native conformation.

The Guinier and Zimm plots of lysozyme in presence of trehalose and PEG are shown in Figure 6, whereas in Table 2 the R_g values obtained for all the investigated mixtures are reported.

Table 2. R_g values obtained for all the investigated mixtures.

Mixtures	*R_g (Å) (ellipsoidal model)*	*R_g (Å) (Guinier)*	*R_g (Å) (Zimm)*
lysozyme/D_2O T=37°C	*11.8*	*11.8*	*11.9*
lysozyme/D_2O T=60°C	*12.7*	*12.6*	*13.0*
lysozyme/D_2O/trehalose T=37°C		*16.2*	*16.9*
lysozyme/D_2O/trehalose T=60°C		*16.4*	*17.0*
lysozyme/D_2O/PEG T=60°C		*19.4*	*20.3*

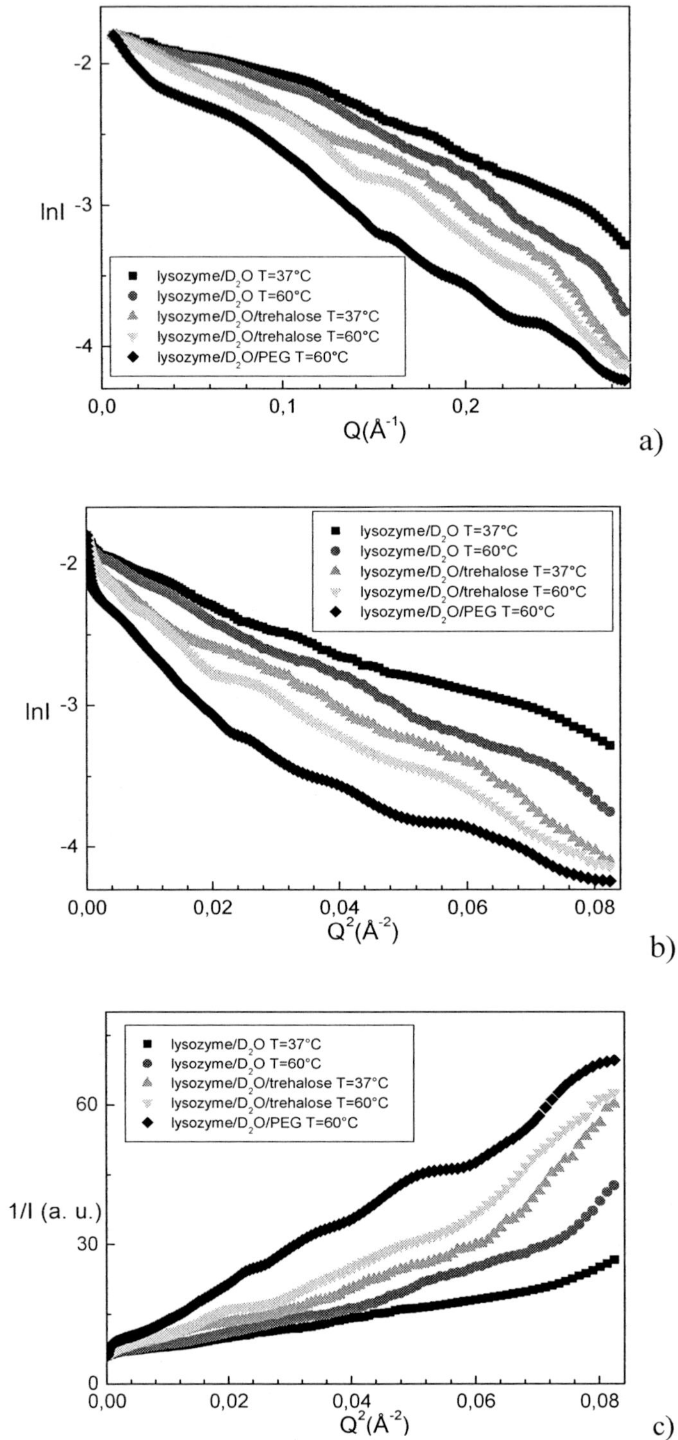

Figure 6. a) SANS intensity profiles as a function Q, b) Guinier and c) Zimm plots for lysozyme/D2O, lysozyme/D2O/trehalose and lysozyme/D2O/PEG solutions for T=37°C and T=60°C.

CONCLUSION

In this work the effects of additives on OCTase stabilization process have been studied as a function of Hofmeister series, ionic strength, preferential exclusion interactions and anion binding reactions [53-55]. The Hofmeister series had origin from the capability of solutions of different salts to precipitate protein. Some salts of the series increase solvent surface tension and decrease the solubility of nonpolar molecules (salting out), enhancing the hydrophobic interaction. By contrast, other salts increase both the solubility of nonpolar molecules (salting in) and the order in water molecules, reducing hydrophobic effect. However, Hofmeister series mechanism is not completely understood. It does not seem to result only from changes in hydrogen-bonding properties of water, but salts could interact directly with proteins (which are charged and have strong dipole moments) and may even bind specifically. Because of their opposite interactions with nonpolar side chain and with the peptide group, Hofmeister ions are protein denaturants or stabilizers.

By taking into account the cosolutes investigated in the present work, □H* and □S* values in presence of each additive are increased in respect to those obtained for native proteins indicating that the enthalpy and entropy of transfer of the denatured proteins from water to additive solution is larger than those for native proteins.

The activation parameters, in the presence of sodium phosphate, NaCl and (NH4+)2SO4, point out the role played by ionic and hydrophobic interactions in the stabilization process of OCT and closely follow the Hofmeister series of anions (phosphate > SO4= > Cl-). A similar trend in stabilization process has been observed by Damodoran [56] and Jensen [57].

More complex is the understanding of OCT stabilization by trehalose, because many different forces seem to be involved. The Ea values vary in the order ornithine > (NH4+)2SO4 > trehalose > phosphate > NaCl. This indicates that the hydrophobic interactions play a crucial role in the stabilization of OCT compared with ionic interactions. The observed Ea values of trehalose and sodium phosphate are very similar, but the stabilization processes are different. Both phosphate and sulphate enhance hydrophobic interaction avoiding the changes in enzyme structure. The same process seem to be involved in the first time of enzyme-trehalose interaction, but after few minutes a change occurs in the secondary enzyme structure. However hydrophobic surface exposure is prevented in trehalose solution and no aggregation process is evident, as it has been shown for NaCl solution where the unfolding process leads to exposure of hydrophobic surface and to their interaction with concomitant loss of OCT activity.

Salts demonstrate very different activity and (NH4+)2SO4 and sodium phosphate has a very strong stabilization activity at very low concentration values, while NaCl became effective only at high concentration values [58].

The kosmotropes used in this work can be divided into two groups which act somewhat differently: i) polyhydroxy compounds (as trehalose) that fit in well with the hydrogen bonding arrangements, stabilize any cluster “flickering” via hydrogen bond rearrangement, as their hydroxyl groups are similarly separated to water-water separations, and ii) zwitterions where the balance among hydrophilicity, hydrophobicity, anionic and cationic characteristics ensures good solubility but only weak net interactions to surrounding water.

Previous findings obtained by neutron scattering experiments have indicated that the dynamical properties of water result drastically perturbed by disaccharides, and in particular by trehalose [36]. These results clearly indicate that disaccharides show a noticeable kosmotrope character, namely the disaccharide-water molecule interaction strength is much higher in respect to that between the water molecules. The high slowing down effect of the diffusive dynamics observed for trehalose is evidently linked to its extraordinary capability to "switch off" the metabolic functions.

The nature of interactions and the mechanism of stabilization by different salt solutions are also evident from CD measurement and UV spectroscopy.

Changes in CD signal reflect changes in the protein structure. The presence of (NH4+)2SO4 and sodium phosphate do not affect the secondary and tertiary structures of the enzyme, and they prevent the loss of protein secondary structure during thermal unfolding as demonstrated by the little changes in the molar ellipticity at 222 nm (proportional to □-helix content). NaCl protects saline bridge and ionic surface interaction, but does not avoid the loss of □-helix. Thus, NaCl stabilizes the side chain interactions, whereas salts that affect hydrophobic interaction are better stabilizers of the backbone conformation.

UV spectroscopy supports this result, showing very little changes in the region between 260-280 nm and indicating the prevention of loss of tertiary structure and of hydrophobic environments in presence of ammonium sulphate, sodium phosphate and ornithine. NaCl does not prevent this loss resulting in the completely loss of tertiary structure, while trehalose shows an intermediate behavior, preventing the loss of the same structure but with a minor effects respect the three compound first mentioned.

Finally, from the observation of the SANS findings, where an increase of the dimensions of the protein in presence of trehalose and PEG is evident, we can conclude that processes originated by the cosolute capability to create a continuous network among the units of lysozyme occur. In addition, by increasing temperature the value of the gyration radius is not changed, this circumstance emphasising the stabilising effect of trehalose and PEG on lysozyme. Therefore trehalose and PEG inhibits significantly the temperature induced swollening of lysozyme.

ACKNOWLEDGMENTS

F. Migliardo gratefully acknowledges UNESCO-L'Oréal for the International Fellowship for Young Women in Life Sciences 2008.

The authors gratefully acknowledge the ISIS facility (Chilton, UK) for the dedicated runs at the LOQ spectrometer.

REFERENCES

[1] Miller, S. L., Lazcano, *A. J Mol Evol* 1995, 41, 689–692.
[2] Forterre, P. *Cell* 1996, 85, 789–792.
[3] Brock, T. D., Brock, K. M., Belly, R. T., Weiss, R. L. *Arch Mikrobiol* 1972, 84, 54–68.
[4] Stetter, K. O. *Extremophiles* 2006, 10, 357–362.
[5] Kashefi, K. Lovley, D. R. *Science* 2003, 301, 934-939.
[6] Arnott, M. A., Michael, R. A., Thompson, C. R., Hough, D. W., Danson, M. J. *J Mol Biol* 2000, 304, 657–668.

[7] Madern, D., Ebel, C., Zaccai, G. *Extremophiles*, 2000, 4, 91-98.
[8] Ginzburg, M., Sachs, L., Ginzburg, B. Z. *J Gen Physiol*, 1970, 55, 187–207.
[9] Ginzburg, M., Ginzburg, B. Z. *In Biomembranes*; Eisenberg, H.; Katchalsi-Katzir, E.; Manson, L. A., Plenum, New York, 1975; 219–251.
[10] Grogan, D. W. *Mol Microbiol* 1998, 28, 1043–1049.
[11] Daniel, R. M., Cowan, D. A. *Cell Mol Life Sci* 2000, 57, 250–264.
[12] Vielle, C., Zeikus, G. *J. Microbiol Mol Biol R* 2001, 65, 1–43.
[13] Britton, K. L., Baker, P. J., Borges, K. M., Engel, P. C., Pasquo, A., Rice, D. W., Robb, F. T., Scandurra, R., Stillman, T. J., Yip, K. S. P. *Eur J Biochem* 1995, 229, 688–695.
[14] Anderson, P. M., Casey, C. A. *J Biol Chem* 1984, 259, 456-62.
[15] Xiong, X., Anderson, P. M. *Arch Biochem Biophys* 1989, 270, 198-207.
[16] Bellocco, E., Di Salvo, C., Laganà, G., Lezzi, U., Tellone, E., Kotyk, A., Galtieri, A. *Physiological research* 2002, 51, 151-8.
[17] Massant, J., Wouters, J., Glansdorff, N. *Acta Crystallogr*. 2003, 59, 2140-2149.
[18] Legrain, C., Villeret, V., Roovers, M., Gigot, D., Dideberg, O., Pierard, A., Glansdorff, N. *Eur. J. Biochem.* 1997, 247, 10461055-10461061.
[19] Bon, C., Lehman, M. S., Wilkinson, C. *Acta Crystallogr.* 1999, D55, 978-984.
[20] Jolles P. *EXS* 1996, 75, 3-12.
[21] Baker, L. J., Hansen, A. M. F., Rao, P. B., Bryan, W. P. *Biopolymers* 1983, 22, 1637-1643.
[22] Svergun, D. I., Richard, S., Koch, M. H. J., Sayers, Z., Kuprin, S., Zaccai, G. *Proc. Natl Acad. Sci. USA* 1998, 95, 2267-2272.
[23] Minezaki, Y., Niimura, N., Ataka M., Katsura, T. *Biophys. Chem*. 1996, 58, 355-363.
[24] Ye, W. N., Combes, D., Manson, P. *Enzyme Microb Technol* 1988, 10, 498–502.
[25] Appleton, B., Gibson, T. D., Woodward, J. R. *Sensors and Actuators: B Chemical* 1997, 43, 65-69.
[26] Baldwin RL. How Hofmeister ion interactions affect protein stability. *Biophys J* 1996; 71: 2056–63.
[27] Donna, B., Wright, D. D., Banks, J. R., Lohman, J. L., Hilsenbeck, L., Gloss, M. *J Mol Biol* 2002, 323, 327–44.
[28] Sola-Penna, M., Meyer-Fernandes, J. R. *Z Naturforsch* 1994, 49C, 327-330.
[29] Sola-Penna, M., Meyer-Fernandes, J. R. *Z Naturforsch* 1996, 51, 160-164.
[30] Sola-Penna, M., Meyer-Fernandes, J. R. *Archives of Biochemistry and Biophysics* 1998, 360, 10–14.
[31] Lee, J. C., Timasheff, S. N. *J Biol Chem* 1981, 256, 7193-7201.
[32] Branca, C., Faraone, A., Magazù, S., Maisano, G., Migliardo, F., Migliardo, P., Villari, V. *Rec. Res. Develop. in Phys. Chem*. 1999, 3, 361-403.
[33] Branca, C., Magazù, S., Maisano, G., Migliardo, F. *Rec Res Develop in Phys Chem* 2002, 6, 35-73.
[34] Magazù, S., Migliardo, F., Ramirez-Cuesta, A. J. *Journal of Royal Society Interface* 2005, 2, 527-531.
[35] Lerbret, A., Bordat, P., Affouard, F., Descamps, M., Migliardo, F. *J Phys Chem B* 2005, 109, 11046-11053.
[36] Magazù, S., Migliardo, F., Telling, M. T. F. *J Phys Chem* B 2006, 110, 1020-1026.
[37] Magazù, S., Migliardo, F., Ramirez-Cuesta, A. *J. Biophys Chem* 2007, 125, 138-142.

[38] Branca, C., Magazù, S., Maisano, G., Migliardo, F., Romeo, G. *J Phys Chem B* 2001, 105, 10140-10145.
[39] Magazù, S., Maisano, G., Migliardo, F., Mondelli, C. *Biophys J* 2004, 86, 3241-3249.
[40] Magazù, S., Migliardo, F. *Chemical Physics Research Journal* 2007, 1, 37-43.
[41] Varga, B., Migliardo, F., Takacs, E., Vertessy, B., Magazù, S., Mondelli, C. *Chem Phys* 2008, 345, 250-256.
[42] Magazù, S., Migliardo, F., Telling, M. *T. F. Food Chemistry* 2008, 106, 1460-1467.
[43] Magazù, S., Maisano, G., Migliardo, F., Benedetto, A. *J Phys Chem B,* 2008, 112, 8936-8942.
[44] De Martines, M. L., Mcintyre, P., Hoogenraad, N. *A. Biochem Int* 1981, 3, 371-378.
[45] Laemmli, U. K. *Nature,* 1970, 227, 680-685.
[46] Bradford, M. M. *Analytical Biochemistry* 1976, 72, 248-254.
[47] Lusty, C. J., Jilka, R. L., Nietsch, E. H. *J Biol Chem* 1979, 254, 10030-10036.
[48] King, S. M. *Using Colette* 2004, Rutherford Appleton Laboratory, Chilton, UK.
[49] Shi, D., Morizono, H., Yu, X., Tong, L., Allewell, N. M., Tuchaman, M. *Biochem J* 2001, 354, 501-509.
[50] Eisenstein, E., Osborne, J. C., Chaiken, I. M., Hensley, P. *J Biol Chem* 1984, 259, 5139-5145.
[51] Chen, Y. H., Yang, J. T., Martinez, H. M. *Biochemistry* 1971, 11, 4120–4131.
[52] ing S. M. *In Modern Techniques for Polymer Characterisation*; Pethrick, R. A.; Dawkins J. V., John Wiley & Sons, UK, 1999; 171-232.
[53] Cacace, M. G., Landau, E. M., Ramsden, J. J. *Quart Rev Biophys* 1997, 30, 241–277.
[54] Record, M. T. J., Zhang, W., Anderson, C. F. *Advan Protein Chem* 1998, 51, 281–353.
[55] Saunders, A. J., Davis-Searles, P. R., Allen, D. L., Pielak, G. J., Erie, D. A. *Biopolymers* 2000, 53, 293-307.
[56] Damodoran, S. *Int J Biol Macromol* 1989, 11, 2-8.
[57] Jensen, W. A., Armstrong, J., DeGiorgio, J., Hearn, M. T. W. *Bioch Bioph Acta* 1996, 1296, 23-24.
[58] Goto, Y., Nishikiori, S. *J Mol Biol* 1991, 222, 679-686.

In: New Trends in Chemical Physics Research
Editor: James N. Lapherty

ISBN: 978-1-61668-535-5

Chapter 7

CHEMICAL PHYSICS OF PHONONS AND SUPERCONDUCTIVITY: AN HEURISTIC APPROACH

Michael J. Bucknum and Eduardo A. Castro*

INIFTA, División Química Teórica, Sucursal 4,
Casilla de Correo 16,
Universidad Nacional de La Plata, La Plata 1900,
Buenos Aires, Argentina

ABSTRACT

Several potential heuristic angles are explored with regard to the parametrization of the superconducting transition temperature, T(c), and its relationship to the chemical physics of bonding. Fruitful angles explored in this paper include the relationship of the gas of Cooper pairs in a superconductor to a van der Waals equation of state for such a fluid and how it may be controlled and exploited through the consideration of such variables as pressure, P, and volume, V, of the fluid of Cooper pairs resident in a superconductor. Other angles explored, in an attempt to get a heuristic handle on the superconducting transition temperature, T(c), for superconducting compositions across the Periodic Table, include the introduction of a Morse's anharmonic chemical bond potential for the Debeye frequency of the phonons as it emerges from the central result for T(c) in superconductors from the Bardeen-Cooper-Schreifer (BCS) formulation of superconductivity. Yet another angle explored in this heuristic reasoning, is the use of a Badger's relationship between chemical bond force constant, k, and equilibrium internuclear distance, r(e), to try to gain some understanding of the nature of the electron-phonon coupling mechanism in superconductors. A relationship between r(e) and A(e) emerges as a result of this line of thought, given as r(e)·A(e) = "critical temperature constant", from which examination of spectroscopic data across the Periodic Table may yield a compositional solution of the appropriate chemical bonding, with the assumption

* Corresponding author: E-mail: eacast@gmail.com

of the proper electron-phonon coupling constant, C, and density of states, N(0), that ultimately may yield more desirable superconductivity transition temperatures.

I. SUPERCONDUCTIVITY AND THE GAS LAWS

Considering our Equation (1) from the microscopic version of the BCS theory [1]:

$$k \cdot T(c) = 1.14 \cdot h\,\omega(D) \cdot \exp(-1/\lambda) \quad (1)$$

one can multiply both sides by NA, the Avogadro constant, to yield a corresponding macroscopic version of this Equation (1), as Equation (2), our macroscopic BCS theory result.

$$NA \cdot k \cdot T(c) = NA \cdot 1.14 \cdot h\omega(D) \cdot \exp(-1/\lambda) \quad (2)$$

Using the identities NA . k . T(c) = R . T, and P . V = R . T, for an ideal gas, we have Equation (3):

$$P \cdot V = NA \cdot 1.14 \cdot h\omega(D) \cdot \exp(-1/\lambda) \quad (3)$$

Assuming the Bose gas of Cooper pairs inside the superconducting lattice behaves as an ideal gas, then there seemingly is nothing we can do with this macroscopic BCS theory expression at all. P and V would thus be the thermodynamic values of the Bose gas of Cooper pairs as their pressure and volume they occupy, for a mole of Cooper pairs constrained within a mole of superconducting material.

But what if the Bose gas of Cooper pairs is non-ideal, what if it obeys a van der Waals (vdW) equation of state for a gas, as Equation (4) suggests:

$$(P + (a^2/V^2)) \cdot (V - b) = NA \cdot 1.14 \cdot h\omega(D) \cdot \exp(-1/\lambda) \quad (4)$$

Therefore, if one could measure the vdW parameters, "a" the attractive force between Cooper pairs, and "b" the finite volume occupied by Cooper pairs, one would have a means of controlling the superconductivity mechanism somewhat and possibly affecting the superconducting temperature onset, T(c) through a control of these thermodynamic variables.

II. SUPERCONDUCTIVITY AND THE ELECTRON-PHONON COUPLING PARAMETER, Λ

An article by the Mao-Hemley group [2] at the Geophysical Laboratory in Washington, D.C., describes measurements of conductivity and superconductivity in B allotropes under pressure. They cite an expression for the electron-phonon coupling of superconductivity, the parameter λ in Equation (1) above. The expression is for s-p metals and superconductors and is given as Equation (5):

$\lambda = ((N(0)\ .\ C) / \langle\omega^2\rangle))$ (5)

The parameter "C" is supposed to be a numerical constant, so we don't worry about that. N(0) is the DOS(Ef) = density of states at the Fermi energy, Ef. And, importantly, $\langle\omega^2\rangle = (k/\mu)$ is the averaged phonon frequency squared.

So we see that the phonon frequencies and ultimately the phonon width at the Fermi energy, play an important role twice in the BCS theory result given in Equation (1) here. In the term, hω(D) , in Equation (1), it seems that the higher the phonon frequencies attained in the BCS mechanism, the higher will be the transition temperature, T(c). But in the electron-phonon coupling parameter, because of the inverse exponential-type function it enters as in Equation (1), it seems that the higher the phonon frequencies contributing to the BCS mechanism, the lower will be the transition temperature, T(c). It is as if the two effects in the BCS theory result of Equation (1) are indeed opposed to each other.

The BCS theory doesn't seem to give a monotonic-type, useful predictive result. It contains parameters that work at cross-purposes to each other. We think, also, judging from the paper we have cited [3], where we note there is yet another expression for the superconducting transition temperature T(c), that indeed there may be a plethora of such expressions for T(c). Indeed, it seems that there is no concensus in the physics community over what the proper physical expression should be for the behavior of T(c) in superconductors. Our earlier result for substituting into (1) the classical Morse potential, U(r), to yield a semiclassical BCS-type theory with maximum contributions, in a continuum of states, from anharmonicity and anharmonic vibrations of the lattice, would seem to work to increase the transition temperature, T(c), were it not for the term in the electron-phonon coupling, given by exp(-1/λ), which suggests such anharmonicity corrections would serve to suppress the superconducting transition temperature, T(c).

Finally, there is the term N(0) which is DOS(Ef). Clearly the bigger the DOS(Ef) the bigger will be the contribution to increasing the transition temperature of superconductivity, T(c). In this instance, and only in this instance, does it seem that a band structure can play a role. And the flatter the bands are at the Fermi energy, the more so will N(0) contribute to a higher transition temperature, T(c).

This seems to be all we can gather from Equation (1) and the other references we have studied thus far. It is odd that the BCS theory has this competition built into it from contributions of the lattice vibrations. It is a conundrum indeed.

III. Superconductivity and the Morse Potential and Badger's Rule

In treating anharmonicity in the BCS expression for superconductivity, one can start[2] with Equation (6), which is the ordinary result from the BCS theory, in which one approximates phonons in a harmonic potential:

$k\ .\ T(c) = 1.14\ .\ h\ \omega(D)\ .\ \exp(-1/\lambda)$ (6)

and thus one can use an anharmonic Morse's potential[4], $U(r) = A(e)(1 - \exp^{(a(r-re)2})$, where the anharmonicity parameter is given by, $a = \sqrt{(k/2A(e))}$, and $A_{(e)}$ is an equivalent atomization energy for the atom of interest in the unit cell of the superconducting lattice for its anharmonic dissociation from the lattice, to solve the Schroedinger equation for the diatomic molecular potential, and thus to replace the phonon energy term in (6), h ω(D) , in order to get an analogous BCS expression for anharmonic phonons as in Equation (7):

$$k\ .\ T(c) = 1.14\ .\ (h\ \omega(D) - (h\omega(D))^2/4D(e))\cdot \exp(-1/\lambda) \qquad (7)$$

In (7) we have thus introduced the effect of the bond strength D(e) into the BCS theory of superconductivity.

And from the point of view of Badger's relation [5], a correlation between bond force constant, k, and bond distance, r(e), developed in the 1930's to relate together various spectroscopic information across the Periodic Table as, $k = a(r(e) - b)^{-3}$, where k and r(e) have their usual definitions, and "a" and "b" are empirical fitting constants of the appropriate dimensions, which is the Badger's relation, and one can look again to the BCS expression, this time to Equation (8) for the electron-phonon coupling constant [6]:

$$\lambda = ((N(0)\ .\ C)/\langle\omega^2\rangle) \qquad (8)$$

and one can replace the term in ω by recognizing the relationship $\omega = \sqrt{(k/\mu)} = \sqrt{(a(r(e) - b)^{-3}/\mu)}$ where μ is the reduced mass of the chemical bond of interest in the electron-phonon coupling phenomena.

Therefore, upon squaring the latter expression for the phonon frequency, in terms of $\omega^2 = (k/\mu) = (a(r(e) - b)^{-3}/\mu)$, we get the following result for the electron-phonon coupling parameter from the BCS theory:

$$\lambda = ((N(0)\ .\ C)\cdot(a(r(e) - b)^3.\ \mu)) \qquad (9)$$

where the electron-phonon coupling constant is in terms of the variables in N(0), the density of states at the Fermi level, the Badger chemical bond parameters a and b, the equilibrium internuclear distance of interest r(e), the reduced mass, μ, of the bond of interest and the electron-phonon coupling constant C.

If we thus introduce all of the changes made above into Equation (6), we would have a BCS expression that introduces both the phonon bond dissociation energy, D(e), and the phonon equilibrium internuclear distance, r(e), explicitly into the overall BCS picture of superconductivity as in Equation (10):

$$k\ .\ T(c) = 1.14\ .\ (h\ \omega(D) - (h\omega(D))^2/4A(e))\cdot \exp\text{-}(1/((N(0)C)\cdot(a(r(e) - b)^3\mu)) \qquad (10)$$

Thus one has effectively transformed the BCS expression, in terms of phonon frequencies, ω(D), into an equivalent, viable expression in terms of phonon properties of dissociation energy, A(e) and equilibrium bond length, r(e), that are in fact more intuitive quantities to deal with and that have been determined spectroscopically quite accurately across the Periodic Table.

Equation (10), and the effects of chemical bond length, r(e), and chemical bond strength A(e), can thus be simply interpreted for the chemical bond that has given rise to the phonon of interest in the given superconductivity mechanism, by making some obvious assumptions. Thus by assuming that the given maximum values for the associated density of states of the material at the Fermi energy, N(0), and simultaneously a maximum in the electron-phonon coupling constant C hold for the purposes of testing out the effect of these chemical bond parameters on the overall critical temperature, T(c), of various superconducting compositions, one would analyze Equation (10) to find that as r(e) gets larger, to a power of 3, the corresponding critical transition temperature gets larger, T(c). By an analogous algebraic analysis, one can see that as A(e) gets larger, similarly T(c) gets larger as well. So one would like to maximize both these chemical bond parameters, r(e) and A(e), in order to maximize the critical superconducting temperature in a given composition. The approach one would take, in such a case, would be of course to scan the spectroscopic data across the Periodic Table, and see which diatomic chemical bond potentials maximize the product function given in Equation (11) below:

$$r(e) \,.\, A(e) = \text{“critical temperature constant”} \qquad (11)$$

It seems to be the case, in this connection, that such expressions as Equation (10) and Equation (11) lead to a more chemically intuitive understanding of the phonons involved in the superconductivity mechanism. One can thus use (10) and (11) to get a more predictive handle on the superconductivity transition temperature, a central result emerging out of the BCS theory, using chemically intuitive concepts such as the spectroscopic chemical bond parameters given by r(e), and A(e).

REFERENCES

[1] J. Bardeen, L.N. Cooper and J.R. Schreiffer, *Physical Review*, 108, 1175 (1957).

[2] M.I. Eremets, V.V. Struzhkin, H-K. Mao and R.J. Hemley, *Science*, 293, 272 (2001).

[3] Ibid [2].

[4] L. Pauling, *The Nature of the Chemical Bond,* 3rd edition, Cornell University Press, Ithaca, NY, 1960, p.594.

[5] Ibid [4], p.231.

[6] Ibid [2].

In: New Trends in Chemical Physics Research
Editor: James N. Lapherty
ISBN: 978-1-61668-535-5

Chapter 8

A REVIEW PAPER ON ERUCAMIDE: ITS SYNTHESIS, PHYSICAL AND CHEMICAL PROPERTIES

Neelam Rani[a], Pallavi Jain*[b] and Mu. Naushad[b]
[a]Department of Chemistry,
Kamrah Institute of Information Technology, Haryana, India
[b]Department of Chemistry, Inst. of Sci. and Tech.,
SRM University, Modinagar, India

1. INTRODUCTION

Naturally occurring triglycerides, oils and wax esters have been used as lubricants for thousands of years and continue to play an important role in today's lubricants industry. Brassicas (Rapeseed-Mustard) is the second most important edible oil seed crop in India. The major fatty acid present in this oil is Erucic acid (40-45%). High Erucic acid Rapeseed oil has immense potential for industrial utilization. Erucic acid 13-docosenoic acid ($C_{22}H_{42}O_2$) (Mol.wt 338) is the precursor to various other chemicals, of which Erucamide is very important, because erucic acid is mono unsaturated, physical properties such as melting point and titre are comparable to oleic acid [1].

1.1. Erucamide

It is one of the most important nitrogen derivatives of erucic acid. Erucamide is used as a slip and anti-block agent for the production of low linear density polyethylene films, molded plastics and poly vinyl chloride resins. Adding 0.05–0.2% of erucamide can reduce friction from 4.55–1.11 between poly propylene films. Erucamide has a high melting point and good

* Dr. Jain palli24@yahoo.com

stability at higher temperature compared with other amides. [2] "Slip" is the action of two surfaces sliding over each other. The higher the slip, the lower the friction between the surfaces. "Block" is the force necessary to pull two surfaces apart such as two sheets of poly ethylene film. Erucamide has advantages over oleamide and stearamide (two other commonly used plastic additives). The mono-unsaturated oleamide is a good slip agent but a poor antiblock agent conversely, the saturated stearamide is a good antiblock agent but a poor slip agent, erucamide bridges the gap between the two providing good performance in both areas. Approximately 100 million pounds of HEAR oil are used annually for this application.

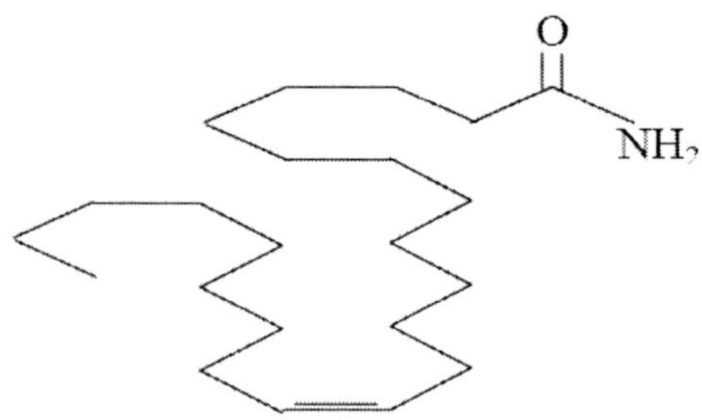

Figure 1. Structure of erucamide.

Brief Description of Erucamide [3]

Derivative of	:	Erucic Acid
Chemical formula	:	$CH_3 (CH_2)_7 CH= CH (CH_2)_{11} CO NH_2$
Mol. Wt.	:	338
Appearance	:	Beige solid
Solubility in Water	:	Insoluble
Stability	:	Stable under ordinary Conditions
Acid Value	:	Less than 0.5
Iodine value	:	72-80
Melting Point	:	78° C
Flash Point	:	230 °C

1.1.2. Principle of Action of Erucamide

Slip effect is a term used to describe the sliding of parallel film surfaces over each other or the sliding of film surfaces over substrates slip agents such as stearamide, oleamide and Erucamide- are added to poly ethylene (PE) or poly propylene (PP) films during the extrusion process to decrease friction, both film–to–film friction and that between the film and production equipment in order to increase the output of the line and improve packaging line operations. Slip effect is measured as COF (Coefficient of friction) [4]

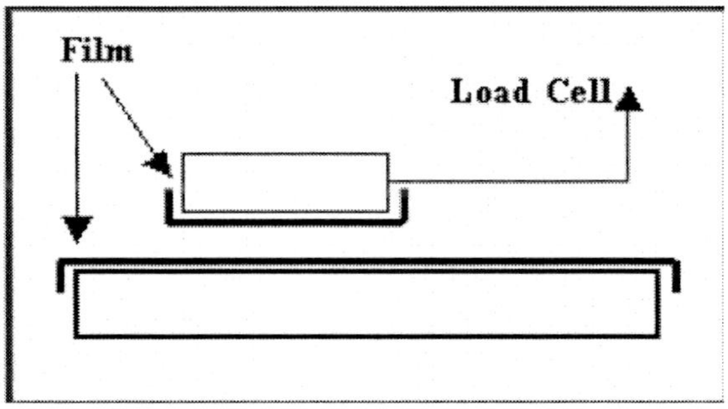

Schematic Representation of COF Measurement.

Decreased COF is a result of the migration of slip additive to the film surface due to the defined incompatibility of the slip additive with the polymer. The migration speed of the slip agent to the surface is largely decided by the additive's chain length compatibility with the polymer and the cyrstallinity of the polymer. The larger the slip agent (more carbon atoms) the more compatible it is with the polymer and the slower it migrates.

Due to migration of the slip agent to the surface of the poly olefins, primary amides are not only used as slip agents for poly olefin films but are also recommended for use as demoulding agents in injection moulding applications.[5]

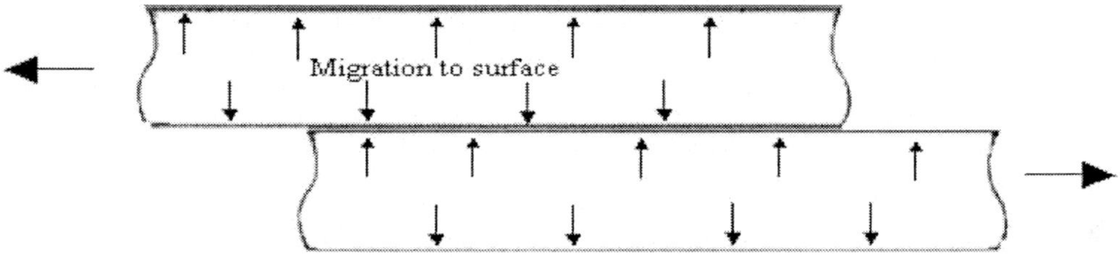

Migration of slip additives to the film surface reducing coefficient of friction

1.1.3. Mechanism of Action

Because slip additives are incompatible with the polyethylene matrix, the majority will exude to the film's surface, preventing direct film–to–film contact and thus reducing film-to-film friction.

The amide molecules orient on the surface in a way that the polar-$CONH_2$ groups are in contact with the polyethylene surface and the hydrocarbon chains are projected outwards. Because the majority of the total slip additive in the polymer matrix will come to the surface, the following should be kept in mind:

At a given film thickness, a higher slip additive concentration will bring more slip on the surface and result in a lower COF. Concentration above a monolayer thickness will not further reduce the COF and may decrease gloss and increase haze.

For the same reason, at a given slip additive concentration, an increase in film thickness will result in a lower COF, since the slip additive migration is not instantaneous, a reduction of the COF will only be observed after time, corresponding to the growing layer of additive on the film's surface. It takes several hours before the equilibrium is reached.

The migration rate depends on the type of amide that is used. Due to the lower molecular weight, oleamide will migrate faster to the surface than Erucamide. This results in a lower initial COF for resins with oleamide. Once the migration is complete, a lower COF will be observed with erucamide since it is more efficient in reducing friction. [6]

1.2. Chemical Methods for Synthesizing Fatty Amides

1.2.1. Dehydration of Ammonium Salts

The simplest method of preparing method of fatty acid amide is thermal dehydration of ammonium salt. [7]

The synthesis of amide from fatty acid involves the transitory formation and subsequent dehydration of the salt.

$$RCOOH \xrightarrow{NH3} RCOONH_4 \xrightarrow[-H2O]{} RCONH_2$$

1.2.2. Ammonolysis of Ester

The ammonolysis of ester is the most direct and possibly the least expensive method for preparation fatty acid amides. With glyceride esters the products are mixed amides of the alkyl component of the parent oil.

$$\begin{matrix} CH_2OCOR \\ | \\ CHOCOR \\ | \\ CH_2OCOR \end{matrix} \quad + \quad 3NH_3 \quad \rightleftharpoons \quad 3RCONH_2 + \begin{matrix} CH_2OH \\ | \\ CHOH \\ | \\ CH_2OH \end{matrix}$$

With fractionally distilled methyl esters obtained by methanolysis of mixed glycerides, relatively pure amides are produced by ammonolysis.

$$RCOOR' \quad + \quad NH_3 \quad \rightleftharpoons \quad RCONH_2 + R'OH$$

$CR' = CH_3$ etc.[8]

The reaction is catalyzed at least in nonaqueous medium of ammonium chloride. [9] Better conversions to amides are carried out with saturated alcoholic ammonia solutions which produce homogeneous reaction media. [10, 11]

The satisfactory conversion of higher fatty ester to amides opportunity requires the use of excess ammonia, 150-500 lb/m^2 pressure and temperature of 100-110 ^{0}C for period of several hours. [12]

1.2.3. Ammonolysis of Acid Chloride

The rapid, exothermic, irreversible reaction of a fatty acid chloride with NH_3 at relatively low temperature affords an amide. [13]

$$RCOCl + 2NH_3 \longrightarrow RCONH_2 + NH_4Cl$$

The initial mole of ammonia reacts with the evolution of hydrogen chloride, which combines, with a second mole of ammonia, converted to an amide. When the reactions are carried out in benzene or diethyl ether, ammonium chloride generally precipitates and the amide is found in the filtrate.

The preparation of fatty acid amides via acid chloride offers a number of advantages.

- Ammnolysis of acid chloride is a rapid reaction, which may be carried out at low temperature. It is reversible and therefore, does not require large excess of ammonia for removal of the products for its completion and no catalyst is required.
- The amide is not exposed to heat for long periods consequently the color of the product depends only on that of the starting material.
- Usually no decolorizing treatments, such as those normally required to obtain low color products from acid through the dehydration of ammonium salt are necessary.

1.2.4. Ammnolysis of Acid Anhydrides

Ammnolysis of fatty acid anhydrides yields amides together with ammonium salt when excess ammonia is used. The exact composition of the product:-

$$(RCO)_2O + NH_3 \longrightarrow RCONH_2 + RCOOH$$

At evaluated temperatures the dehydration of the intermediate ammonium salt may occur only the amide is produced

$$(RCO)_2O + 2NH_3 \longrightarrow 2RCONH_2 + RCOOH$$

Relatively little use has been made of the syntheses of amides from acid anhydrides. [14]

1.2.5. Inter Charge Reactions

Several inter charge reactions between urea or low homolog amides and fatty acid have been used in synthesizing higher fatty acid amides.

$$RCOOH + H_2NCONH_2 \longrightarrow RCONH_2 + CO_2 + NH_3$$

$$RCOOH + H_2NSO_2NH_2 \longrightarrow RCONH_2 + SO_3 + NH_3$$

$$RCOOH + HCONH_2 \longrightarrow RCONH_2 + HCOOH$$

$$RCOOH + CH_3.CONH_2 \longrightarrow RCONH_2 + CH_3COOH$$

Example: When an intimate mixture of oleic acid and urea are heated rapidly and maintained at 230°C for 2sh and the reaction mixture extracted with petroleum oleamide (48% yield) was obtained on crystallization somewhat better conservation to amide are claimed to resulting by heating the acid and urea with diammonium hydrogen phosphate as catalyst[15]

1.2.6. Addition of Water to Nitriles

Generally, amides are dehydrated to nitriles, but if water is added across the nitrogen carbon triple bond of a nitrile the amide may be regenerated, this reaction has been accomplished by heating the nitrile with concentrated hydrochloric, sulfuric and glacial acetic acid at low temp and by heating with hydrogen in alkaline solution.[16]

$$RCN + H_20 \longrightarrow RCONH_2$$

This method is not frequently applied for the preparation of amides.

1.3. Properties of Fatty Amides

1.3.1. Physical Properties

The amides of the fatty acids exhibit many anomalous physical properties. Various explanations have been proposed to account for the structure of amides which has been assumed to correspond to either the ketonic [17-22] $RCONH_2$ or to an enolic configuration [23-27] RCNHOH more recently dynamic tautomeric equilibrium between the two forms have received considerable acceptance largely on the basis of pertinent ultra violet and infrared absorption data.[28-29]

1.3.1.1. Hydrogen Bonding

Some X-ray data have been interpreted to indicate that bonding in the solid state amides consisting of polymers of symmetrical cyclic groupings of monomers linked through weak hydrogen bonds, but the high dielectric constant and dipole moments of these substances are not in accordance with this structure. [30]. Correlation of the specific heats and electrical conductivities with temperature in the solid state has indicated that transition occurs at 56^0C and 640 C respectively for palmitamide and stearamide. [31]

1.3.1.2. Melting and Freezing Point Behaviour

Replacement of the hydroxyl of carboxyl group of a fatty acid with an amino group increases the melting point appreciably. The melting points of amides exhibit alternation between the even and odd carbon series, and marked increase occur between successive members at the lower end of both series.

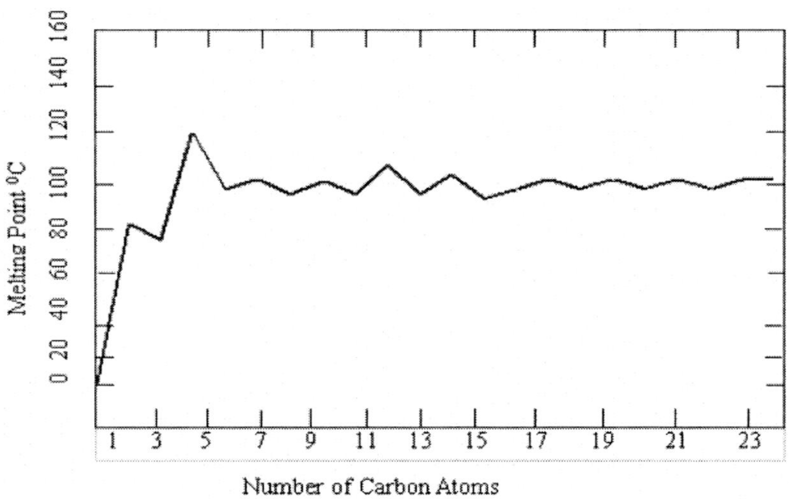

Melting Point of Fatty Amides

Binary freezing point data for mixtures of the C_{18} – amides, such as oleamide, linoleamide, elaidamide and steramide, indicate that the reparation and purification of these amides by solvent crystallization methods would be quite difficult [32].

1.3.1.3. Solubilities

The solubilities of fatty amides in many solvents appear to constitute further evidence of unusual stretches of these compounds. In some non polar solvents there is little order in the solubility pattern for the homologous series. [33] In general, the amides are less soluble in hexane than in benzene or cyclo hexane, and there appears to be a direct relationship in this solvent between the solubilities and the melting point of the solutes.[34]

In polar solvents although the solubility decreases with increasing molecular wt. the intervals between the solubility curves are not uniform.

1.3.2. Chemical Properties

Amides of carboxylic acids were generally considered to be neutral, relatively inert compounds, and of limited utility for further chemical reactions later investigations have demonstrated that the amide functional group is inherently quite reactive.

Some of the chemical properties are as follows.

1.3.2.1. Hydrolysis

Hydrolysis of amides is catalyzed by acids and bases generally the reaction is sluggish because of the insolubilities and high melting points of the higher amides.

$$RCONH_2 + H_2O \longrightarrow RCOOH + NH_3$$

High boiling solvents and pressure are generally employed to hydrolyze amides under alkaline conditions. The reaction is molecular and the acid catalyzed reaction has a large temperature coefficient rate of hydrolysis can be used to locate the branching in high molecular weight amides. [35, 36]

1.3.2.2. Alcoholysis and Ester-Ester Interchange Reactions

The alcoholysis reaction is the reverse of the ammonolysis of esters, and proceeds whenever the reaction conditions are favourable in presence or excess of alcohol and removal of ammonia. Reaction usually occurs to a limited extent when R= CH_3, C_2H_5, etc., and the product obtained from a low homolog acid can be fractionally distilled and removed at temperatures sufficiently low to prevent benzoylation of the amide[37].

$$RCONH_2 + R'OH \rightleftharpoons RCOOR' + NH_3$$

$$RCONH_2 + R'OH \longrightarrow RCOOH + R'NH_2 \rightleftharpoons RCOONH_3R'$$

$$RCONH_2 + R'NH_2 \longrightarrow RCONHR' + NH_3$$

$$RCONH_2 + C_6H_5COOH \longrightarrow RCOOH + C_6H_5CONH_2$$

$$RCONH_2 + Br_2 + 2NaOCH_3 \longrightarrow RNHCOOCH_3 + 2NaBr + CH_3OH$$

1.3.2.3. Salt Formation with Mineral Acids

Amides particularly the lower number, may titrated in acetic anhydride solution with perchloric acid. [38]

$$RCONH_2 + HClO_4(CH_3COOH) \longrightarrow RCONH_3^+ClO_4^-$$

1.3.2.4. Reaction with Nitrous Acid

With Nitrous acid amides generally give acids. [39]

$$RCONH2 \longrightarrow RCN + H_20$$

$$2RCONH2 \longrightarrow RCN + RCOOH + NH_3$$

1.3.2.5. Dehydration to Nitriles

The industrial production of fatty nitrous is based on reaction. [40]

$$RCONH2 \longrightarrow RCN + H_20$$

$$2RCONH2 \longrightarrow RCN + RCOOH + NH_3$$

1.3.2.6. Hofmann Degradation Reaction

The hofmann degradation reaction involves the convolves the conversion of an amide to an amine containing one less carbon atom. [41, 42]

$$RCONH_2\ Br_2 \longrightarrow RCONHBr + HBr$$

$$RCONHBr + NaOH \longrightarrow RNCO + NaBr + H_2O$$

$$RNCO + H_2O \longrightarrow RNH_2 + CO_2$$

$$RNCO + RNH_2 \longrightarrow RCHCONHR$$

The amide is usually dissolved in a slight excess of cold aqueous hypohalite solution followed by rapid warming. This technique is suitable for amides upto C_8, but above this chain length it is preferable to dissolve the amide in methanol containing sodium methoxide, followed by the addition of Bromine which generally results in smooth conversions to methyl carbamates.

$$RCONH_2 + Br_2 + 2NaOCH_3 \longrightarrow RNHCOOCH_3 + 2NaBr + CH_3OH$$

On saponification with alkali the latter are continued to amines.

1.3.2.7. Reduction to Primary and Secondary Amines

High pressure hydrogenation of higher homolog amides with copper-chromium oxide – type catalyst yields variable mixtures of primary and secondary amines.

$$RCONH_2 + 2H_2 \longrightarrow RCH_2NH_2 + 2H_2O$$

$$2RCONH_2 + 4H \longrightarrow (RCH2)_2NH + 2H_2O + NH_3$$

The reduction of amides to primary amines in good yields and with little formation of secondary amines appears possible with metallic hydrides; Palmitamide and steramide have been converted to hexa decyl and octa decyl amine, respectively by treatment with Lithium Aluminum hydride. [43]

1.3.2.8. Reaction with Formaldehyde

A number of industrially important chemical intermediates are prepared by reaction of fatty amides with formaldehyde or its polymers. [44, 45]

$$RCONH_2 + HCHO \longrightarrow RCONH_2OH$$

$$2RCONH_2 + HCHO \longrightarrow RCONHCH_2NHCOR + H_2O$$

$$2RCONH_2 + 3\ HCHO \longrightarrow RCONCH_2OCH_2 - \overset{\overset{\displaystyle COR}{|}}{N} - CH_2 + 2H_2O$$

The reaction products of fatty amides and formaldehyde have considerable utility per se, and, as chemical intermediates. Textile auxiliaries, wetting agents, water proofing compounds, dyeing assistants, waxes, detergents and other products can be prepared from them.

1.3.2.9. Ethhoxylation with Ethylene Oxide

Fatty amides react with oxide to give a series of products ranging from fluid liquid and soft pastes to high melting waxy materials, depending upon the molar preparation of ethylene oxide employed. [46]

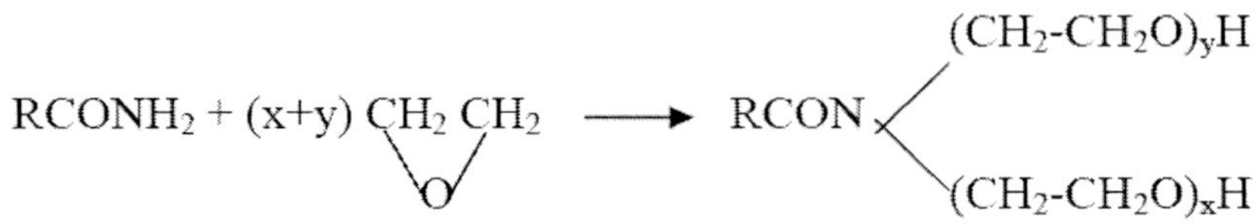

1.3.2.10. Acylation

Two processes are available for acylating amides with reactive acid chlorides, one which comprises, heating a mixture of the amide and the appropriate chloride in vacuo until all the hydrochloric acid has been evolved. [47]

$$RCONH2 + R'COCl \longrightarrow RCONHCHOR' + HCl$$

1.3.2.11. Alkylation: Conversion to Hydroxamic Acids

The reaction of acetamide with hydroxylamine hydrochloride gives acethydroxamic acid. [48]

$$CH_3CONH_2 + {}^{+}NH_3OHCl^{-} \longrightarrow CH_3CONHOH + NH_4Cl$$

1.3.2.12. N-Halogenation

N-halogenation of fatty amide (where the halogen is chlorine or bromine) may be accomplished by treating the amide with the free halogens at low temperature

$$RCONH2 + X2 \longrightarrow RCONHX + HX$$

X = Br, Cl usually [49]

1.3.2.13. Formation of Alkali Metal Derivatives

Alkali metal derivatives of fatty amides can be formed under suitable conditions e.g. reaction with finally divided metals with anhydrous, inert, organic solvents such as toluene. The alkali metal derivatives are never isolated but serve as intermediates for the preparation of other derivatives.

$$RCONH_2 + M \longrightarrow RCONHM + H$$

M = K, Na etc.

1.3.2.14. Sulfonation

Much of the interest in these products resides in the fact that the water soluble products possess detergent, surface active, and wetting properties. Sulfonation amide may be prepared by the acylation of amino sufonic acid which may be prepared by the acylation of amino sulfonic acids such as methyl amino methane sulfonic acid with acid chloride. [50]

$$CH_3NHCH_2SO_3H + RCOCl + NaOH \longrightarrow CH_3\text{-}N(COR)\text{-}CH2SO_3Na + NaCl + H_2O$$

1.3.2.15. Nitration

Nitration of fatty amides is relatively unimportant because of the difficulty in preparing and handling nitramides.

$$RCONH_2 + HNO_3 \longrightarrow RCONHNO_2 + H_2O$$

1.3.2.16. Reaction with the Grignard Reagent

Primary and secondary amides react with Grignard regents to give mono and di substituted derivatives respectively. [51]

$$RCONH2 + R'MgX \longrightarrow RCONHR' + MgXH$$

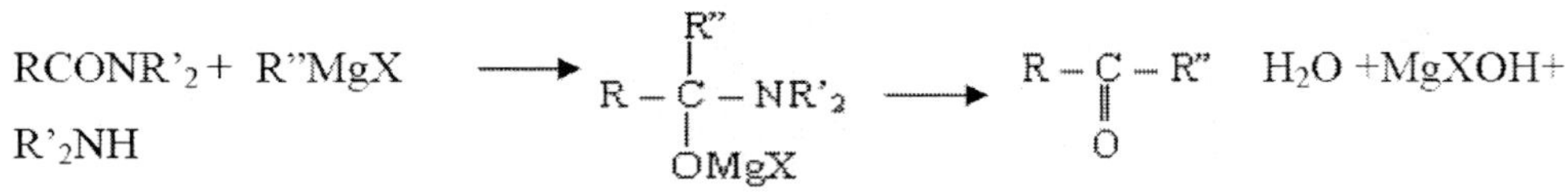

1.4. General Characteristics of Erucamide

The general characteristics of erucamide as compared with other amides can be discussed as follows:

- Estimated mechanism for developing lubricity
- Amounts of erucamide added and amounts of amide on LDPE
- Melting point behavior
- Effect of film thickness on coefficient of friction

1.4.1. Estimated Mechanism for Developing Lubricity

Erucamide comes into contact with a resin, than transfers to the resin surface during the formation process and finally exhibits lubricity.

1.4.2. Amounts of Erucamide Added and Amounts of Amide on LDPE

Erucamide transfer to the film surface until a particular concentration is reached. This feature allows the substance to offer excellent lubricity.

1.4.3. Melting Point Behaviour of Erucamide

Regardless of the carbon number, the melting point of saturated fatty acid amides falls in a range of $100^0 - 110^0$ C in contrast, the melting point of unsaturated fatty acid amides is significantly affected by the carbon number of such fatty acid amides; there melting point fall in a range of 70^0-85^0C relative to the carbon number in a range of 18 to 22. Usually, as a lubricant a fatty acid amide whose carbon number is in a range of 18 to 22 is used.

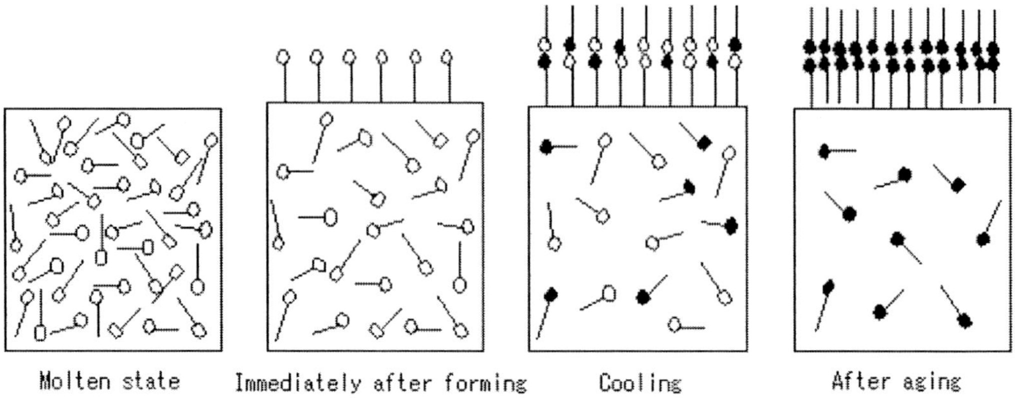

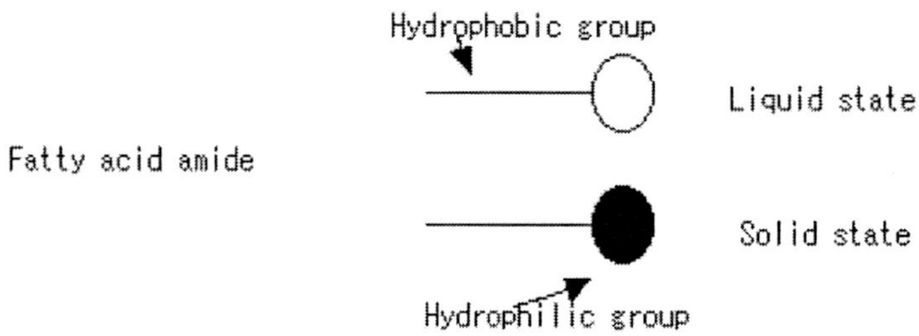

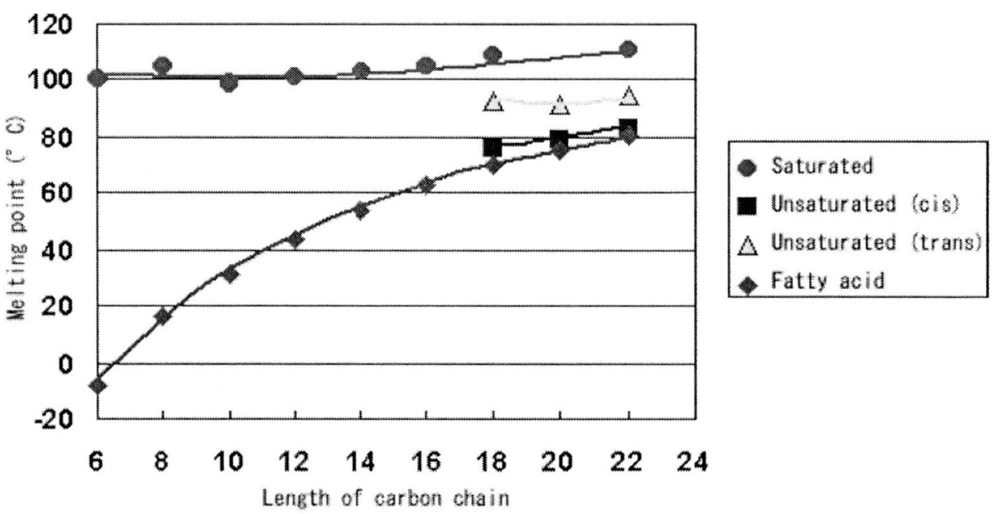

1.4.4. Effect of Film Thickness on Coefficient of Friction

Film thickness affects the lubricity affects of fatty acid amides. In a film thickness range of 20 to 30 f Êm, the amount added must be adjusted. [52]

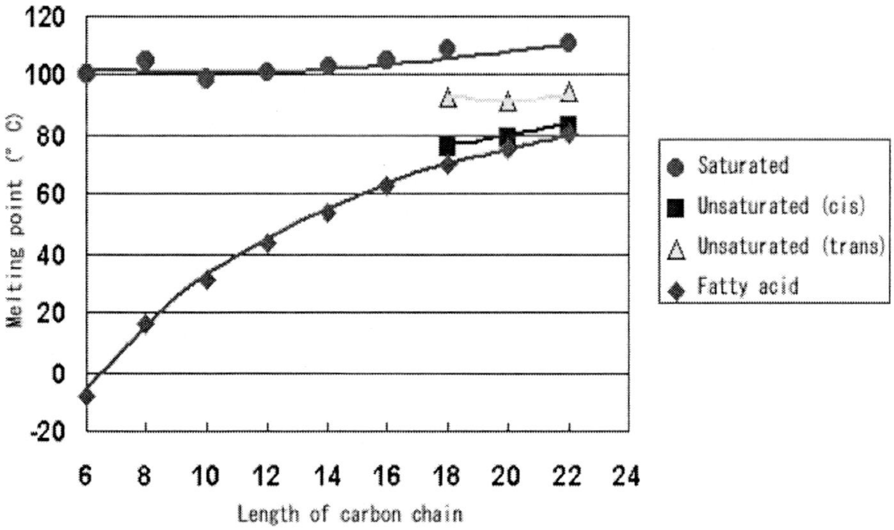

1.5. Comparison between Erucamide, Oleamide and Stearamide

Erucamide has advantages over Oleamide and Stearamide, two other commonly used plastic additives. The rate of diffusing of slip additive to the surface of the polymer film varies as a function of chemical structure. The lower molecular weight oleamide migrates to the film surface more rapidly and is often referred to as a "fast blooming" slip. Although "slow blooming" erucamide develops its lubricating effect at a slower rate initially, after 4-6 h. it produces lower COF values than equal quantities of oleamide.

Antiblocking performance is another key characteristic of amide slips "Blocking" is the tendency for polyethylene film sheets to adhere to one another as they are separated vertically; antiblocking is therefore the ability to reduce inter-sheet adhesion characteristics. In most cases, the effectiveness of amorphous and crystalline silica, diatomaceous earths and talc antiblocks far outweigh the marginal antiblock properties provided by amide slip agents. However, because inorganic antiblocks increase haze levels and can negatively affect other optical and physical film properties, some high clarity film applications rely exclusively on the slip agent to provide the necessary reduction in blocking force. In these cases, erucamide provides superior antiblocking compared to equal levels of oleamide.

Stearamide (another fatty acid amide) provides even better antiblocking, but since it does not function effectively in reducing COF, it is usually used in combination with oleamide or erucamide to achieve a desired balance of slip and antiblock characteristics. [53]

When comparing oleamide with erucamide (both unsaturated fatty acids), the initial slipping property of oleamide is better than of erucamide while the heat resistance and anti blocking properties of erucamide are better than those of oleamide.

Property	Advantage
Initial slipping property	Oleamide > erucamide
Final slipping property	Erucamide = Oleamide

Transparency of film	Erucamide = Oleamide
Anti - blocking property	Erucamide > Oleamide
Color stability	Erucamide > Oleamide
Heat resistance	Erucamide > Oleamide
Sublimation resistance	Erucamide > Oleamide

1.6. Applications

1.6.1. Polymer Industry

One of the important uses of erucamide is as an anti-sticking lubricant release and antistatic agent of polyethylene (PE) and polypropylene (PP) film. Adding 0.05 – 0.2% erucamide can reduce friction from 4.55 to 1.11 between poly propylene films. It can reduce sticking of the plastic to itself by minimizing and facilitate processing on high speed packaging equipment. It finds application in manufacturing resin compositions with good lubricity for cap lines and caps used to seal a PET bottle resulting in good sealing property and no odor from no water stored in it. It is also used in it.

It is also used in manufacturing oxygen – absorb in multilayer thermoplastic resin films used in packaging containers. A three layer film comprise of anti blocking layers containing micro particles with melting point higher than that of ethylene polymers, middle layers containing fatty acid amides lubricants and surface layers containing lubricants and optionally the micro particles.

It is also used in manufacturing high frequency welded olefin polymer bags with zippers containing ethylene vinyl acetate co-polymer (EVA) fatty acid amide type slip agents, inorganic anti blocking and inorganic neutralizes. Erucic acid amide is used in a mixture containing HDPE, Et. acrylate co-polymer red P and some additives and extruded to give a label sheath having good fire resistance, mechanical strength and lubrication.[54]

1.6.2. Paint Industry

Erucic acid amide is used in manufacturing thermoplastic elastomer containing 90:10 isotactic propylene-atactic polypropylene mixture and 1000 ppm erucic amide. It was gravure printed with pigmented PUA ink and used in manufacture of weather resistant decorative building exterior material.

1.6.3. Textile and Surfactant Industries

Erucic amide along with ethylene – Et acrylate co-polymer, Hybrar (Styrene based thermoplastic elastomer) is used in manufacturing good crease and scratch resistant soft cloths with excellent softness low noise on bending.

High erucic acid rapeseed (HEAR) oil surfactants have good solubility property. These are also mild to skin, non toxic, and biodegradable. Erucamide can also be used as foam stabilizer and antiseptics of sulfonate detergents.

1.6.4. Other Applications

Erucamide, a nitrogen derivative of erucic acid is also used in the manufacturing of printing inks as a dispersing agent. It is also used in high grade lube oils, dye dispersants, waterproof and antifog agent for paper making and textile industries, foam stabilizer, metal-wire drawing lubricant agent etc. Erucamide has low vapor pressure and is less volatile, making it more suitable for higher temperature processings. Lower volatility means that more of the slip will stay on the film surface and not end up venting off as smoke or depositing on metal die surfaces. In addition, erucamide is less prone to thermal oxidation during processing and contributes less color (yellowing) and odor to the final product. The chemical stability of erucamide also translates into lower rates of oxidation, rancidification and color formation during product storage. [55]

2. Review of Literature

2.1. Compositions

M. Asuka [56], in a study of strong and transparent antistatic and antiblocking Polyolefin compositions and films and bags made from them reported that the compounds comprise (A) Polyolefin resins 100, (B) vinylaromatic type elastomers 2-10 (C) Spherical polymer particles having average diameter 0.5-8μm and f value (i.e., trueness of the spheres as determined by f $=[M/(\pi/4)]^{0.5}/N_{max}$ where M =any given cross- section area of the particles in mm^2; N_{max} = maximum diameter of the cross section in mm) >0.8, 0.03-0.5, (D) higher falty acid amides 0.03-04 and (E) carboxylic acid partial esters of 4-6 valent polyols, 0.05 to1.5 –parts. Thus emulsion polymn. of Me methacrylate 90 with butanediol diacrylate 5 and $BuMe_2SiO(SiMe_2)_{60}SiMe_2\ (CH_2)_3\ OCOCMe{:}CH_2$ 5 parts gave a copolymer which was spray dried and milled in a jet mill to give spherical particles with n 1.48, diameter 3mm and f value 0.91. Blending a propylene co-polymer (MFR 6.0 g/10 min) 100 with Hybrar HVS 3(Hydrogenated triblock isoprene- Styrene rubber) 2, the spherical particles 0.35, erucic acid amide 0.18 and diglycerin mono stearate 0.24 part at 210° for 10 min, role pressing into a film, folding and heat sealing an 2 sides gave a bag with haze 1.9% friction coefficient 0.38 on outer and inner surface and good antiblocking and antistatic properties.

M. Asuka, [57] studied polyolefin compositions containing polymer micro particles and antistatic, transparent and blocking- resistant films and bags therefrom and stated that these compounds are useful for packaging materials, containers, etc., comprise polyolefins, 0.01-0.3 phr spherical resin mocroparticles with average diameter (Φ) 0.5-8 μm and sphericity (f) ≥0.8, $f=[M/(\pi/4)]^{0.5}/N_{max}$; $M(mm^2)$=cross-sectional area of the microparticles;, N_{max} (mm) = maximum diameter of the cross-section], 0.01-0.3 phr higher fatty acid amides,and 0.05-1.5 phr partial esters of 4- to 6-valent polyols with aliph. mono carboxylic acids. Thus propylene co polymer (MFR 6.0 g/10 min), 0.08 phr microparticles [n=1.48; Φ 3μm; f 0.91; prepared from Me methacrylate 90, butanediol diacrylate 5, and $C_4H_4SiMe_2O(SiMe_2O)60SiMe_2\ (CH_2)_3\text{-}OCOCMe{:}CH_2]$ 5 parts, 0.25 phr erucic acid amide and 0.24 phr diglycerol monostearate were kneaded and molded into a film showing haze 1.2%, dynamic friction coefficient 0.16, and surface resistivity $6.2X10^{13}\ \Omega$.

M. Totani et al. [58] reported block coplymer compositions for providing transparent heat shrinkable films with rigidity, impact resistance, blocking resistance, and improved printability, comprise (a) 20-100 parts of an aromatic vinyl compound to conjugated diene block copolymer having weight ratio of aromatic vinyl compound to conjugated diene 60-90 :

40-10 and no. –average mo1ecular weight 40,000- 500,000 (b) 0-80 parts of an aromatic vinyl polymer selected from (i) aromatic vinyl hydrocarbon polymers, (ii) co polymers of aromatic vinyl hydrocarbon and (meth) acrylic acid (iii) Co-polymers of aromaticvinyl Hydrocarbon and (meth) acrylic acid esters, and (iv) a rubber modified styrene polymer, (c) 0.01-0.2 phr a fatty acid mono amide and (d) 0.01-0.3phr of a fatty acid bisamide. A 60-μm biaxially stretched transparent film was obtained from a compn. comprising 100 parts of a A-B-B- type butadiene-styrene block co-polymer, 0.08 phr of erucic acid amide, and 0.04 phr of N,N'-ethylene bis stearylamide and had tensile elasticity 17,000 kg/cm^{2}. Izod impact strength 2.0 kg. cm/cm, and heat shrinkability> 20%.

K. Oka et al. [59] while studying the heat-resistant monodispersed methacrylate polymer particles and blocking-resistant transparent film stated that methacrylate polymer particles containing 0.001 –1 phr antioxidants, satisfy average diameter(ϕ) 0.1-1000 μ m, variation coefficient [vc; (std. deviation) /-(averagediameter)X100] 1-30, and decomposition initiation temperature (T_D) 250-300^0.The components contain 100 parts polymers selected from polypropylene, propylene- olefin co-polymers, polyethylene, polyesters, polyamides, PVC, nylon and polystyrene and 0.1-3 parts of the particles. Thus 96 parts methyl methacrylate was emulsion seed polymerised with 4 parts ethylene glycol dimethacrylate in PVC and treated with p-methoxy phenol to give particles showing OH value 19, ϕ 3.2 μm, vc 6 and T_D ~258^0

M. Nakano et al. [60] white studying the polyolefin based transparent stretch labels with good ink fixability and blocking resistance reported that these labels contain ethylene – hexane co – polymers (hexane content 4-15%) prepared in the presence of metallocene – based catalysts 100, higher aliph acid amides having unsatd. double bonds 0.01 – 0.1, and cross linked poly (methyl methacrylate) beads with average grain size 1-10 μm 0.05 – 0.3 part, and satisfy thickness 40-200 μm and haze ≤ 5%. Thus, 100 parts evolve SP 1540 (Ethylene–1-hexene co-polymer; MI 4g / 10 min, d.0.915 g/cm^3) was blended with 0.1 part erucic amide and 0.2 part Epostar MA 1002 (I; ϕ 2μm) extruded, and corona treated to give a 100- μm thick stretchable film, which was printed from one side, heat-sealed at side edge, and cut to give a stretch label for bottle showing residual stretch 0.8%, haze 4.0% static frictional coefficient 0.10 and excellent fixability of printing inks.

M. Asuka et al. [61] studied the polyolefin based compositions for films with good blocking resistance, transparency, slippage, and less dynamic friction coefficient and reported that these compounds comprise (A) 100 parts polyolefins, (B) 0.01 – 0.3 part spherical polymer particles [prepared by copolymg. R_3 SiO (R_2SiO)n SiR_2X [R=C_{1-2} (un)substituted monovalent hydrocarbyl ; ≥ 50% of R=Me; X=org.group containing radically polymerizable and non cross linkable monomers,and radically polymerizable and crosslinkable monomers; refractive index (RI) 1.45 – 1.52; average particles size(APS) 0.5 – 8 μ; sphericity (f) ≥ 0.8 [f=[M/$(\pi/4)^{0.5}$/ N max; M= cross section of fine particles (mm^2) ;N max =maximum major axis of cross section(mm)], and (C) 0.01 – 0.3 part higher fatty acid amides. Thus a compound containing propylene copolymers 100, spherical polymer particles [prepared from Me methacrylate 90, butanediol diacrylate 5, and $BuMe_2SiO(Si\ Me_2O)_{60}S$ $iMe_2(CH_2)_2OCOCMe$: CH_2 5 parts ; RI=1.48, APS=3μm; f=0.91] 0.08, and erucamide 0.28 part was kneaded and formed into a film which showed haze 0.9, dynamic friction coefficient 0.19 and good blocking resistance.

M. Zajac et al.. [62] while studying the processing of plastic waste containing poly vinyledene chloride (PVDC) reported that these waste comprising poly propylene (PP) and ≥

15% PVDC is blended with 5-40 % mixture containing 0.1-5 % erucamide, 1-10 % higher fatty acid alkaline earth metal salts and 85-98.9 % Ca, Al or Mg carbonates and/or hydroxides . For example, a biaxially oriented waste PP film containing 40 % PVDC was combined with 20% mixture containing 4% erucamide, 16% Ca stearate and 80% $CaCO_3$ and subjected to melt – blending hot -pressing without decomposition.

H. Ooe. et al. [63] studied polypropylene polymer heat-shrinkable packaging films which are biaxially oriented have good transparency and slipping properties and have haze ≤ 5% and static friction coefficient At 50^0 ≤ 1.5, comprise (A) crystalline propylene polymers having xylene soln fraction at 20% ≤ 10% 100, (B) antioxidants having static friction coefficient At 50^0 ≤ 1.5, 0.01-1.0 antiblocking agents having average particle diameter1-10 μm, 0.05-0.6, (D) slipping agents comprising C_{22-26} fatty acid amide 0.01 –0.3, and (E) fatty acid metal salts 0.005 –0.2 and /or (F) synthetic hydrotalcites 0.005-0.5 parts . Thus a biaxially oriented film prepared from ethylene –propylene co-polymer having ethylene content 3.0 % MFR 2.5g/10 min., and xylene sol. fraction 2.4% 100 BHT antioxidants, 0.2 synthetic silica ,antiblocking agent 0.2, erucic acid amide slipping agent 0.08, and calcium stearate 0.1 parts had haze 4.0 % and static friction coefficient 0.65, compared with 4.0 and ≥2.0, resp. for film prepared using oleic acid amide slipping agent instead of erucic acid amide.

D. Bedford et al. [64] in the study of antistatic hard surface cleaning composition especially suitable for cleaning glass and glossy or shiny surfaces, reported that these compounds comprises (i) water, (ii) 0.01-5% of one or more alkyl or alkenyl ethoxylated amides. preferably oleyl ethoxylated amide and /or erucic ethoxylated amide, (iii) 1-10% of at least one member selected from the group comprising glycol ethers such as butoxypropanol or methoxyisopropanol and C_6-C_{11} alcs., and optionally (iv) 0.005-5 % of an antistatic agent. The antistatic agent may be a quaternized hydrolyzed keratin or a quaternary ammonium compound containing a polypeptide coupled to fatty acid moieties. Thus a composition comprising butoxy propanol 2.1, methoxyisopropanol 1.8, polyoxyethylene oleylamide 0.3, Trideceth wetting agent 0.006, quaternized hydrolyzed keratin 0.1, and water 100 % gave lower smearing level for glass mirror tile and high antistatic properties for glass and poly carbonates.

P.T.Alder et al. [65] studied biaxially oriented multilayer polyolefin films containing fatty acid amide and antiblocking agent with good slip properties and stated that these films comprises a propylene polymer base layer containing ≥ 1 fatty acid amide and a printable outer layer including a blend of a crosslinked silicone antiblocking agent and an inorganic antiblocking agent. Thus a three-layer film was prepared by co-extruding a polypropylene base layer containing 0.16% erucamide and 0.07% bis ethoxylated amine, propylene/ethylene co-polymer (95.3/4:7) first outer layer containing 0.15% crosslinked silicone and 0.165% silica, and second outer layer containing a 40:60 blend of propylene/ethylene/1-butene copolymer(83/2/15), containing 0.15% crosslinked silicone and 0.20% silica, by stretching 4.5 times in the extrusion direction and 10 times in the transverse direction and corona discharge treating, having good slip properties and printability.

P.T.Alder et al. [66] while studying biaxially oriented multilayer polyolefin films containing fatty acid glyceryl monoester and silicone and inorganic antiblocking agents with improved slip properties reported that these films having good hot and cold slip properties for treated and untreated film surfaces, comprise a propylene polymer base layer containing ≥1000 ppm fatty acid glyceryl monoester(based on the wt. of the base layer), and a printable outer layer (on the base layer) containing a blend of a crosslinked silicone antiblocking agent and an inorganic antiblocking agent. Thus a three-layer film (40 μm thick) manufactured by coextruding a polypropylene base layer containing 0.128% glyceryl monostearate, 0.03% bis-ethoxylated amine, 0.064% erucamide and pigments, with outer layers of propylene/ethylene/1-butene co- polymer (91.4/3.8/4.7) containing 0.2% Tospearl 120, 0.1% Tospearl 145 (both are crosslinked silicone) and 0.125% silica and by stretching 4.5 times in the extrusion direction,and 10 times in transverse direction had coefficient of friction for the corona discharge –treated face to metal 0.19 at 40^0 compared to 0.23 for a propylene ethylene co-polymer (96.3:3.7) film without silicone antiblocking agent.

P.T.Alder et al. [67] studied biaxially oriented multiplayer olefin polymer films heat sealble at high packaging speeds and stated that polyolefin films having a low heat seal threshold combined with heat sealing at high speeds for use on packaging machinery, comprise a first outer polymeric layer (≥ 0.7 μm thick) containing an antiblocking agent and a 1-butene-propylene co-polymer and a second outer polymeric layer having a coefficient of friction 0.1-0.35. Thus a three-layer film with 30 μm thickness was prepared by co-extrusion of first outer layer of 71:28:1 propylene/1-butene/ethylene terpolymer containing 0.1% fine silica and 0.1% crosslinked silicone spheres, an~95.5 propylene/ethylene co-polymer second outer layer containing 0.1% fine silica and 0.1% erucamide, and a polypropylene core layer containing 0.16% glycerylmonostearate, 0.08% erucamide and 0.04% bisethoxylated amine antistatic agent, stretching 4.5 times in the extrusion direction and 10 times in the transverse direction, and corona - charged treating the second outer layer to surface energy 40 m J/m^2 showing heat sealed threshold of first outer layer 82^0 and good heat seals at 70 m/min on Versaflow seal machine at 145^0.

J. Kuroki et al. [68] in study of Transparent, heat-sealable propylene copolymer films with no loss of lubricity after hot lamination, noticed that the title film comprise seal layer and support layer in ≥10: ≤ 90 thickness ratio, comprising (A) 100 parts propylene-α-olefin copolymer (DSC fusion peak at $\leq 140^0$) comprising propylene –ethylene random co-polymer (3.8 % ethylene) and/or propylene –ethylene –1 butene random co-polymer (0.5-5%ethylene; 3-25% 1-butene); (B) 100 parts propylene -α-olefin copolymer (DSC fusion peak at $\leq 150^0$) comprising propylene-ethylene-random co-polymer (2-8% ethylene) and/or propylene-ethylene-1-butene random copolymer (0.5-5% ethylene; 3-25%1- butene); (C) 0.01-1 part blocking preventer chosen from silicones; polystyrene and PMMA spherical microparticles with average particle diameter0.5-8 μm and sphericity ≥ 0.8 as measured as $[M/(\pi/r)]^{0.5}/N_{max}$ (M=particle cross section area in mm^2; N_{max} = longest diameterof the particle cross section); (D) 0.01-0.2 part slip agent comprising erucamide and behenamide in 0.2 –5:1 ratio where in the seal layer comprise at least the components A,C and D and the support layer at least the components B and C. A 5 μm thick seal layer comprised 100 parts 1-butene-ethylene-propylene copolymer (Fusion peak at 138^0), 0.15 part silicone spherical particles (average diameter 5 μm) , 0.04 part erucamide , 0.08 part behenamide, and 0.15 part silica and a 20 μm –thick support layer comprised 98:2 propylene-ethylene copolymer (fusion peak at 152^0), 0.06 part erucamide, 0.04 behenamide, and 0.20 part silica. The two layers are coextruded, treated with corona discharge, and dry laminated on a biaxially stretched polypropylene film using a poly urethane adhesive.

K. Kanamori et al. [69] while studying poly-olefin based multilayer films with good optical and antiblocking properties reported that these films are useful for packaging of food, pharmaceuticals, industrial parts etc. comprise at least (a) layer of an ethylene-C_{3-12} olefin copolymer (I; eg.ethylene-1-hexene copolymer, 1-butene ethylene co-polymer) with d 0.905-0.945 and melt flow rate 0.5-30g/10 min containing 0.01-5.0 phr. unsatd. fatty amide with m.p 65-90^0 (eg. erucamide, ethylene bisoleamide), and (b) a layer of I with d 0.900-0.940 and MFR 0.5-3.0 g/10 min contg 0.01-0.20 phr. unsatd. fatty amide with m.p. 65-90^0 and 0.2-2.0 phr. antiblocking agent (eg. powd. crosslinked PMMA, Me methacrylate-styrene copolymer, diatomite, synthetic aluminosilicate, Mg silicate) with wt.-av diameter 8-20 μm and n_D 1.45-1.58, providing the d of (a) layer is ≥0.005 higher than that of (b) layer.

Vinyl chloride polymer and polyurethane compositions for cable sheath and air hose with high flexibility and low frictional properties was studied by N. Fujio et al. [70] and he observed that the compounds, useful for cable sheath esp. in operational parts such as robotic arms of mobile machines, comprises (1) poly (vinyl chloride) or vinyl chloride copolymers 100, liq. plasticizer 55-90 and fatty acid amide 0.2-3.0 parts, or (2) thermoplastic polyurethane 100 and fatty acid amide 0.2 – 3.0 parts. Thus, poly (vinyl chloride) 100 was mixed with DOP liq. plasticizer 60, Armoslip E (erucyl amide) 0.5, tribasic lead sulfate 6, lead stearate 1 and calcium carbonate 50 parts, and molded to give a test specimen showing cable bending resistance 50,000 – 60,000 cycles and good processability.

A. kohara et al. [71] studied multilayer polyethylene laminated films for liquid packaging. These title films, with good low–temperature sealability and useful for high speed liq. packaging process, comprise (a) an ethylene- C_{4-10} α - olefin copolymer (I) layer with only 1 DSC peak at < 100^0, mol. wt. distribution 1.8 – 3.2, and containing unsaturated fatty amides (eg. erucamide), unsaturated fatty bisamides (eg. ethylene bis -oleamide) and polymer particles with decomposition temperature > 250^0 (b) an 1 layer with DSC peak at 110 – 125^0 and optionally (c) other layer (e.g., of Bonyl) on the b layer side.

H. Taniguchi et al. [72] while studying Propylene polymer composition for automobile interior parts, stated that these compounds contain 100 parts propylene block copolymers (structural units derived from ethylene 2-20 mol %; ambient – temperature – sols. in n-decane 6-20%; melt flow rate 10 – 50g /10 min; isotactic fraction of boiling heptane insols. 0.960 – 0.995) and 0.05 – 2.5 parts stearamide, oleamide or erucamide. The compounds are resistant to whitening caused, e.g. by ejector pins, and have good mech. properties and dimensional stability.

K. Ooki et al. [73] studied functional agent– containing multilayer packaging films with durable functionality, comprise a shielding layer (e.g., of plastics, deposited film, metal foil, a foil Tec Barrier), a functional agent containing layer (e.g., containing erucamide, antislipping agent, stearyl monoglyceride antistatic agent) a sealant layer (e.g., Mirason 16P) and optionally a base layer (e.g., of paper) on the shielding layer side.

K. Kobayashi, [74] in a study of thermo – plastic alefin elastomer compositions, with good recycling properties and scratch resistance, no toxic gas emission in burning, and no tacky surface, reported that these compoundscomprise 20- 85 parts crystalline polyolefins, 15-80 parts (hydrogenated) block copolymer comprising styrene (deriv.) polymer block and isoprene polymer or isoprene buta-diene copolymer block with overall 1,2- or 3,4 bonding of isoprene unit ≥ 40%, and 0.1–10 phr. higher fatty amides. An injection-moldable composition comprised polypropylene 50, hydrogenated styrene – isoprene – styrene block copolymer (20% styrene; 1, 2 or 3,4 bonding content 55%) 50, erucamide 3.0 and hydrogenated styrene -

isoprene – styrene block co-polymer (30% styrene; 1,2- or 3,4 – bonding content 7%) 10 parts.

A powder coating for the interior of containers was studied by L. Kiriazis et al. [75] he was of the opinion that the title coatings, which have good adhesion and elasticity and can be used at thicknesses <15μm contain epoxy resins. [Epoxy equivalent weight (EEW) 300 – 5500], phenolic curing agents [OH equivalent weight (HEW) 100–500], and ≤2% lubricants, and have average particle size (D) 5–20 μm. A mixture of epoxidized novolak (DER 642U, EEW 500) 580, bis phenol A- based curing agent (DEH 82, HEW 850) 270, flow modifier 5, fine SiO_2 $_{143}$, and fluidizing aid 2 parts, containing 0.5% erucamide (lubricant), D 9 μm, was coated on the inside of cans and baked at 220^0 for 3 min. to give coatings with good resistance to H_2O and 5% lactic acid at 128^0 (30 min.).

H. Benkhoff et al. [76] while studying sealable and printable oriented polyolefin multilayer films stated that, Title films esp. those sealable on both sides and printable on 1 side have poly (dimethyl siloxane) (I) lubricant in an outer layer that has not been treated with corona, F, flame, or plasma and a combination of I and erucamide as the lubricant in another outer layer that has been treated with corona, F, flame or plasma.

M. Hashi et al. [77] studied polypropylene packaging bands. He was of the opinion that Bands contain 0.05–2 phr fatty amide slip agents and have good moving properties on an automated packing machine. Bands are drawn lengthwise and embossed on the surface and / or rear side and have thickness 0.12–24 mm before embossing. Thus a band contained 0.1 phr erucamide, $CaCO_3$, pigments etc.

T. Takahashi et al. [78] reported Proplyene polymer heat–bandable fibers with soft handle and nonwoven fabrics there from and told that these fibers are prepared by melt. Spinning compounds containing polypropylene (I) and 0.01 – 3.0% silica derivatives and/or fatty acid amides and exhibiting temperature of crystisation $<-(0.094CE^2 - 4.12\ C_E + CE + 127)^0$ (C_E = ethylene content in %; $0 \leq C_E \leq 8$). The nonwovens are useful for diapers and sanitary napkins. A composition containing I and 0.1% erucamide was pelletized, melt spun, crimped, made into a carded web, and bonded at 142^0 to give a nonwoven fabric showing softness rating (A best, D worst) A.

D.G Needham et al. [79] investigated a novel coating process for making dustless color concentrate granules and stated that the title granules having improved dilution and dispersion characteristics comprise individual resin particles coated with pigment and binder. The process comprises preparing a mixture of a pigment (TiO2, Carbon black, diazo yellow, phthalocyanine green) particulated resin, specifically HDPE, and a suitable binder, specifically erucamide having a freeze point $\geq 100^0$ F, where the particulated resin has a higher m.p than the binder. A processing aid, specifically Zn stearate is used for preventing agglomeration of the pigment and the granular color conc. The mixture is agitated in a ribbon blender and heated by frictionto a temperature of $< 200^0$ F (near the m.p of the binder) to cause the binder to become tacky and/or to melt, the resulting coated resin particles are cooled and the title concentration recovered.

T. Kawaguchi et al. [80] studied nonpolluting thermoplastic elastomer compositions with improved surface properties and stated that these compounds comprise 50–80 parts block copolymer having vinyl aromatic compound polymer block (A) and conjugated diene-vinyl aromatic compound random copolymer block (B) 20-50 parts ethylene-propylene copolymers; and 0.2 – 5.0 phr higher fatty acid amides. The compounds containing AB'–

ABC- , or ABA–type co-polymers [B' = conjugated diene polymer or conjugated diene-vinyl aromatic compound random copolymer block; C=vinyl aromatic compound-conjugated diene tapered block (vinyl arom. compd. is gradually increased)] as the block co-polymers are also claimed. Thus, a blend of Dynaron 1320 P (hydrogenated SBR rubber) 65, erucic acid amide (I) 0.90, a dispersant 0.05, and ethylene – propylene copolymer 35 parts was melt -kneaded pelletized and injection – molded to give a test piece showing hardness (ASTM D – 2240) 72, 100% tensile stress 31, tension set 11,dynamic sliding resistance 4.2 and good blocking and scratch resistance and surface touch, vs 72, 31, 11, 61 and poor, resp., for a control prepared without I and the dispersant.

H. Maciejewska et al. [81] while studying poly propylene compositions reported solvent-resistant polypropylene; compounds contain isotactic poly propylene 96–98, oleamide, erucamide, or 1: (1–1.5) oleamide- or erucamide – silica mixtures 0.25–0.6; and optionally stabilizers such as mixtures of antioxidants and halogen absorbers 0.2 – 0.5%.

T. Minaini et al. [82] in a study of Thermal recording material reflected that these material comprise a support with a thermal recording layer containing (a) colorless or leuco basic dye, (b) ≥ 1 a color developer selected from 4- hydroxy – 4' – isopropoxy diphenyl sulfone, 2, 4' – dihydroxydiphenyl sulfone, 4,4'– dihydroxydiphenyl sulfone, bis(3– allyl – 4 – hydroxylphenyl) sulfone, 1,3 – bis (α – methyl - α - (4' – hydroxy phenyl) – ethyl benzene and (c) erucic amide. The material shows good whiteness and recording sensitivity.

T. Matshmura et al. [83] studied isotactic polypropylene compositions for fibers; polypropylene fiber and their manufacture and reported that title compounds comprise 100 parts cryst polypropylene (I) with isotactic pentad ratio (LPF) ≥ 96% and (wt. Average mol. wt.) / (no. average mol. wt.) (Mw / Mn) = 2–15 and 0.01 – 1 part fatty acid amides are spun and drawn to give title fibers giving clear color phase in blending with pigments and drawing at high ratio, i.e., preventing whitening as a result of crystn. Thus, 100 parts I (LPF 98.0%, Mw/ Mn 3.0) was mixed with erucamide 0.10, sumilizer GA80 0.05, Irgafos 168 0.10; and Ca-stearate 0.05 part; spun; and drawn to give title fiber showing improved spinnability and gloss.

Bing Lin Chen et al. [84] studied Maleic acid monoamides in polyolefin compositions with improved slip properties and adhesion of water – based ink and reported that, Polyolefin compounds are claimed, comprising an ethylene or propylene (co) polymer blend with 0.1 – 0.5 phr. of ≥1 alkenyl monoamide of a dicarboxylic acid $RNHCOZCO_2H$ (R= C_{12-24} alkenyl; Z=hydrocabylene), specifically N – erucyl – monoamide of maleic acid.

M. Isoda et al. [85] while studying polyolefin compositions and films thereof stated that the compounds contain arom. Phosphonites I [sic]

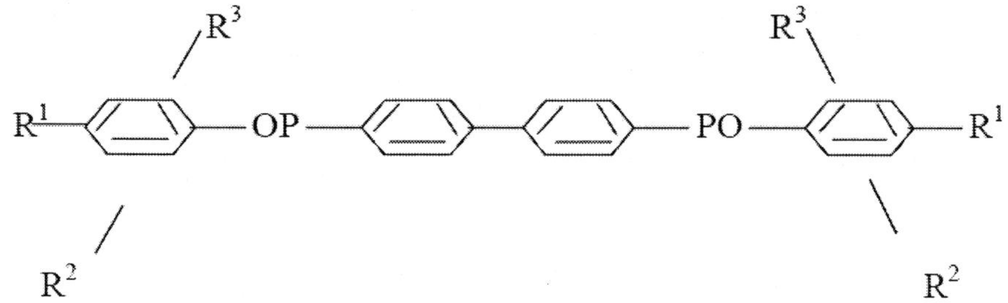

(R^{1-3} = C_{1-4} hydrocarbyl, hydrotalcites, antiblocking agents, and lubricants and / or surfactants and are molded into films with good thermal stability, blocking and discoloration resistance, and anticorrosive, antistatic, antifogging, and heat – sealable properties. Thus 1 butene – ethylene copolymer was blended with I (R^1, R^3= CMe_3; R^2 = Me) 0.1, hydrotalcite 0.05, silton JC 300 (amorphous aluminosilicate) 0.45, and erucamide 0.12% and press molded to give a sheet showing yellowness index (JIS – K 7103)-2.0.

R. Edwards et al. [86] studied polyolefin extrusion coating compositions having good coatability, good adhesion to the substrate, and good chill roll release and stated that polyethylene composition is a blend of a polyethylene component having melt index ~2-100 dg/min. at 190^0 79.93 – 97.97, a specific quantity of a defined tackifying resin having ring and ball softening point 100-140^0, 2-20 and a very specific minute quantity of a unique fatty acid amide, erucamide 0.03 – 0.07%. Polyethylene containing the tackifier Nirez 1135 was modified with 0.025 –0.1% erucamide and extrusion coated to a paper substrate at an extrudate temperature 243^0; showing the noise level of chill roll release lowest (95-97 dB) at 0.04-0.06%.

Antiblocking agents and polyolefin compositions and films containing the agents was studied by M. Mita et al. [87] and he noticed that antiblocking agent powders with powdering degree [i.e. average particle size (P) 0.1-10 μm] comprise basic Mg carbonate optionally assocd. with ≥ 1 of zeolite A, silica gel, talc, feldspar, apatite, $BaSO_4$, and amorphous aluminosilicate as a result of acid – treatment on zeolite A, (2) compounds containing polyolefins 100, the antiblocking agents 0.01 – 1, and higher fatty acid and/or their derivs. 0.01–1 part, and (3) films comprising the compounds Thus linear low – d. 90:10 ethylene – 1 butene copolymer containing 0.4 phr basic Mg. Carbonate (P 3.1 μm) and 0.2 phr erucic amide was inflation – molded to give films, 2 of which were laminated and pressed for 12 h to give a test piece showing peeling strength 12 g/100 cm^2 and haze 5.2%.

O. Berbert et al. [88] while studying PVC – free thermoplastic elastomeric compositions for stoppers of beverage container covers reported that the title compounds comprise olefin or styrene (co) polymers or elastomers (e.g. polyethylene, polypropylene, EPR, EPDM) modified by (a) viscosity reducing agent, e.g. olefin wax stearamide 1-15, or paraffin or naphthenic oils 5-30, (b) lubricants, e.g. silicates, carbonates, sulfates, amorphous or crystalline oxides 0.1 –2, and (d) synthetic zeolites as deodorants 0.1–3%.

A.G. Hoechst [89] studied heat sealable, oriented polyolefin laminates and stated that the title laminates, with low friction coefficient (C_5) and good processability, comprise base layers of polypropylene (I) containing tertiary aliphatic amines and amides of H_2O-sol. C_{8-24} carboxylic acids, and sealable outer layers containing SiO_2, and are free of silicone oils. A laminate (overall thickness 20μm) comprising a base layer of I containing 0.11% N, N-bis (2 hydroxy ethyl) – C_{12-28} alkyl amine (Armostst 300) and 0.14% erucamide between two 0.7-μm outer layers of 8:3:88.5 1 – butene – C_2H_3 – C_3H_6 co-polymer containing 0.33% carboxylic acid – treated SiO_2 (average particle size 4μm) was prepared by coextrusion and biaxial orientation and corona treated. This laminate had good printability C_f (after 14 days) 0.20, and Cf (after 14 days and printing and coating) 0.16; vs. good, blooking and bloking resp., in the absence of SiO_2.

Multilayer ethylene polymer films containing lubricants and antibloking agents for lamination was studied by S. Harauchi et al. [90] and stated that the title films comprise 3 layer containing mainly ethylene-α-olifin. co-polymer with 1 surface 1 layer contg ≥ 0.1 %

silica as an antibloking agent-and 0.04-0.30 % mixt of lubricants m. < 100^0 and ≥ 100^0 [i.e., erucamide m 84° and N,N' – dioleoyl-1, 6-Hexanediamine (I) m 118°] and the other surface layer containing 0.02-0.25 % lubricant mixture A film (i.e. of a blend contg 20 % LDPE comprised 1 surface layer containing erucamide 0.06, I 0.06 , and silica 0.32%,another surface layer containing0.06% erucamide and an intermediate layer containing 0.06 % erucamide and 0.06 % I. The film was dry laminated with a polyether – adhesive coated polyamide film, showing adhesion 400 g /15 mm. initially and after 6 mo at 40°.

H. Morli. et al. [91] while studying plastic bottles reported that the single or multi–ply title bottles showing adequate slip properties during manuf. filling, transporting, and packaging contain ≥2 lubricants (in outer layer in case of multi–ply bottles) of different m.p. Ethylene- Proplylene copolymer was used as inner and outer layers with Eval. EP-F core and the outer layer contained 1000 ppm oleamide and 1500 ppm erucamide.

A. Tokutake et al. [92] studied polyolefin compositions and blocking resistant films. Thereform and reported that the anti corrosive heat and weather resistant polyolefin compounds

Q =

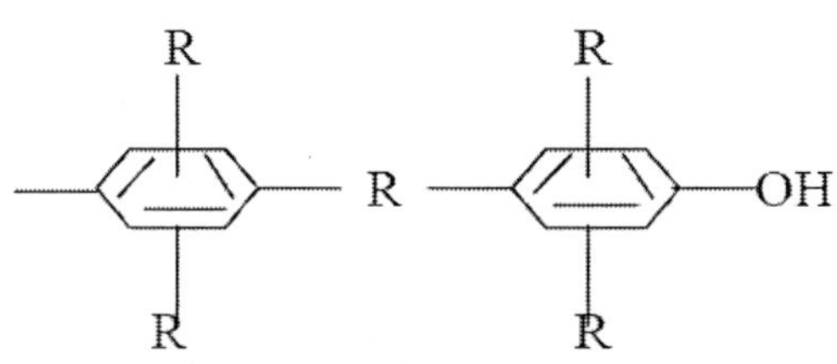

contain P compd (s) chosen from P (OR^1) (OR^2) (OR^3) and P (SR^4) n (SR^5)3-n [R^{1-3} = H, hydrocarbyl, O-contg hydrocarbyl with at least one of R^{1-3} being Q cylohexylphenyl, $(C_pH_{2p}O)_qZ^3$, Z^4 $OCrH_{2r+1}$; Z^3 =H, alkyl, cycloalkl, aryl aralkyl, alkaryl; Z^4 = (un)substituted arylene ; p = 2-9; r= 1–10; R^4, R^5 = alkyl, cycloalkyl alkoxycarbonylalkyl; n = 0–3] hydrotalcite, and blocking preventers and optionally surfactants and lubricants. A 96:4 ethylene –1 butene copolymer compns contain tris (4-cyclohexylphenyl) phosphite 0.1, hydrotalcite 0.05, Silton JC 30, 0.45 and erucamide 0.12 %

T. Yokomizo et al. [93] in the study of multilayer polyethylene bottles for shampoos and liquid detergents noticed that the title bottles with repelling property surfacant-containing shampoos and liq. detergents comprise an inner layer containing repelling materials (e.g.4000-10,000 ppm erucamide or 1-5% silicone oil.

R. Christopherson et al. [94] studied antistatic, printable, sealable, polypropylene film and stated that Biaxially oriented, three – layer polymeric films having a core layer comprising a propylene homopolymer, a 1st heat sealable layer or 1 surface of the core layer and including an ionic hydrocarbyl sulfonate and non –silica, non – migratory slip and/or antiblock agent and a 2nd heat – sealable layer on the other surface of the core layer and including a non migratory slip and/or antiblocking agent but substantially no ionic hydrocarbyl sulfonate have good antistatic properties, combined with good slip/antiblock properties. A 20 μ m-thick (total) stretched title film was prepared by co-extrusion of polypropylene core containing0.16% erucamide and 0.08% bis-ethoxylated C_{12-24} alkylamine with a 1st (0.8-μ m-thick) layer of an ethylene/propylene random copolymer (I) (4%ethylene) containing 2% alkyl sulfonate (mean mol. wt. 328) and 0.3% nylon 12, and a 2nd (0.8-μm-thick) layer of I containing0.1% silica.

Multilayer plastic films having good slip properties were studied by T. Hoshino et al. [95] and he observed that film comprise a seal layer containing 90-99% C_2H_4-C_3H_8 copolymer (I, 2-10% C_2H_4),1-10% HDPE,and 0.03-0.25 phr.slip agents m.100-125^0 and a poly propylene layer. Thus a seal layer contained 0.04 phr. erucamide slip agent and an interlayer contained 0.04 phr. behenic acid amide slip agent, and the film had good slip properties after manuf., heat treatment, and lamination and heating.

T. Hoshino et al. [96] studied propylene polymer films with improved slipping property and stated that the title films, useful for packaging materials, contain (A) 100 parts mixts.of 90-99% ethylene-propylene copolymer (I) with ethylene content (x) 2-10% and 1-10% high-d. polyethylene (II) 0.03-0.25 part slipping agents with m.p 100-125^0 Thus 100 patrs mixture containing 95% I (x 3.6) and 5 parts II; 0.05 part erucamide (m.p.83.5^0), and 0.05 part behenamide (m.p.112.0^0) were melt kneaded and extruded to give a film with friction coefficient 0.81 initially and 0.17 after heating and dry-laminating and good bleeding-out property.

A. Tokutake et al. [97] studied heat-resistant storage stable polyolefin compositions for antiblocking films and stated that the compounds comprise a polyolefin, ≥1 a P compd., hydrotalcite (I), and an additive selected from an antiblocking agent, a lubricant, and/or a surfactant. A film from 1 butene-ethylene copolymer containing $P(OH)_2$ $OZnOP(OH)_2$ 0.1 I 0.05 zeolite 0.45, and erucamide 0.12% showed blocking 0.49 kg/cm^2, haze 5%, and good heat resistance.

O. Shimomura et al. [98] made an investigation for mechanical properties of PVC-talc binary composite systems. Separation and identification of organic compounds in talc as an organic filler and stated that the yield strength of PVC filled with ≤50 phr talc, mica, kaolin, and $CaCO_3$ was studied with increasing amt. of talc or mica, the yield strength of the filled PVC decreased. Erucamide (I), palmitic acid, docosane, bis (2-ethylhexyl) phthalate, di-Bu-phthalate, 1–docosanol, bis (2-ethylhexyl) adipate, and metallic salts of lactic acid, formic acid, and AcOH were isolated from talc and detected I possibly reacted with PVC, and its presence in the PVC-mica system alleviated the problem of poor adhesion at the interface. Newly developed fiberous materials interconnecting PVC and mica were observed using SEM. The presence of aliph.amide with long alkyl chain such as behenamide or lauramide could greatly increase the yield strength.

K. Soejima et al. [99] while studying easily Peelable laminate sheets and manufacture there of reported that the title sheets easily peelable without possibility of readhesion to the original state, useful in pressure–sensitive recording materials contain two or more thermoplastic resin layers of different adhesive properties between two substrates and the member with lower adhesive strength is peelable. Erucamide (900 ppm) – containing LDPE and neat LDPE (lower layer) were coextruded on wood free paper, where in before adhesion the neat LDPE side was treated with ozone and the lower paper treated with corona discharge. The top paper was easily peelable.

K. Azuma et al. [100] in a study of reversible heat sensitive recording media and recording there with reported that the title media comprise a substrate, reversible heat – sensitive recording layer, and elec. Heater layer. A transparent polyester film was coated on one side with an epoxy resin containing carbon fiber to form a 20 μm thick elec. heater layer with surface resistance 10^3 Ω/cm, coated on the other side with a compn. from N – stearyl erucamide 20, vinyl acetate vinyl – alc.- vinyl chloride co-polymer 80, and THF 800 parts to form a 15 μm thick recording layer,and printed on the recording layer a UV – curable urethane – acrylic protective layer.Turbid images are formed on the recording layer by a thermal head (0.5 m J/dot), and the image can be erased by heating the recording surface to 90^0 by the heater layer.

Henning Hans Hub et al. [101] studied thermoplastic polyolefin molding compositions containing hexamethylenebiserucamide and stated that molding compounds with good lubricity and transparency and low blocking contain 50-99.5:50-0.5 EVA, C_2H_4-(meth) acrylic acid copolymers (optionally neutralized), or C_2H_4- (meth) acrylate ester copolymers 80-99.99, N,N'-hexamethylenebiserucamide (I) 0.01-2,and other additives 0-19.99% EVA(97:3) containing 750 ppm I had blocking force 14 N with scattering value 8.1%; vs 150 and 9.4 resp., with 1000 ppm erucamide in place of I.

T. Yamaguchi et al. [102] I n a study of heat and impact-resistant ethylene polymer compositions for lining of container caps reported that The title compounds showing good stress crack resistance, contain low-density polyethylene 30-75, ethylene-C_{3-10} α –olefin copolymers 20-70, and ethylene-propene copolymers or hydrogenated block copolymer elastomers 5-30% as well as 0.01-1 phr lubricants. A mixture of petrothene 203.70, Nipolon L-F 20R 20, JSR-EP 02P 10, and erucamide 0.1 part was molded inside a cap shell as a lining showing good adhesion and impact strength and heat – resistant sealing properties.

M. Yamanaka et al. [103] while studying in –mold labels with good lubricity reported that the title labels contain adhesive layers containing 0.5- 5% C_{12-19} aliphatic amides and / or fatty acid metal salts as lubricants. Thus a multilayered label, which contained an adhesive layer prepared from a 0.5% erucamide containing ethylene – Me acrylate copolymer, showed good adhesion to HDPE blow molded containers.

N. Myata et al. [104] studied polypropylene compositions containing fatty acid amides and stated that the heat and chemical resistant title compounds with improved moldability, useful for packaging foods, comprise propylene polymers 100, neutralizing agents 0.01-5, sorbitol -based nucleating agents 0.05 – 0.5, and fatty acid amides 0.05 – 0.5 parts. Thus R 740 (ethylene – propylene copolymers) 100, Ca stearate 0.15, dibenzilidine sorbitol 0.3, and erucamide 0.2, and Irganox 1010 0.10 part were mixed, dry blended, pelletized, and subjected to continuous injection molding to show no corrosion on molds nor bleeding.

Bing Lin Chen et al. [105] in a study of N–substituted lactamides for improving the slip of polyolefins and the adhesion of water based inks to polyolefins and stated that olefin polymers contain lactamide MeCH(OH)CONHR (R=C_{12-22} alkenyl), optionally a finely divided inorg. material low-d polyethylene containing 500 ppm N-erucyl lactamide (preparedby reactionof equimol amts of erucylamine and Et lactate) and 1500 ppm Microken 801 was blown into films having coefficient of friction 0.39, ink adhesion (0= no adhesion; 10 = 100% adhesion) 10, and good printability.

K. Shishikura et al. [106) worked on polyethylene compositions with good drawing and extrusion properties and reported that the title compounds suitable for uniaxial elongation comprise (A) polyethylene [I; melt index (MI) 0.1 – 1.5 g /10 min, d. 0.950 – 0.965 g / cm^3]100, (B) basic Mg 12 – hydroxystearate (II) 0.01 – 1.0, (C) hydrotalcite 0.01 – 1.0, and (D) fatty amides 0.01 – 1.0 part. A typical composition comprised I [MI 0.72 g / 10 min, d.0956 g /cm^3]100, II 0.15 hydrotalcite 0.05, and Air flow P10 0.10 part.

John. A. Torkington et al. [107] while studying antiblocking and slip agents in extrudable polyolefin compositions stated that the title additives for extrudable polyolefin compounds comprise precipitated or fumed silica (antiblocking agent), a slip agent, and a polyalkylene polyether specifically polyethylene glycol (PEG). The presence of PEG reduces the amt. of the slip agent (e.g. erucamide) needed. Thus a blow–molded film of an ethylene–4methyl–1–pentene co-polymer containing erucamide 1000, DHT–4A (anion exchanger, surface active agent treated hydrotalcite) 250, Syloblock 47 (pptd. silica) 1000, and PEG (average molecular weight 3200) 670 ppm had friction coefficient 0.16 and induced blocking force 0.45 N; vs. 0.23 and 1.2, respectively, for a similar film containing no PEG.

Vinyl based bilayer coatings and their application was studied by S. Hassl et al. [108] and he reported that Antifriction, water- resistant coatings, useful for plastic foam articles such as those in the sports industry, comprise a waterborne primer containing a vinyl chloride copolymer, a plasticizer; and a slow evaporating coalescent and a solvent borne toplayer containing vinyl acetate–vinyl chloride copolymer plasticizer and erucamide.

K. Tsubushi et al. [109] studied liquid developers for electrostatic image containing erucic acid amide and stated that the liq. developers, prepared by dispersing toner particles containing coloring agents and resins in a high resistance and low dielectric constant carrier liquid made from a mixture of perfluorinated hydrocarbon and aliph hydrocarbon solvents and / or silicon type solvent, contain erucic acid amide (I). The developers show improved antioffset properties and generation of hydrocarbon gases is decreased. Thus a fleshed toner prepared from ammonium fumate (Sic) #44 (carbon black), Epolene E-15 (Paraffin wax), amide I lauryl Methacrylate, Methyl methacrylate. Methacrylic acid glycidyl methacrylate copolymer and Isopar H (solvent) were ball milled and diluted with a mixture of fluorinert FC.72 (Perfluoro carbon solvent) and KIF 96 (silicone oil) to give liquid developer.

R. Kuwayama et al. [110] while studying weather resistant chlorosulfonated polyethylene compositions stated that the title compounds, readily processable on rolls and useful for sheet products contain amides of akenoic erucic and / or ricinoleic acids. A mixture of chlorosuflonated polyethylene 100, MgO_5, TiO_2 35, $CaCO_3$ 65 pentaerythritol 3, BHT 1, and oleamide 2 parts was processed on rolls and showed good weather resistance.

J. Pohrt. et al. [111] in a study of antiblocking thermoplastic molding compositions pointed out that the title compounds contain non elastomeric star block co-polymer of 60-95% styrene and 5-40% C_{4-5} conugated diene prepared using an anionic organolithium catalyst along with $\leq$ 25% std. polystyrene of viscosity no. 70-85 cm^3 / g and 0.01-0.15% steramide and erucamide lubricating agent The combination of amides provides improved antiblocking effect and transparency. For example a composition containing 50% butadiene–Styrene block co-polymer and 50% high–impact polystyrene had transparency value 85.1%, when 2% of 1:1 amide mixture was present compared to 84.6% while using Acrawax. The force required to slide on extrudate using the amide was 56% of that needed when Acrawax was present.

G. Miller et al. [112] reported calendrable styrene – diene block copolymer molding compositions for film production and told that the title compounds comprise (a) non elastomeric 60-95: 5–40 styrene – C_{4-5} conjugated diene block copolymer (prepared using an anionic org. Li initiation) $\geq$ 95, (b) erucamide or stearamide antiblocking agent $\leq$ 0.2 (c) hindered phenol stabilizer 0.05 – 2, and (d) Sn, Ca or alkali metal stearate lubricant 0.05 – 2%. In addn., 0.2 – 10% polystyrene of viscosity no. 70 –85 cm^3 /g may be added. The compounds do not stick on calandering rolls and films obtained from them show better thickness uniformity than extruded films.

H. Iwase et al. [113] studied preparation of transparent and slippery packaging films and stated that these films useful for packaging syringes etc. are prepared by melt laminating slipping agent and antiblocking agent containing heat – sealable resin on transparent films and forming protuberances of 0.5 – 10 μm high on the sealable resin side. Thus LDPE containing 700 ppm erucamide and 1000 ppm powdered silica (diameter 5-20 μm) was melt laminated 50μm on a 12-μm PET film, pressed on the LDPE side on a cooling roll with 3-μm depth concaves and wound 2000 m on a roll showing tan θ (θ = angel of starting slipping of 2 films with 800 g load) 0.1, 0.11 and 0.12 for LDPE – LDPE contact and 0.2 , 0.21 and 0.20 for PET – PET contact, as prepared after 10 days at outer most part and 100 m from center, resp. vs 0.1, 0.095 and 0.222 and 0.2, 0.22 0 and 0.096, respectively, for a laminate containing 1000 ppm erucamide and no. silica.

R. Bal.lons et al. [114] while studying multilayer heat sealable polypropylene films for packaging use reported that these films exhibit good machinability and good bonding with water based-adhesives, consist essentially of (A) on outer heat sealable layer containing ≥ 1 slip additive which is incompatible with I and (c) an outer layer consisting essentially of isotatic I in the substantial absence of slip additives. A typical film consisted of an outer layer of 100% isotactic I containing 3000 ppm kaopolite and 2400 ppm slipernat 44 antiblock particles; a core layer of isotactic I containing 400 – 700 ppm erucamide, and an outer heat sealable layer of 90% propylene – ethylene – butene – 1 copolymer, 10% microcrystal wax and slip and antiblocking agents. The film had sarah lamination strength 80-150 g / in and coefficient of friction 0.25 – 0.35.

H. Furuya et al. [115] in a study of styrene resin compositions containing amides for good melt fluidity reported that these compounds contain 0.01 – 5 phr amide $RCONR^1R^2$ (R= C_7-C_{21} alkyl or alkenyl; R1 = C_8 – C_{22} alkyl or alkenyl R^2 = H, C_{8-22} alkyl or alkenyl; R and / R^1 = alkenyl) Denka ABS – HH containing 1 phr N – lauryl erucamide (I) showed spiral flow 77 cm in injection molding vs 65 with erucamide instead of I.

J. Pinter et al. [116] studied Isotactice polypropylene sheets with low friction coefficient good gloss and high strength in welded joints and stated that H (C_2H_4O) x NR (C_2H_4O)y H (I; R = C_{8-20} alkyl; x,y = 1- 25) HOCH = (CH_2OH) $CH_2O_2CR^1$(R^1=C_{12-22} hydrocarbyl) and C_{16-22} fatty acid amides are used at 0.05 – 1, 0.01 – 0.5 and 0.02 – 0.25%, resp., as antistatic agents in isotactic polypropylene (II) containing Ca stearate (III) and hindered phenols to produce sheets with the title properties. Thus a 0.020 mm II sheet containing penta erythritol tetra kis [3 – (3,5 – di – tert – butyl – phenol 0.25 III 0.88 , I 0.15 erucyl amide 0.1 and glyceryl mono stearate 0.075% exhibited sheet – to – sheet static friction coefficient 0.35 (DIN – 53375) surface resistivity 10^3 Ω (CSN – 341382) before and after 14 days aging and gloss 100% (ASTMO 2457 – 70).

A. Mcintosh et al. [117] studied a method of treating extrudable polymers and observed that unsaturated fatty amides are migratory agents for control of surface friction in polymers when extruded. Woven tapes prepared from extruded polypropylene containing 0.1% erucamide (I) had width loss on weft package change over 0.2% vs 0.9 – 1 without I.

K. Tsubushi et al. [118] while studying electrostatographic liquid developers stated that these electrostatog. liquid developers compound of a toner family containing a dye and a resin dispersed in an aliph hydrocarbon as the carrier liquid contain an erucic amide derivative. The dye is prepared by flashing method using the erucic amide deriv. The developers show antioffset property and are capable of high speed – heat – roller fixing and flash fixing. Thus a mixture of NH4 fumate, C black, Epolene E15 and erucic amid was kneaded at 120^0, dried and pulverized to give a dye. The dye was then combined with glycidyl methacrylate-lauryl methacrylate–methacrylate acid. Methyl methacrylate copolymer in Isoper H, to give a liquid developer that showed excellent fixability.

K. Masahiro et al. [119] in a study of propene polymer films containing slipping agents observed that the title films contain a slipping agent m 100 - 125^0 and a slipping agent m 70-90^0 and have froiction coefficient ≤ 0.7 initially and ≤1.0 often heating at $60^{0,}$ i.e. excessive loss of slipperiness during heating is prevented. An extruded film of a 6.0:2.1: 91.9 1 – butene – ethene – propene copolymer containing 0.09% behenamide and 0.03% erucamide had friction coefficient 0.35 initially and 0.30 after 24 h at $60^{0.}$

T. Fujii et al. [120] while studying antifogging antiblocing and transparent polyolefin compositions reported that these compounds comparise on polyolefin 100, an inorg. filler (having av particle diameter (d) [2 – 10 μm] 0.05 – 1.0, an unsatd. fatty acid amide (having m.p ≤ 110) 0.02 – 1.0 and polyhydric alcohol monoester of a fatty acid (monoester content ≥ 60%) 1 – 5.0 parts a compn. comprising polyethylene (melt index 3 g / 10 min, d 0.931 g / cm^3)100 synthetic silica (d 4 μm, sp area 300 m^2 /g) 0.10 erucamide (m.p 80^0) 0.10, diglycerin monosterate (monoester content 75%) 0.5 part had excellent antifogging properties, degree of antiblocking 30g / $100cm^2$ and haze 4.0%.

Y. Azuma et al. [121] studied ethylene polymer - based film compositions and stated that the title compounds with good mech. properties, transparency and heat sealability and useful for packaging, contains (A) 100 parts low – d copolymer of C_{3-4} α olefin unit 0.4 – 9.5, C_{6-12} α olefin unit 0.4 – 9.5 , and ethylene unit 90 – 99 mol% (B) 0.01 – 2 parts aliph amide with m.p, 60 – 120^0 and / or silica gel with av diameter 1 – 20μ and pore vol. ≤ 2.0 mL / g and (c) 0.05 – 3.0 parts antifogging agents. Thus a co-polymer of ethylene unit 96, 1 – butene unit – 2, and 1 – octene unit 2 mol% (a 0.920; melt index 2.0 g / min) 100, erucamide 0.2,silica gel (av diameter 4.0μ pore vol 0.44 m L / g) 0.2 and stearic monoglyceride 1.0 part were kneaded at 200^0 and extruded at 190^0 into 30 - μ tubular film, which showed low haze, low static coefficient of friction higher blocking resistance and high tear strength.

K. Kane et al. [122] studied rubber belts and told that friction – resistance belts, useful for acoustic devices are prepared by molding rubber containing C_{9-24} fatty amides and / or arom amides, which bloom on the belt surface compounded neoprene containing N- octa decyl erucamide (I) was vulcanized to give a belt with friction coefficient (vs an ABS pulvy) 0.3 vs 0.9 without I.

K. Koyakota et al. [123] reported antiblocing rubber compositions and stated that rubbers containing bis amides RCONH $(CH_2)n$ NHCOR' (R,R' = C ≥ 10 aliph. group; n ≥ 1) have good antiblocking properties. A mixture of NBR, carbon black 50, stearic acid 5, S 0.5, vulcanization accelerators (TT and CZ) 3.0 and 1,2 -erucamide ethane (I) 10 parts had adhesion to Cu ~ 0 kg. / cm^2 (vs ~ 0.9 without I) and gave vulcanizates having handness 67, tensile strength 157 kg/ cm^2 and elongation 320% (vs 65, 180 and 310 resp. without I).

2.2. Synthesis

F. Suo.et al. [124] synthesized N, N-ethylene biserucamide, in this synthesis euric acid (I) was prepd through the pressurizing and heating hydrolysis of repesed oil with zinc oxide catalyst. N, N- ethylene biserucamide (II) was synthesized by using the reaction of I with ethylenediamine (III) (III/I) mol.ratio.0.55/1 at 170° far 3h. The final yield of II calcd according to I was ≥ 90%. II is a good lubricant for processing plastics.

Z. lan et al. [125] synthesized erucamide at normal preassure in which erucic acid was treated with NH_3 at atm. Pressure in the presence of (iso $Pro)_4$ Ti catalyst to give erucamide.

The suitable reaction conditions were:

Reaction time	7-8 h
Temp	170-190°
Catalyst amt	1%

K. Ekpenyong et al. [126] studied selectivity for high value products in process Bio-technology and stated that eruic acid and NH_3 react to perform predominantly erucamide in a second order reaction at < 200°. At higher temperature, selectivity in product formations favours eruconitrile. Mechanistically eruconitrile is formed consecutively from erucamide rather than in a parallel reaction step. Minimization of product coloration resulting from nitrile formation requires that amide production reaction can be carried out < 200° (ie 185-190°)

2.3. Properties

W.C Peloso et al. [127] characterized the degradation of the polymer slip additive erucamide in the presence of inorganic antiblock agents and stated that the antagonism between five com. grades of erucamide and seven com. grades of inorganic antiblock agents used in LDPE film formulations is investigated. Mixtures of erucamide and antiblocking agents were thermally aged in air at 90^0 and the increase in discoloration of the mixtures was quantified by measuring the optical absorbance at 440 nm.. The formation of colored degradation products of erucamide appears to be auto catalytic and increase with increasing pH of the antiblock agents. The experimental results suggest impurities in com. antiblock agents such as transition metal oxides and alk. earth metal oxides as well as surface silanol groups in silica antiblocks proposed to account for the observed variation in the extent of antagonism exhibited by the different antiblock agents.

G. Quijada et al. [128] studied the desorption of erucamide vapor in vacuum from Erucamide / Isotactic polypropylene films and isotactic polypropylene (i-PP) were measured micro gravimetrically under vacuum. Experimental Data were used for testing a theoretical description of diffusion process in the erucamide / i-PP system, a partially inhomogeneous binary and incompatible system, and for the calcn. of the diffusion parameters. For films containing 4.6% and 8% erucamide the kinetic curves show good agreement with the Fickian Law. However, for films contg, 24.0% and 44% erucamide, the experimental results were explained by assuming two terms for the overall transport process, a diffusional term which describes the release amorphous regions of spherulties, and a term describing the release from the globules to the amorphous regions of spherulties. An equation was derived to describe experimental data for these high erucamide content blends (24.0% and 44.0%)

R. Barales et al. [129] studied the solubility of erucamide (13 – cis – docosenamide) in isotactic polypropylene and thermal behaviour of their blends and stated that DSC studies of blends prepared by mixing in the melt demonstrated that crystn. Of erucamide / isotactic polypropylene (i-PP) blends results in separating crystalisation of the two components rater than cocrystn. For low content erucamide blends (< 29%), two melting peaks for erucamide were obsd. The higher melting erucamide crystals – are those situated in the external surfaces of iPP; the lower melting crystals are those forming crystalline micro domains located either with in the amorphous region of spherulites and/or in the amorphous regions which surround the spherulties. This behaviour is in agreement with theoritical predictions for crystalisation in immiscible blends. Additional evidence of incompatibility is the poor interfacial adhesion which has been made clear from the large voids left or the fracture surface where the erucamide globules have sepd. from the matrix and from the smooth surfaces of the exposed erucamide globules. A model is proposed for erucamide / i-PP blends which are able to explain quant. in detail all the thermal and morphol. experimental data.

Y. Pei et al. [130] developed the special master batch for high tenacity pylen filament yarn and told that a special master batch for high tenacity pylen filament yarn was manufactured from polypropylene impregnated with a pigment and compound with erucic amide and stearic acid (as lubricants) by mixing and extrusion pelletization. The effects of the type and content of pigment, matrix resin and the content of the mixed lubricant on the master batch were ≤ 25 and the pigment fraction in the pylen filament yarn was 2-3%. Spinning tests of the yarn showed no broken filaments and the tight luster of the yarn was homogeneous and bright.

G. Quijada et al. [131] studied diffusion of erucamide (13 – cis – Docosenamide) In isotactic polypropylene and observed that diffusion Coefficient Of erucamide (13 – cis – docosenamide, $H_3C\ (CH_2)_2\ CH{:}CH\ (CH_2)7\ CONH_2$ eru) into isotactic polypropylene (i-pp) were determined under different isothermal conditions; viz, 323,333 and 353 K ≤ 15 days some other experiments were performed for several predate time intervals (5.10 and 15 days) at a constant temperature (343 K). The diffusion all which was used to establish eru conc. Profiles of diffusant with in thepolymer matrix is similar to that described and used by Roe et al. The concentration Profiles of additive in film stacks were analyzed after soxhlet extn. By capillary gas–liquid chromatography. These profiles were compared with theoritical curves based on SOINS of fick's diffusion equation for thebest fit with the appropriate boundary conditions. The measured concentration Profiles do not show a good agreement with the fickian law values of the diffusion coefficient D in the from 10^{-9} to 10^{-11} $cm^2\ 5^{-1}$ were obtained probably due to a dual mechanism and / or some other characterstic effects of the diffusion of erucamide through i-pp is to be expected that the diffusion process itself may be more complex than that predicted by a simple fiction model. The diffusion or migration of erucamides through i-pp by a lemiemprirical dwal mechanism is given in which 2 simultaneous fiction process perhaps with 2 types of diffusionmols with diffusion coeffs. D1 D1 and D2 on temp followed arrhenious type relationships over the temp range investigated. The activation energies were calculated as Ed(D) = 117, Ed(D1) = 100 and Ed (D2) = 130 KJ mol^{-1} , respectively. The values of diffusion coefficients and activation energies are in the range found for some other additives. The solubility of erucamide in i-pp as a function of temperature was determined directly by extrapolation of diffusion data to penetration depth equal to zero from a plot of these soly parameters as a function of temp, it is very easily under stood that erucamide is insol. (incompatible) in i-pp for temps. Below 313 K, a good correlation between diffusion coefficient and free volume fraction established by the Williams – Landel – Ferry equation was found.

F. Kitchel et al. [132] while studying the differential coefficient of friction in coextruded blown films pointed out that erucamide is routinely used in the manufacturing of blown low-d polyethylene film to modify the film surface coefficient of friction (COF) slip migration in extruded blown film to produce a film having distinctly different COF values from one side of the film to other was investigated. Factors tested included slip placement (in the multilayer structure) film gauge, layer ratio and slip concentration The greatest differential COF was achieved when slip was concluded in the outside skin layer of the co–extruded film.

2.4. Applications

H. Benkhoff et al. [133] while studying sealable, printable biaxially oriented propylene polymer multilayer films using different lubricants in the two outer layers pointed out that the title multilayer films contain a polydimethylsiloxane lubricant in the outer layer not treated with corona, fluorine, flame, or plasma and an erucic amide lubricant in the corona-, fluorine-, flame- or plasma treated outer layer. The transparent films exhibit good machine processability, transparency, sealability (storagte over 100 days did not result in any significant deterioration of the seal seam), and printability.

W. Dye et al. [134] have studied erucic acid containing additives for increasing the lubricity of well fluids in petroleum drilling and pointed out that an additive for increasing the lubricity of a drilling fluid, suitable for use in a no.of types of drilling fluids, consists of a fatty acid or fatty acid deriv. (preferably erucic acid or an erucic acid – containing fatty oil or deriv.), or if the fluid preferably a seflonated petroleum product. High erucic-acid rapeseed oil is the preferred fatty acid deriv. Addnl. component can include a non –water sol. alc. and a hydrotrope (e.g. a sulfonated xylene salt). Suitable non-water solution alcohols include a poly glycol (esp. poly-propylene oxide or poly glycerol). The method of reducing the lubricity of a well fluid by contacting the fluid with the above - described additive is also described

Donald. P et al. [135] studied lubricious radiation – stable polyolefin compositions and articles therefrom and stated that erucamide (I) or its derivatives are useful as radiation – stable lubricants for compounds based on polyolefins with crystallinity 20 – 90% and poly dispersity ≤ 6 and containing liq. Mobilizers (Such as aliph hydrocarbon oil) and are useful for radiation – sterlizable, lubricious medical goods. Thus polypropylene (poly dispersity 2.8) moldings containing aliph hydrocarbon oil 4.7 Tinuwin 770 0.0012 and 0.003 parts I or erucyl erucamide (II) exhibited lower coefficient of friction after 3 Mrads Co^{60} irradn. than a similar molding containing oleamide instead of I or II.

S. Okai et al. [136] introduced solid artistic drawing materials stated that these materials with good release properties and storage stability contain pigments, waxes and $C_{20\text{-}28}$ fatty acid amides. thus a rod molding containing J2522 (poly ethylene) PEWK2 (poly ethylene wax), paraffin wax, carmine 6B, Zn stearate, and erucamide showed good mold release and storage stability.

K. Wakamatsu et al. [137] had claimed Angiogenesis enhancer and told that Angiogenesis enhancer RA [R = (glycolated or epoxidized) higher aliph hydrocarbon residue; A = H, CO_2H, OH (substituted) carbamoyl [substituted or quanternized, amino] is prepared and formulated for treatment of isochoric disease trauma burn or alopecia. Erucamide enhanced blood vessel elongation in a rat cornea test and in a mouse dorsal air sac test and had an ED_{50} of 2.13 μg in a chorioallantoic membrane assay. An ointment contained erucamide 1 and white veseline 50g N – N - Erucoproline Me ester was prepared by converting erucic acid to the acid chloride and then reacting with L- proline Me - ester.

N. Shamoto et al. [138] introduced cable carrying pipes and stated that pipes used in conduit lines with reduced hanging tension are managed from plastic containing lubricants.

M. Goto et al. [139] studied antifriction additives to lubricating oils and stated that condensation products of C_{22-24} unsaturated fatty acids and alkanolamines reduce friction on the metal surface when added to crankcase oils and gear oils. Thus, 105g $NH(CH_2CH_2OH)_2$ was added over 30 min. to a sol^n of 339 g erucic acid in 200 ml MEK at room temperature stirred for 30 min. to stripped off the MEK at 110^0 , and heated at 165^0 for 2h to give a product mainly consisting of N, N-(2-hydroxy ethyl) erucamide (I) crankcase oil containing 1% I produced good antifriction effects and showed full consumption 98.5 compared to 100 for the base oil.

H. Gebler et al. [140] while studying decreasing friction in extruding cartridge tubes from high density polyethylene stated that internal lubricants for the title process are selected from C_{14-40} mono or poly carboxylic acid or their esters, amides, or salts with group IA, IIA, or II B metals or alkali metal salts of C_{8-26} alkyl Ca montanate or erucamide was extruded and drawn 5.7 – fold lengthwise to give 12 gauge cartridge tubes without problem in several hours operation, vs tearing in 10-30 min. during drawing without lubricants.

K. Takeda et al. [141] claimed Air – tight adhesive tapes and told that title tapes prepared, by coating adhesives on coarse surface closed – cell foam rubber tapes have high friction resistance and are useful for sealing connected parts of communication cables. Thus a coarse surface ethylene – propene rubber foam sheet with Asker hardness 80 was coatd with an unvulcanized butyl rubber adhesive to give a sealing tape having good heat cycle resistance (alternating -20^0 and $+60^0$ cycles).

2.5. Analysis

Thermodynamic analysis and experimental verification of surface segregation of erucamide in LLDPE films was studied by S.Sakhal.akar et al. [142] stated that thermo dynamic model is presented to predict surface segregation of additives in semi crystal polymer. The model incorporates the effect of surface tension and morphology differences between surface and bulk on the segregation phenomenon. Experimental results correlate well with the model predictions.

B. Nitin joshi [143] analysed the diffusion of erucamide in linear low-density polyethylene films using FIIR microspectroscopy.

B. Nitin Joshi [144] analysed erucamide by evaluating bulk-to-surface partitioning of erucamide in LLDPE films using FIIR microspectroscopy. The objective of this research is to experimental Characterize the spatial distribution of and additive across the thickness of a single layer polymer film conch. Distribution data are then used to quantity the amt. of additive that diffuses from the bulk of the film to the film surface as a function of time. These results are important because a film's surface property, such as coefficient of friction, depends on the additive surface concentration.The concentration profiles of an erucamide slip agent in linear low-d. polyethylene (LLDPE) files were measured with fourier transform IR micro spectroscopy this pa0per describe in detail the procedures used to microtome specimens and map the concentration profiles for the erucamide. LLDPE system data show that the additive concentration profile is uniform across the thickness of the film immediately after it has been formed. After aging at 60° for 4.5 h (not yet to equil) approx 40% of the additive accumulates at the film surface.

I. Quijada Garrido et al. [145] in solid state NMR studied of structure and dynamics of erucamide / isotactic poly (propylene) blends. Stead that solid state. NMR spins diffusion expts. And two dimensional (2D)' H- 13C NMR experiments were used to steady the phase structive, mobility and domain sizes of mixtures of isotactic poly propylene (iPP) and erucamide (ERU) a low-molecular weight slop and antiblock additive for iPP films. The steady of the proton line shapes in pure erucamide and iPP together with their mixtures reveals the mobility of the components. The mobility is higher in the mixtures than in erucamide and iPP when they are considered sep. In and, 2D wide line separation (WISE) NMR spectroscopy indicated that erucamide mols. Show both low and high mobilities depending and the erucamide content in the mixture. By measuring proton spin diffusion with H detection, the microdomain dimensions have been estel quantized. The domain sizes range from approx. 13 nm to 60 nm

A. Moltby et al. [146] while studying modifications of surface properties of polyethylene plastomers by the use of additive (ii) analyzed that monolayer film of a metallocene catalyzed polyethylene elastomers were prepared containing erucamide slip agent at 200 ppm and with various concentrations of inorganic antiblocking agents. Coefficient of friction and blocking tests performed on these films and natural silica gave the best performance in combination with the slip agent. Amide leaves of 2000-6000 ppm in combination with natural slica at 1000 ppm were evaluated.

T. Nishhioka et al. [147] stead that the slip reducing, mechanism in poly propylene laminated films steadied by FTIR. Fourier transform (FT) IR spectroscopy was used to analyze the slip- reducing mechanism in cast and laminated films of poly propylene containing erucamide amide and behenic amide lubricants.

G. Frutos et al. [148] studied the uses of the Latin square design to study the influence of the experimental parameters on erucamide differential scanning colorimetric curves. A Latin square design was used to evaluate the effect of the sample mass and the heating and cooling rates on the melting of Erucamide, which is commonly used as an additive in the production of poly olefin films. The application of the latin square revealed that the rate heating is the major factor that influences the erucamide differential scanning calorimetric curve.

R.A. Fidler et al. [149] analysed erucamide by online FIIR spectroscopic measurement of erucamide and slica in molder LDPE. On line FIIR spectroscopy is used as a quality and process control tool in loco-d-poly ethylene manually. The design of online FFIR analyzer and its application in characterization of anti-blocking and lubricating agent's erucamide and silica is described.

T. Nishioka et al.. [150] analysed the slip reading mechanism in polypropylene laminated films of FIIR and XPs and stated that FIIR. Spectroscopy and XPs were applied to examine the slip reducing mechanisms of poly proplylene laminated films. The relationship between the slip and the amount of lubricants was established on the basis of quantitative analysis of lubricants on the cast and laminated poly propylene film surfaces.

These results suggested that erucie amide used as a lubricants for the cast film migrated from the film surface into the adhesive layer, probably due to inter mol.interactions chydrogen bonds) with anisocyenurate. Ether polyol used as an adhesive for laminating the films. This led to an understandiong of the slip reducing mechnism in the laminated films.

B. Vigerust et al. [151] studied quantitative analysis of additives in low density poly ethylene using infrared spectroscopy and multivariate calibration and stated that multivariate calibration methods were used to establish a new method for measuring 3 selected additives in LDPE. The detection of the concentration of SiO_2, erucamide and BHT is based on IR spectroscopy and a calibration model. The concentration is between 20 and 1100 wt/ppm. Compared with traditional chem. analysis, the new method is both time and cost effective and is more precise. The method has potential for quality control of polyethylene.

K. Wakamatsu et al. [152] studied isolation of fatty acid amide as an angiogenic principle from bovine mesentery and stated that erucamide (13-docosenamide) was found to be the major bovine mesentry angiogenic lipid as assessed by chorioallantoic membrane (CAM) assay. 2μg of this lipid caused antiogenesis in the assay. Antiogenic activity of this naturally occurring lipid was also found by rat corneal micropocket and mouse dorsal air-sac assays. Specificity of the chem. structure, which elicited activity, was low how ever. The mechanism of angiogenic activity is unknown and this lipid does not promote proliferation of endothelial cells of induce inflammatory effects.

REFERENCES

[1] Inform, Vol.6, No. 9 (September, 1995).
[2] Erucic acid and erucamide –chemicals from rapeseed oil in industry and cooking.(www.Google.com)
[3] Chemical Land 21.com
[4] www.Special chem. For polymers.com
[5] Uniqema storefront@SpecialChem.
[6] *Slip and Anti-Block Additives in Polyethylene film* (Exon Mobil Chemical) www.Google.com
[7] J. Dumas, Ann. chim.etphys.[2], 44,29-144 (1830).(Fatty acids, their their chemistry, properties, production and uses by Klare. S. Markley)
[8] L. Meyer, *Ber.*, 22, 24-27 (1889).
[9] V. F.Balaty, L. L. Fellinger and L. F.Audrieth, *Ind. Eng. Chem.*, 31, 280-282 (1939)
[10] A.W. Hofmann, *Ber.*, 15, 977-984 (1882).
[11] M.Carlet, *Bull. Soc. Chim.* France [1], 1, 76-81 (1859).
[12] A.I. Shatenshtein, Russian Pat.50, 964 (Apr.30, 1937).
[13] L. Henry, *Ber.,* 2, 494-495 (1869).
[14] D.D. Coffmann, N. L. Cox , E. L. Martin, W. E. Cochel , and F. J. Van Atta, *J. Polymer Sci.,* 3, 85-95 (1948).
[15] E.T. Roe, J.T. Scanlon, and D. Swern, *J. Am. Chem. Soc.*, 71, 2215-1218 (1949).
[16] B. Radziszewski, *Ber.*, 18,355-356(1885).
[17] L. Claisen, *Ann.*, 287, 360-371(1895).
[18] Hantzsch and E. Voegelen, *Ber.*, 34, 3142-3163(1901).
[19] K. Auwers, *Ber.,* 34, 3558-3559 (1901).
[20] Hantzsch and F.E. Dolfus, *Ber.*, 35, 226-265 (1902).

[21] Schmidt, *Ber.*, 36, 2459-2482 (1903).
[22] E. Fawsitt, *Proc. Roy. Soc.* Edinburgh, 25, 51-60 (1904).
[23] A.M Buswell, W. H RodeBush, and M.F. Roy, *J. Am. Chem. Soc.*, 60, 2444-2449 (1938).
[24] M.Freymann and R. Freymann, *Bull. Soc. Chim.* France [5], 4, 944-950 (1937).
[25] J.A. Wankelyn and R.Schenk, *J. Chem. Soc.*, 21, 31-32 (1868).
[26] B. Rathke, *Ber.*, 14, 1774-1780 (1881).
[27] K.Auwers, Z. *physik. Chem.*, 12, (689-722) (1893); 15, 33-55 (1894); 23, 449-468 (1897).
[28] L.Claisen and L.Meyerowitz, *Ber.*,22, 3273-3281 (1889).
[29] Pinner, *Ber.*, 25, 1434-1438 (1892).
[30] E.Henderson, *Proc. Roy. Soc.* Edenburgh, 48, 20-27 (1928).
[31] T.Sakurai, *Bull. Chem. Soc.* Japan, 28, 660-664 (1995).
[32] R. R. Mod, F. C. Magne, and E.L. Skau, *J. Chem. Eng. Data*, 5, 478-479 (1960).
[33] N.Sperber, D. Papa, and E.Schwenk, *J.Am.Chem.Soc.*, 76, 3039-3040 (1948)
[34] C.W. Hoerr and H.J. Harwood, *J. Org. Chem.*, 16,779-791 (1951)
[35] J.C.Crocker, *J. Chem.Soc.*, 91, 593-612 (1907).
[36] J.C.Crocker and F. H. Lowe, *J. Chem. Soc.*, 91, 952-962 (1907).
[37] Darapsky and D.Hillers, *J. Prakt. Chem.*, 92, 297-341 (1915).
[38] D.C.Wimer, *Anal. Chem.*, 30,77-80 (1958)
[39] G.Heyl and V, Meyer, *Ber.*, 28,2776-2793 (1895).
[40] P.E Gagnon, J.L.Boivan, and P.A. Boivan, *Can.J. Research*, 28B, 207-212 (1950).
[41] A.W. Hoffmann. *Ber.*, 14, 2725-2736 (1881); 15, 407-416 (1882) ; 15, 752-762 , 762-787 (1882) , 17 1406-1412 (1884).
[42] E. Jeffreys, *Ber.*, 30, 898-901 (1897).
[43] F.Wessely and W.Swoboda, Monatsh., 82, 621-627 (1951).
[44] A.W.Baldwin, E.E. *Walker, and Imperial Chemical Industries*, Ltd., British pat. 475,170 (Nov. 15, 1937).
[45] T, Yoshizaki, *J.Chem. Soc.* Japan, *Ind. Chem. Sect.*, 54, 777-779 (1951).
[46] Armour and Company trade names: Ethomids HT-15, -25 and -60.
[47] Turinsky and R. A. Reck (to Armour and co.), U.S. Pat. 2,672,470 (March 16, 1954).
[48] Hofmann, *Ber.*, 22, 2854-2856 (1989).
[49] Bittner, PB report 63969; *Bibliog. Sci. Ind.Rept.*, 5, 570-579 (1947).
[50] F. Guenther, F. Munz, and H. Haussmann (to I. G. Farbenind A.-G.), U.S. Pat. 1,932,177 (Oct. 24 1933).
[51] V. Paraskova, *Compt. rend.*, 198, 1701-1703 (1934).
[52] Industrial Chemical Division, Evaluation of lubricity with poly olefins. (www.Google.com).
[53] Biochemical process chemicals for plastic manufacturing. (www.Google.com).
[54] *Indian journal of plant Genetic Resources* 11(2): 251-257.
[55] Erucamide 90%, Erucamide 95% ,Erucamide 97%.(www.Google.com).
[56] Asuka, Masahiro (Sekisui Chemical Co., Ltd, Japan) Jpn.Kokai Tokkyo Koho JP 11, 181, 165 [99 181, 165] (Cl.C08L23/00), 6 Jul 1999, Appl. 1997/357, 397, 25Dec.1997; 13pp. (Japan). Chem. Abstr. 131:74609a, No.6, 1999.

[57] Asuka, Masahiro (Sekisui Chemical Co., Ltd. Japan) Jpn. Kokai Tokoyo Koho JP 11 193, 339 (99 193, 339) (Cl. C08L23/02), 21 Jul 1999, JP Appl. 1997/297, 338, 29 Oct 1997; 11 pp. (Japan) *Chem. Abstr.* 131:103190q, No.8, 1999.

[58] Totani, Hideki; Nakasato, Masayoshi, Suzuki, Hiroshi (Denki Kagaku Kagyo K.K., Japan) Jpn. Kokai Tokkyo Koho JP11 228,783[99 228,783](Cl.C08L53/02),24 Aug. 1999,Appl.1998/32,707, 16 Feb 1998; 11 pp. (Japan) *Chem.Abstr.*131:171064b, No.13 1999.

[59] Oka, Kenji (Tocero K.K.Japan) Jpn, Kokai Tokkyo Koho JP 11 12,327 (99 12,327) (Cl. C08F220/14, 19 Jan 1999, Appl. 97/187, 504,26 Jun 1997;7pp.(Japan) *Chem.Abstr.*130:126105j, No.10,1999.

[60] Nakano, Masaki; Tsuruta, Rintaro (mitsui chemicals Inc., Japan) Jpn. Kokai Tokkyo Koho JP 11 52, 865 [99 52,865] (Cl. G09F3/06) 26 Feb 1999, Appl. 97/214,753, 8 Aug 1997; 7 pp (Japan). *Chem.Abstr.*130:238527e, No. 1999.

[61] Asuka, Masahiro (Sekisui Chemical Co., Ltd. Japan) Jpn.Kokai Tokkyo Koho JP 11 100,467 [99 100,467] (Cl.C08L23/00), 13 Apr 1999, Appl. 97/261,496, 26 Sep 1997; 7 pp,(Japan) *Chem.Abstr.*130:312624, No.23, 1999.

[62] Zajac, Marian Pinter, Jozef; Grom, Jan; Durca, Milan (VUSAPL, a. s., Slovakia) Slovakia SK 278,593 (Cl.C08L27/08), 5 Nov 1997, Appl. 6,257, 6 Nov 1989; 2 pp. (Slo). *Chem.Abstr.*129:290913w, No.22, 1998.

[63] Ooe, Hisashi; Inukai, Akihiro (Grand Polymer K.K., Japan) Jpn,. Kokai Tokkyo Koho JP 10 168,252 [98 168,252] (Cl.C08L23/10), 23 Jun 1998,Appl. 96/333,807, 13 Dec 1996; 7pp (Japan). *Chem. Abstr.* 129:123471u, No.10, 1998.

[64] Ueno, Akira; Hojo, Keiji; Okano, Tomomichi; Nishida, Shigeo (Lion Corp., Japan) Jpn Kokai Tokkyo Koho JP 10 273,699 [98 273,699] (Cl C11D10/02), 13 Oct 1998, Appl. 97/95,354,28 Mar 1997; 5pp. (Japan). *Chem.Abstr.*129:332495u, No.25, 1998.

[65] Alder, Paul Thomas (Hoechst, Trespaphan Gmbh, Germany) Eur. Pat. Appl. EP 774,347(Cl. B32B27/18), 21 May 1997, GB Appl. 95/23,350, 15 Nov 1995; 5 pp.(Eng.) *Chem.Abstr.*127:5928j, No.1,1997.

[66] Alder, Paul Thomas; King Dawn Janine (Hoechst, Trespaphan Gmbh, Germany) Eur. Pat. Appl. EP 774,346 (Cl. B32B27/18), 21 May 1997, GB Appl. 95/23, 312, 15 Nov 1995; 5 pp (Eng). *Chem.Abstr.*127:5930d, No.1, 1997.

[67] Alder, Paul Thomas; Gilbert Timothy James (Hoechst Trespaphan Gmbh, Germany) Eur. Pat. Appl. EP 774,350 (Cl. B32B27/32), 21 May 1997, GB Appl. 95/23,349, 15 Nov 1995; 5 pp.(Eng). *Chem.Abstr.*127:5930d, No.1, 1997.

[68] Kuroki, Junichi, Tomita, Hiroto (Dainippon printing co., ltd., Japan) Jpn. Kokai Tokkyo Koho JP 09 76,433 [97 76,433] (Cl.B32B27/32), 25 Mar 1997, Appl. 95/262, 051, 18 Sep. 1995; 7 pp. (Japan).*Chem.Abstr.*127:18870j, No.2, 1997.

[69] Kanamori, Kazunori, Chikanari, Kenzo, (Sumitomo Chemical Co., Ltd., Japan) Jpn. Kokai Tokkyo Koho JP 09,136,390 [97,136,390] (Cl. B32B27/32), 27May 1997, Appl. 95/297, 133, 15 Nov 1995; 13 pp. (Japan).*Chem.Abstr.*127:82519d, No.6, 1997.

[70] Fujio, Nobuhiro, Ishihara, Akihiro (Tatsuta Electrical Wire and Cable Co., Ltd., Japan). Jpn Kokai Tokkyo Koho JP 09,251,811 [97,251,811] (Cl. H01B3/44), 22 Sep 1997. JP Appl. 96/3, 034, 11 Jan. 1996, 9 pp.(Japan).*Chem.Abstr.*127:279257, No.20,1997.

[71] Kohara, Atsushi; Shirai, Takafumi; Moryama, Tamio; Ichiki, Makoto (Kojin Kk, Japan)Jpn. Kokai Tokkyo Koho JP 08,252,894 [96,252,894] (Cl. B32B27/32), 1 Oct

1996, Appl. 95/82, 098,15 Mar 1995; 9 pp. (Japan). *Chem.Abstr.*126:32478y, No.3, 1997.

[72] Taniguchi, Hiroshi; Hirayama, Katsura; Toki, Akio (Mitsui Petrochemical Ind; Toyota Auto Body Co., Ltd. Japan) Jpn Kokai Tokkyo Koho JP 08,302,149 [96,302,149] (Cl. C08L53/00), 19 Nov 1996, Appl. 95/106, 268, 28 Apr 1995; 8 pp. (Japan). *Chem.Abstr.*126:90217y, No.7, 1997.

[73] Ooki, Kazuhiro; Katayama, Shinya (Tokyo Paper Mfg, Japan) Jpn. Kokai Tokkyo Koho JP 08, 300, 573 [96,300,573] (Cl. B32B27/18), 19 Nov. 1996, Appl. 95/128, 806, 1 May 1995; 5 pp. (Japan). *Chem.Abstr.*126:118754, No.9 1997.

[74] Kobayashi, Kyoko; Ito, Juichi; Uchama, Akira (Mitsui Petrochemical Ind, Japan) Jpn. Kokai Tokkyo Koho JP 09 48, 882 [97 48, 882] (Cl. C08L23/02], 18 Feb 1997, Appl. 95/202, 609, 8 Aug 1995; 17 pp. (Japan).*Chem.Abstr.*126:252313f, No.19,1997.

[75] Kiriazis, Leonidas (Basf Lacke + Farben Ag, Germany) Ger.Offen.DE 19,531,585 (Cl. C09D5/46), 6 Mar 1997, Appl. 19,531, 585, 28 Aug 1995; 8 pp. (Ger) *Chem.Abstr.* 126:252473h, No.19, 1997.

[76] Benkhoff, Hermann (Wolff Walsrode Ag, Germany) Ger. Offen. DE 19,534,625 (Cl. B32B27/32), 20 Mar 1997, Appl. 19,534,625, 18 Sep 1995; 9 pp.(Ger).*Chem. Abstr.*126:294399u, No.22, 1997.

[77] Hashi, Masaaki; Niikura, Junji; Asano, Hiroyuki (Sekisui Jushi Kk, Japan) Jpn. Kokai Tokkyo Koho JP 08,175,563 [96,175,563] (Cl. B65D63/10), 9 Jul 1996, Appl. 94/324,246, 27 Dec 1994; 4 pp. (Japan). *Chem.Abstr.*125:223953, No.18, 1996.

[78] Takashashi, Tetsuya; Nakamura, Akira; Kotani, Terumitsu (Showa Denko Kk, Japan) Jpn. Kokai Tokkyo Koho JP 08,246,230 [96,246,230] (Cl. D01F6/46), 24 Sep 1996, Appl. 95/80,761, 13 Mar 1995; 8 pp. (Japan). *Chem.Abstr.*125:331447k, No.26, 1996.

[79] Needham, Donald G. U.S.US 5,455,288 (Cl. 523– 205; C08K9/02), 3 Oct 1995, Appl. 188, 447, 26 Jan 1994; 5 pp. (Eng). *Chem.Abstr.*124:10241x, No.2, 1996.

[80] Kawaguchi, Takao; Kumagai, Katsushi; Ishikawa, Tetsuo; Sunakawa, Masaro (Hitachi Ltd) Jpn Kokai Tokkyo Koho JP 07,233,306 [95,233,306] (Cl. C08L53/02), 5 Sep 1995, JP Appl. 93/334, 931, 28 Dec 1993; 9 pp. (Japan). *Chem.Abstr.*124:31802w, No.4, 1996.

[81] Maciejewska, Hanna; Uhniat, Maria; John, Roman; Dolasinskl, Czeslaw; Jaskola, Konard; Tyburski, Miroslaw; Sep., Karol M.; Kurkowska, Maria; Urbanski, Janusz (Instytut Ciezkiej Syntezy Organicznej "BLACHOWNIA"; Mazowieckie Zaklady Rafineryjne i Petrochemizne) Pol. PL 161,935 (Cl. C08L23/12), 31 Aug 1993,Appl. 282,974, 21 Dec 1989; 4 pp.(Pol). *Chem.Abstr.*124:57918p, No.6, 1996.

[82] Minami, Takehiro (Shinoji Seishi Kk, Japan) Jpn Kokai Tokkyo Koho JP 07,290,835 [95,290,835] (Cl. B41M5/30], 7 Nov 1995, Appl. 94/91, 952, 28 Apr 1994; 8 pp. (Japan).*Chem.Abstr.*124:216233e, No.16, 1996.

[83] Matsumura, Tooru; Hata, Toshikuni (Tonen Kagaku Kk, Japan) Jpn. Kokai Tokkyo Koho JP 08 13,238 [96 13,238] (Cl. D01F6/46), 16 Jan 1996, Appl. 94/163,148, 22 Jun 1994; 6 pp. (Japan). *Chem. Abstr.* 124:263311p, No.20, 1996.

[84] Chen, Bing Lin (Witco Corp.) U.S. US 5,393,814 (Cl. 524 – 219; C08K5/20), 28 Feb. 1995, Appl. 153,779, 17 Nov 1993; 5 pp. *Chem.Abstr.*123:34490h, 1995.

[85] Isoda, Masayoshi; Ishibashi, Tadayoshi; Tokutake, Atsuo (Nippon Petrochemical Co. Ltd.) Jpn. Kokai Tokkyo Koho JP 07 62,162 [95 62,162] (Cl. C08L23/02), 07 Mar 1995, Appl. 93/243,483, 25 Aug. 1993; 10 pp. *Chem.Abstr.*123:201364, 1995

[86] Edwards, Ray, Foster, Bruce W. (Eastman Chemical Co.) U.S. US 5,387,630 (Cl. 524-210; C08K5/20), 07 Feb 1995 Appl. 115,271, 31 Aug 1993; 8 pp. *Chem.Abstr.*123:59045e, 1995.

[87] Mita, Muneo; Takatsu, Shozo; Hiramatsu, Tsunenosuke; Yamasaku, Takashi; Umemoto, Keigo (Nippon Chemical Ind; Kosei Kk; Eishin Kasei Kk) Jpn Kokai Tokkyo Koho JP 06,256,589 [94,256,589] (Cl. C08L23/00), 13 Sep 1994, Appl. 93/65,938, 03 Mar 1993; 6 pp. *Chem.Abstr.*122:107659m, 1995.

[88] Berbert, Otacilio Teixeira; Soares Joao Alfredo de Almeida (Itap S/A) Braz. Pedido PI BR 92 02,713 (Cl. C08L23/04) 01 Feb 1994, Appl. 92/2,713, 13 Jul 1992; 17 pp. *Chem.Abstr.*122:11748p, 1995.

[89] Hoechst A. – G. Ger. Offen. DE 4,306,153 (Cl. B32B27/32), 01 Sep 1994, Appl. 27 Feb. 1993; 9 pp. *Chem.Abstr.*122:162899q, 1995.

[90] Harauchi, Shigeaki; Tanaka, Katsuhiro; Murai, Toshuki; Haneda, Kenji (Okura. Industrial Co. Ltd) Jpn.Kokai Tokkyo Koho JP 06 93,250 [94 93,250] (Cl. C09K3/10), 05 Apr 1994, Appl. 92/270,971, 14 Sep 1992; 5 pp. *Chem.Abstr.*121:232316c, 1994.

[91] Morii, Hitoshi; Iso, Yukio; Tomita, Toshiaki (Kureha Chemical Ind Co Ltd; Kureha Purasuchitsukusu Kk) Jpn. Kokai Tokkyo Koho JP 06 72,422 [94 72,422] (Cl. B65D1/00), 15 Mar 1994, Appl. 92/245, 889, 21 Aug. 1992; 5 pp. *Chem.Abstr.* 121:111004r, 1994.

[92] Tokutake, Atsuo; Sakuma, Hisao; Shibayama, Mctoyuki; Yoshikawa, Toshitsune; Noguchi, Katsutoshi; Satu, Fumio (Nippon Petrochemical Co Ltd) Jpn. Kokai Tokkyo Koho JP 06 80,830 [94 80,830] (Cl. C08L23/00), 22 Mar. 1994, Appl. 92/274,767, 02 Sep. 1992; 15 pp. *Chem.Abstr.*121:135580b, 1994.

[93] Yokomizo, Takayuki (Toppon printing Co Ltd) Jpn. Kokai Tokkyo Koho JP 06 99,481 [94 99,481] (Cl. B29C49/22), 12 Apr 1994, Appl. 92/249,146, 18 Sep 1992; 4 pp. *Chem.Abstr.*121:135904k, 1994.

[94] Christopherson, Roy (Courtaulds Films (Holdings) Ltd.) Eur. Pat. Appl. EP 573,225 (Cl. B32B37/32), 08 Dec. 1993, GB Appl. 92/11,938, 05 Jun 1992; 5 pp. *Chem.Abstr.*120:246957d, 1994.

[95] Hoshino, Toshio; Matsuo, Hironobu; Maruyama, Muneyoshi (Tonen Kagaku Kk) Jpn. Kokai Tokkyo Koho JP 06 00,928[94 00,928] (Cl. B32B27/32), 11 Jan 1994, Appl. 92/186,168, 19 Jun 1992; 6 pp. *Chem.Abstr.*120:247034n, 1994.

[96] Hoshino, Toshio; Matsuo, Hironobu; Maruyama, Muneyoshi (Tonen Kagaku Kk) Jpn. Kokai Tokkyo Koho JP 06 01,894 [94 01,894] (Cl. C08L23/16), 11 Jan 1994, Appl. 92/186,250, 19 Jun. 1992; 5 pp. *Chem.Abstr.*120:247101, 1994.

[97] Tokutake, Atsuo; Sakuma, Hisao; Shibayama, Motoyuki; Yoshikawa, Toshitsune; Noguchi, Katsutoshi; Sato, Fumio (Nippon Petrochemicals Co Ltd) Jpn. Kokai Tokkyo Koho JP 06 80,829 [94 80,829] (Cl. C08L23/00), 22 Mar 1994, Appl. 92/274,766, 02 Sep. 1992; 17 pp. *Chem.Abstr.*120:272333s, 1994.

[98] Shimomura, Osamu; Goto, Kanio; Takai, Yoshio; Sawada, Masami; Takahashi, Shigetoshi (Dep. Appl. Chem., Osaka Inst. Technol., Osaka, Japan 535). Nippon Gomu Kyokaishi 1993, 66 (5), 335 – 43 (Japan) *Chem.Abstr.*120:219417w, 1994.

[99] Soejima, Keiichi; Kohama, Juji; Kawaguchi, Koji (Jujo Paper Co Ltd) Jpn. Kokai Tokkyo Koho JP 05, 307,362 [93,307,362] (Cl. G09F3/10), 19 Nov 1993, JP Appl. 92/16,387, 31 Jan 1992; 6 pp. *Chem.Abstr.*120:219871q, 1994.

[100] Azuma, Kensaku; Fukaya, Kazuhiko; Endo, Takashi; Iida, Tsutomu (Tomoegawa Paper Co Ltd) Jpn. Kokai Tokkyo Koho JP 05,155,148 [93,155,148] (Cl. B41M5/36), 22 Jun 1993, Appl. 91/349,838 10 Dec 1991; 5 pp. *Chem.Abstr.*119:252343, 1993.

[101] Hub, Hans Henning (BASF A. - G.) *Eur. Pat.* Appl EP 521,426 (Cl. C08L23/08), 07 Jan 1993, DE Appl. 4,122,313, 05 Jul 1991; 07 pp. *Chem.Abstr.*119:204611k, 1993.

[102] Yamaguchi, Takao; Sawayama, Matsuo (Shibasaki Seisakusho Kk; Tosoh Corp) Jpn.Kokai Tokkyo Koho JP 05,163,389 [93,163,389] (Cl. C08L23/04), 29 Jun 1993, Appl. 91/329,145, 12 Dec 1991; 10 pp. *Chem.Abstr.*119:205169j, 1993.

[103] Yamanaka, Masatsuki; Yamabe, Takashi (Oji Yuka Goseishi Kk) Jpn. Kokai Tokkyo Koho JP 05 80,702 [93 80,702] (Cl. G09T3/10), 02 Apr 1993, Appl. 91/35,622, 04 Feb 1991; 8 pp. *Chem.Abstr.*119:140853f, 1993.

[104] Myata, Naoaki; Iida, Makoto; Ootani, Yasuo (Idemitsu Petrochemical Co) Jpn. Kokai Tokkyo Koho JP 05 17,637 [93 17,637] (Cl. C08L23/10), 26 Jan 1993, Appl. 91/193,461, 08 Jul. 1991; 5 pp. *Chem. Abstr.* 119:51063z, 1993.

[105] Chen, Bing Lin; Barker, James A. (Witco Corp.) U.S. US 5,206,273 (Cl. 523 – 300; C08J3/28), 27 Apr 1993, Appl. 791,562, 12 Nov 1991; 7 pp. *Chem.Abstr.*119:118729, 1993.

[106] Shishikura, Katsuhiro; Maehara, Hiroyuki (Chisso Corp.) Jpn. Kokai Tokkyo Koho JP 04,142,350 [92,142,350] (Cl. C08L23/04), 15 May 1992, Appl. 90/263,947, 03 Oct. 1990; 6 pp. *Chem.Abstr.*118:40127h, 1993.

[107] Torkington, John Alexander (BP Chemicals Ltd.) Eur.Pat Appl. EP 526,117 (Cl. C08L23/02), 03 Feb 1993, GB Appl. 91/16,498, 31 Jul 1991; 8 pp. *Chem.Abstr.*118:235245c, 1993.

[108] Hassl, Scott; Kenow, Mark (PDI, Inc.) U.S. US 5,192,608 (Cl. 428-319.7; B32B27/00), 09 Mar. 1993, Appl. 781,861, 24 Oct 1991; 6 pp. *Chem.Abstr.*118:236074h, 1993.

[109] Tsubushi, Kazuo; Kuramoto, Shinichi; Umemura, Kazuhiko; Uematsu Hidemi (Ricoh Co., Ltd.) Jpn. Kokai Tokkyo Koho JP 03,200,266 [91,200,266] (Cl. G03G9/12), 02 Sep 1991, Appl. 89/342.245, 28 Dec 1989; 6 pp. *Chem.Abstr.*116:95773e, No.10, 1992.

[110] Kuwayama, Rikiji; Kaji, Shinji; Ono Yasuharu Sato, Hisashi (Showa Denko Dupont K.K.; Hitachi Cable Ltd.) Jpn. Kokai Tokkyo Koho JP 03 227,349 [91,227,349] (Cl. C08L23/34), 08 Oct 1991, Appl. 90/18,957 31 Jan 1990; 5 pp. *Chem.Abstr.*116:61393r, No.8, 1992.

[111] Pohrt, Juergen; Jenne, Helmert; Walter, Hans Michael; Gausepane, Hermann; Benedix, Franz (BASF A-G) Eur.Pat Appl. EP. 390,019 (Cl. C08L53/02), 03 Oct 1990, DE. Appl, 3,910,259, 30 Mar 1989; 6 pp. *Chem.Abstr.*114:83252e, No.10,1991.

[112] Pohrt, Juergen; Hatzmann, Guenter (BASF A-G) Ger. Offen DE 4,009,065 (Cl. C08L53/02), 04 Oct 1990, DE Appl. 391,040, 24 Mar 1989; 4 pp. *Chem.Abstr.*114:123889b, No.14, 1991.

[113] Iwase, Hiroshi; Imaizumi, Atsushi (Toppan printing Co., Ltd) Jpn. Kokai Tokkyo Koho JP 02,184,436 [90 184,436] (Cl. B32B27/18), 18 Jul 1990 Appl. 89/4,200 11 Jan 1989, 5 pp. *Chem.Abstr.*113:233005 xs, No.26, 1990.

[114] Balloni, Ricardo; Donovan Kevin M; Keung, Jay.k (Mobil Oil Corp.) U.S. US 4,956,232 (Cl. 428-349; C09J7/02); 11 Sep 1990, Appl. 290,131, 27 Dec. 1988, 5 pp. *Chem.Abstr.*113:213587k, No.24, 1990.

[115] Furuya, Hirnobu; Takahashi, Hideki (Nippon Oil and Fats Co., Ltd) Jpn. Kokai Tokkyo Koho JP 01,201,352 [89,201,352] (Cl. C08L25/04), 14 Aug 1989 Appl. 88/25, 695, 08 Feb 1988; 3 pp. *Chem.Abstr.*112:78739y, No.10, 1990.

[116] Pinter, Jozef; Saraz, Jozef; Kondrkova, Dagmar Kopernicky Ivan; Kasovic, Vratko; Simko, Karol; Schmidt, Gejza Czech.CS254 945 (Cl. C08L23/12), 15 Nov 1988, Appl. 86/3,684, 21 May 1986, 4 pp. *Chem.Abstr.*111:8461g, No.2, 1989.

[117] Mcintosh, Alastair; Simpson, George (Don and Low (Holdings) Ltd) Brit. UK Pat. Appl. GB 2,204,043 (Cl. C08K5/20), 02 Nov 1988, GB. Appl. 87/4,187 23 Feb 1987; 6 pp. *Chem.Abstr.*111:76829n, No.10, 1989.

[118] Tsubshi, Kazuo Kuramoto, Shinichi; Minamitani, Toshiki; Umemura Kazuhiko; Nagai Kay Oko (Ricoh Co. Ltd) Jpn. Kokai Tokkyo Appl. 87/301,378, 28 Nov 1987; 5 pp. *Chem.Abstr.*111:244283, No.26, 1989.

[119] Masahiro, Kakugo; Seiichiro; Ima (Sumitomo Chemicals Co. Ltd) Eur. Pat. Appl. EP 249; 342 (Cl. C08L23/10), 16 Dec 1987, JP. Appl. 86/110,231, 13 May 1986; 14 pp. Chem.Abstr.108:168788k, No.20, 1988.

[120] Fujii, Toshio; Hasegawa, Koichi (Mitsubishi Chemical Ind. Co. Ltd) Jpn. Kokai Tokkyo Koho JP 63,145,350 [88,145,350] (Cl. C08L23/02), 17 Jun 1988, Appl. 86/291,739, 08 Dec 1986; 7 pp. *Chem.Abstr.*109:232185m, No.26, 1988.

[121] Azuma Yasushi; Tanaka; Yoshikatsu (Idemitsu Petrochemical Co.,Ltd) Jpn, Kokai Tokkyo Koho JP 03 06,038 [88 06,038] (Cl. C08L23/08), 12 Jan 1988 Appl. 86/150,029;26 Jun 1986; 5 pp. *Chem.Abstr.*108:222819m, No.26, 1988.

[122] Kanai, Kazumi (Hokushin Kogyo; Inc.) Jpn. Kokai Tokkyo Koho JP 62,147,141 [87,147,141] (Cl. F16G1/00), 01 Jul 1987 *Appl.* 85/285,476, 20 Dec 1985; 3 pp. *Chem.Abstr.*107:156140j, No.18, 1987.

[123] Koyakata Keisuke (Fujikura Rubber, Ltd.) Jpn. Kokai Tokkyo Koho JP 62 72,734 [87 72,734] (Cl. C08L21/00), 03 Apr. 1987, Appl. 85/210,989 26 Sep 1985; 3 pp. *Chem.Abstr.*107:200122f, No.22, 1987.

[124] Suo, Fuxi (Dep. Chemistry, Anyang Teacher's school Anyang, Peop. Rep. China 455000) Huaxue Shijie 1997, 38(2), 80-82 (Ch.), Shanghaishi Huaxue Huagong Xuehui. *Chem.Abstr.*127:222199b, No.16, 1997.

[125] Lan. Zhongwei; YuXiaoqi; Wang; Qinghua; Zhao; Jihong (Dep. Chem.Sichuan Univ, Chengdu Peop.Rep. China) Huaxue Shijie 1990, 31(8),350-1 (Ch) *Chem.Abstr.*114:187923r, No.20, 1990.

[126] Ekpenyong, Kieran I; Walls, J.Ray (Dep. Chem, Univ Jos, Jos Nigeria) *Ind. Eng. Chem. Res.*1990, 29(I), 121-5(Eng). *Chem.Abstr.*112:22724 xs, No.4, 1990.

[127] Peloso, Christopher W.; O'Connor, Michael J.; Bigger, Stephen W; Scheirs, John (School of Life Sciences and Technology , Victoria University of Technology, Melbourne, 8001 Australia). *Polym. Degrad. Stab.* 1998, 62(2), 285-290 (Eng), Elsevier Science Ltd. *Chem.Abstr.*130:111022f, No.9, 1999.

[128] Quijada-Garrido, I.; Barrales- Rienda, J.M.; Alejo Espinoza, L.; Fiero, J.L.G. (Departmento de Quimica – Fisica de Polimeros Instituto de Ciencia y Technologia de Polimeros C.S.I.C., Madrid Spain E-28006) Macromolecules 1996, 29(27), 8791-8797 (Eng), American Chemical Society. *Chem.Abstr.*126:32161q, No.3, 1997.

[129] Quijada-Garrido, I.; Barrales-Rienda, J.M.; Perena, J.M.; Frutos, G. (Departmento de Quimica – Fisica de Polimeros Instituto de Ciencia y Technologia de Polimeros

C.S.I.C., Juan de la Cierva,Madrid Spain E-28006). *Polymer* 1997, 38(20), 5125-5135 (Eng), Elsevier. *Chem.Abstr.*127:248685g, No.18, 1997.

[130] Pei, Yuntong; Li, Hengtai (Research Institute, Luoyang Petrochemical Complex, Luoyang, Peop. Rep. China 471012). Suliao Gongye 1997, 25(4), 85-87(Ch), Suliao Gongye Bianjibu. *Chem.Abstr.*127:319978t, No.23, 1997.

[131] Quijada-Garrido, I.; Barrales-Rienda, J.M.; Frutos, G. (Departmento de Quimica – Fisica de Polimeros Instituto de Ciencia y Technologia de Polimeros Madrid, Spain E-28006). Macromolecules 1996, 29(22), 7164-7176(Eng). *Chem.Abstr.*125:223037t, No.18, 1996.

[132] Kitchel, Frank. T. (USI Div. Quantum Chem. Corp., Rolling Meadows. IL 60008 USA). Tappi J. 1988, 71(12), 156-9 (Eng). *Chem.Abstr.*110:116138v, No.14, 1989.

[133] Benkhoff, Hermann; Brandt, Rainer; Schwarz, Willi (Wollff Walsrode Ag, Germany) Ger. Offen. DE 19,534,623 (Cl. B32B27/32), 20 Mar 1997, Appl. 19,534,623, 18 Sep 1995; 9 pp. (Ger). *Chem.Abstr.*126:278626k, No.21, 1997.

[134] Dye, William; Clark, David E.; bland, Ronald G. (Baker-Hughes Incorp.) Eur. Pat. Appl. EP 652,271 (Cl.C09K7/00), 10 may 1995, US Appl. 150,217, 10 Nov 1993; 27 pp. *Chem.abstr.*123:118055h, 1995.

[135] Payne, Donald D. Jr.; Waller, Jess M.; Clarke, Richard P.; Titus George R.; Martin, David A. (Becton, Dickinson and Co.) *Eur. Pat. Appl.* EP 476,401 (Cl. C08L23/02), 25 Mar 1992, US Appl. 576,750, 04 Sep 1990; 11 pp. *Chem.Abstr.*116:2368435, No.24, 1992.

[136] Okai; Sadahiro (Pentel Co. Ltd.) Jpn. Kokai Tokkyo Koho JP 03 35,069 [91 35,069] (Cl. C09D13/00), 15 Feb 1991, Appl. 89/170,460, 30 Jun 1989; 5 pp. *Chem.Abstr.*115:94607w, No.10, 1991.

[137] Wakamatsu Kaori Kondo, Koichi; Sudo Katsuichi (Takeda Chemical Industries Ltd) Eur.Pat. Appl. EP 314,105 (Cl. A61K31/16), 03 May 1989 JP. Appl. 87/274,461 29 Oct 1987; 17 pp. *Chem.Abstr.*112:70939t, No.9, 1990.

[138] Shamato, Naoki; Suzuki Hideo; Sugawara, Yasuyuki (Fujikara Ltd) Jpn. Kokai Tokkyo Koho JP 01,110,013 [89,110,013] (Cl. H02G9/06), 26 Apr 1989. Appl. 87/263; 724, 21 Oct 1987; 2 pp. *Chem.Abstr.*112:140776p, No.16, 1990.

[139] Goto, Masahisa; Nagagawa, Hiroshi; Ooshima Masataka (Idemitsu Kosan Co. Ltd) Jpn. Kokai Tokkyo Koho JP 63,180,000 [88,180,000] (Cl. C10M159/12), 23 Jul 1988, Appl. 87/10,073 21 Jan 1987; 4 pp. *Chem.Abstr.*109:233991h, No.26, 1988.

[140] Gebler, Horst; Pfahler, Gerhard; Schindler, Arnold (Hoechst A-G) Eur.Pat, Appl. EP 254,264 (Cl. C08L23/06), 27 Jan 1988, DE Appl. 3,625,258 25 Jul 1986; 4 pp. *Chem.Abstr.*109:74544x, No.10, 1988.

[141] Takeda, Kazutoki; Kuboba, Yoshinori; Sasaki, Atsushi; Yanaginara, Morio (Nippon Telegraph and Telephone Public Corp.) *Jpn. Kokai Tokkyo Koho* JP 63 23,984 [88 23,984] (Cl. C09J7/02), 01 Feb 1988, Appl. 86/165,629, 16 Jul 1986; 4 pp. Chem.Abstr.108:188155e, No.22, 1988.

[142] Sakhalkar, Sachin S.; Hirt, Douglas E. (Department of Chemical Engineering, Clemson University, Clemson, SC 29634-0909 USA). *Annu. Tech. Conf. - Soc. Plast. Eng.* 1998, 56th (Vol. 3), 2826-2831 (Eng). *Chem.Abstr.*130:39300a, No.4, 1999.

[143] Joshi, Nitin B. (Clemson Univ., Clemson, SC USA). 1998. 212 pp. (Eng). Avail UMI, Order No. DA9833473, From Diss. Abstr. Int. B 1998, 59(5), 2318. *Chem.Abstr.*130:82205k, No.7, 1999.

[144] Joshi, Nitin B.; Hirt, Dougles E. (Department of Chemical Engineering, Center for Advanced Engineering Fibers and Films, Clemson University, Clemson, SC 29634-0909 USA). *Appl. Spectrosc.* 1999, 53(1), 11-16 (Eng), Society for Applied Spectroscopy. *Chem.Abstr.*130:210334r, No.16, 1999.

[145] Quijada-Gerrido, Isabel; Wilhelm, Manfred; Spiess, Hans Wolfgang; Barrales-Rienda, Jose Manuel (Max –Planck Institut Polymerforschung, D-55021 Mainz, Germany). Macromol. Chem. Phys. 1998, 199(6), 985-995 (Eng), Huethig and Wepf Verlag. *Chem.Abstr.*129:16621w, No.2, 1998.

[146] Maltby, Adam; Marquis, Richard (Croda Universal Ltd, Hull, UK HU6 7PH). Polym., Laminations, *Coat. Conf.* 1997, 2, 423-426 (Eng). *Chem.Abstr.*129:123137q, No.10, 1998.

[147] Nishioka, T.; Tanaka, Y.; Teramae, N.; Gohshi, Y. (Polym. Res. Lab., Idemitsu Petrochem. Co., Ltd., Ichihara, Japan 299-01). *Proc. SPIE-Int. Soc. Opt. Eng.1993, 2089(9th INTERNATIONAL CONFERENCE ON FOURIER TRANSFORM SPECTROSCOPY,* 1993), 472-3 (Eng). *Chem.Abstr.*122:33088a, 1995.

[148] Frutos, G.; Quijada, I.; Barrales J.M. (Fac. Farm., Univ. Complutense, Madrid, Spain 28040). *Analyst* (Cambridge, U.K.) 1994, 119(7), 1547-51 (Eng). *Chem.Abstr.*112:213591x, 1994.

[149] Fidler, R.A. (Polym. Process Control Div. Automatik Mach. Corp., Charlotte, NC 28273 USA). *Annu. Tech. Conf.-Soc. Plast. Eng.* 1992, 50th (Vol. 1), 211-14 (Eng). *Chem.Abstr.*120:272074h, 1994.

[150] Nishioka, T.; Tanaka. Y.; Kume, K.; Satoh, K.; Teramae, N.; Gohshi, Y. (Polym. Res. Lab., Idemitsu Petrochem. Co., Ltd., Ichihara, Japan 299-01). *J.Appl. Polym. Sci.* 1993, 49 (4), 711-17 (Eng). *Chem.Abstr.*119:118568v, 1993.

[151] Vigerust, B; Kolset, K; Nordenson S; Henriksen, A; Kleveland, K (Cent. Ind. Res., 0314 Oslo, Norway). *Appl.Spectrosc.* 1991, 45 (2), 173-7 (Eng). *Chem. Abstr.*114:144706e, No.4, 1991.

[152] Wakamatsu, Kaori; Masaki,Tsuneo; Iton, Fumio ; Kondo, Koichi; Sudo, Katsuichi (Tsukuba Res. Lab, Takeda Chem Ind. Ltd, Tsukuba, Japan 300-42). *Biochem. Biophys. Res. Commun.* 1990, 168(2) 423-9(Eng). *Chem.Abstr.*113:18371h, No.3, 1990.

In: New Trends in Chemical Physics Research
Editor: James N. Lapherty

ISBN: 978-1-61668-535-5

Chapter 9

COMPUTER MOLECULAR MODELLING WITH EXPERIMENTAL STUDIES, OF PYROLYSIS CHEMISTRY OF LOW-RANK COALS CONTAINING INORGANIC COMPLEXES AND CATALYTIC STEAM GASIFICATION

George Domazetis[*,1,2]***, Bruce D. James***[†2] ***and John Liesegang***[3]
[1]Clean Coal Technology Pty. Ltd., Australia.
[2]Chemistry Department, La Trobe University, Victoria, 3086, Australia
[3]Physics Department, La Trobe University, Victoria, 3086, Australia

ABSTRACT

Metal-mediated coal pyrolysis chemistry precedes catalytic steam gasification, which may produce hydrogen-enriched gases with CO_2 capture, and remove sulphur, nitrogen, and ash pollutants. Catalytic species are formed during the pyrolysis, which are derived from aqua-inorganic species added to low-rank coal; the addition of these inorganic species must be done with a careful adjustment of the pH, otherwise a mixture of various inorganic compounds are physically added to the coal, which may not form catalytically active species. Inorganic compounds (Fe, Co, Ni, Na, K, Ca, Mg, Ba, Al, Cu, Pb, and Zn) have often been added to low-rank coals using various methods, resulting in coal samples containing poorly characterized inorganic mixtures. Metal-mediated pyrolysis chemistry is extremely complex and studies at a molecular level require both experimental and high-level molecular modelling computations. In this paper, molecular models of low-rank coal containing inorganic species have been constructed and optimised to ground state using semi-empirical quantum mechanics computations (SE). These models have been used to develop char models containing metal species; SE and *ab initio* density functional theory computations are used to examine metal-mediated pyrolysis chemistry and reaction routes to CO_2, H_2O, CO and H_2 formation. Insights from experimental and

[*] Corresponding author. †deceased 2009. E-mail: G.Domazetis@latrobe.edu.au

molecular modelling studies include: the nature of aqua-metal complexes added to coal, transformations of metal species in char molecules, and reaction mechanisms, particularly for the formation of CO_2, CO and H_2. Reactions leading to the formation of H_2 and CO from steam and char containing metal clusters and the nature of active sites have been examined using computational molecular modelling. Experimental and modelling data indicate that catalytic steam gasification of low-rank coals containing Fe and Ni significantly increases the yield of H_2.

Keywords: Low-rank coals; computational molecular modelling; pyrolysis chemistry; catalytic gasification.

1. Introduction

An understanding of pyrolysis of low-rank coals containing metal complexes is critically important for progress in catalytic steam gasification, which can produce hydrogen-enriched gases with CO_2 separation and capture and the removal of sulphur, nitrogen, and ash pollutants. Low-rank coal is an heterogeneous substance and its utilisation involves very complicated physico-chemical processes. The extremely complicated nature of pyrolysis chemistry necessitates a reliance on: experimental and computer molecular modelling studies of inorganic species added to low-rank-coals, the impact of these on the distribution of pyrolysis products and their transformations during heating. Modelling of coal pyrolysis requires a molecular representation of the low-rank coal molecule to provide a molecular basis for reaction routes of metal-mediated pyrolysis and catalytic gasification.

Experimental and molecular modelling studies of metal-mediated pyrolysis face a number of difficulties. Experimental studies need to ensure reproducibility of the data using well mixed coal samples, to avoid the problems inherent in the heterogeneous nature of coals and the variations in data observed from different apparatus and conditions. It has been demonstrated over a lengthy period that pyrolysis experimental results are dependent on a number of factors, including the nature of the coal sample, the apparatus and particularly changes of temperature with time. The variation in the characteristics of each coal and its uses, have resulted in a large number of apparatus and analysis techniques. Pyrolysis experiments have been carried out using slow to extremely high heating rates, and the major goal has been to measure the amounts of the products formed with temperature. Sophisticated experimental techniques have also been utilised to obtain data on the transformations of inorganic species in coal, and these have been useful for computational molecular modelling studies of the chemical pathways of metal-mediated pyrolysis. As a result of this variation in experimental apparatus and method, experimental results often differ; it is often difficult, therefore, to correlate experimental data with molecular modelling results.

Pyrolysis encompasses disruption of hydrogen bonds, loss of oxygen functional groups as CO_2, CO, and H_2O, bond breaking that forms molecular fragments (depolymerisation) and cross-linking (bond-formation), loss of gaseous hydrocarbons, transport of tars, and formation of char with an increase in the relative proportion of aromatic hydrogen. Pyrolysis of low-rank coal is often categorised as low-, or high-temperature. Pyrolysis has been thought to consist of primary and secondary processes, and also of various stages subjected to a given temperature-time ramp [1, 2]. Kinetic treatments of pyrolysis, which are phenomenological models, cannot deal with the molecular transformations of inorganic species, or predict the nature of the catalysts that may form during metal-mediated pyrolysis; these require high-level coal molecular modelling studies [3]. Molecular modelling of coal pyrolysis requires a molecular structure of low-rank coals that adequately encapsulates the properties of these coals; the complicated nature of the chemistry, and the stringent requirements for calculations at a molecular level, necessitates the use of semi-empirical quantum mechanics (SE), and where possible, *ab initio* computations, such as Density Functional Theory (DFT). Theoretical studies are difficult because coal is an heterogeneous substance without a distinct molecular structure. Instead, coal molecular models have been developed using experimental data obtained from well mixed samples that reflect the averaged properties of the coal. Molecular structures of coal have been based on experimental data obtained from high-rank coal [4] and from low-rank coals [5]; experimental techniques have been important in elucidating many coal structural features [6].

Computational studies of molecular models of coal have often used low-level techniques, such as molecular mechanics; such treatments are inadequate for mechanistic studies of coal pyrolysis. Insights into chemical reaction mechanism require high-level techniques, such as SE and *ab initio* DFT computations; these computations are rigorous, and any ineffectual chemical structures can be rejected if, for example, a self-consistent field result could not be obtained. Additionally, computational packages that optimise molecular structure to a ground state provide a great deal of information to enable a judgement of the suitability of the molecular model (e.g., bond lengths and angles, hydrogen bonds, total energy, calculated heats of formation, partial charges, and bond orders). The size of the molecular model, however, is limited by the capabilities of the SE computational package and computer hardware, and a molecule needs to be created after a detailed assessment of the similarity of the measured and calculated properties of the suitably sized model [7,8] A 3-D molecular structure of low-rank coals must encapsulate measured properties, such as the elemental composition, distribution of functional groups, and the ratios of aromatic and aliphatic carbon and hydrogen. The model must also be comparable with experimentally measured fundamental aspects of low-rank coals, such as for example hydrogen-bonded and bulk water in coal, inclusion of aqua-ionic and aqua-transition metal species and thermal changes leading to char formation with inorganic species. The model should also provide insights into specific reactions leading to major products during pyrolysis. Such modelling effort improves our understanding of the chemical reaction routes for the formation of the major products (H_2O, H_2, CO and CO_2) during pyrolysis and for studies of catalytic steam gasification.

This paper discusses the chemistry of metal-mediated pyrolysis of low-rank coals using SE and DFT molecular modelling computations of coal and chars molecular structures to elucidate major events during pyrolysis, and shows these are consistent with experimental data; these progress our understanding of metal-mediated pyrolysis and the chemistry of catalytic steam gasification.

2. Experimental and Computational Methods

Brown coal with low levels of silica and clay, acid washed (aw) to very low amounts of ash was obtained from the La Trobe Valley in Victoria, Australia (Yallourn and Loy Yang

Open Cut mines) and Germany (Rheinbraun GBT) for these studies. Coal samples were prepared for pyrolysis and steam gasification experiments by: (i) removing all inorganic species in the coal samples, as described previously [10], and (ii) adding selected inorganic aqua-species to the aw coal using a step-wise adjustment of the solution pH of the metal salt mixed with the coal sample [9,10]. Initial experiments were carried out by adding salts of Fe, Ni, Co, Al, Ca, Mg, and various combinations of these, to aw coal samples. Detailed pyrolysis experiments were carried out using coal samples containing aqua-iron species, and aqua-cobalt species; steam gasification was mainly carried out using samples of brown coal containing iron species. All chemicals used were AR grade.

2.1. Pyrolysis

Details of pyrolysis apparatus and methodology are found in [11]. Briefly, samples consisted of aw coal and coal with known amounts of added inorganic species were pyrolysed using a quartz tube reactor placed inside a Lindberg furnace with a constant flow of nitrogen, at specific temperatures. Two sets of experiments were performed to: (i) determine the total weight loss of the coal sample at specific temperatures, and also the CO_2 and CO concentrations at these same temperatures, and (ii) measure the CO_2 and CO concentrations for a number of specific time-temperature ramps. For (i) the concentration of gases was monitored periodically using Fourier transform infra-red (FTIR); thus at each temperature of the experiment, the total weight loss, and the total concentrations of CO_2 and CO formed, were measured. For (ii) a known weight of coal heated to predetermined settings and CO_2, CO concentrations were measured at given periods using the FTIR. Once steady state conditions had been achieved, the furnace setting was changed to the next temperature; this was repeated with temperature increases usually of 50 °C, to a final temperature between 300 and 500 °C.

Char samples were cooled under nitrogen and stored in a desiccator for elemental analysis, scanning electron microscope – energy dispersive X-ray microanalysis (SEM-EDX) and X-ray diffraction (XRD) studies. Inorganic species on the surface of coal and char samples were examined using X-ray Photoelectron Spectroscopy (XPS). Small amounts of volatiles produced when heating the coal to >600 °C condense as an oily substance on the cooler walls of the quartz tube; when sufficient amounts of the condensate had formed, they were collected by washing the tube with ethanol, and the FTIR spectrum recorded for a thin film of the oils formed on a KBr disc under vacuum.

2.2. Gasification

The apparatus used for pyrolysis was also used for experiments in which the coal samples were heated under an atmosphere of helium and steam [12] Coal samples were aw coal, and the same coal containing a known amount of added iron hydroxyl complexes. Experiments were performed by placing a porcelain boat holding the coal sample into the quartz reactor at a set temperature between 200 °C and 900 °C, and helium gas passed at an accurate rate through a round bottom flask containing gently boiling water, the vapour of which then flowed through the quartz reactor. The product gases exiting the quartz reactor were passed through a water trap cooled in an ice bath, and analysed using FTIR and gas chromatography (GC); the GC analysis was performed using a double column Shimadzu GC-4B PTF, with a silica gel 60/80 packed column, specific for H_2, CO, CO_2, and CH_4, and a thermal conductivity detector. Data were recorded with an eDAQ Powerchrom 280. The GC was calibrated using standard gas mixtures of H_2, CO, CO_2, and CH_4 in helium, supplied by Linde.

2.3. Analytical Procedures

The methods used for the analysis and characterization of coal samples have been discussed previously [9- 13]; briefly, elemental composition of coal samples, including oxygen and chloride, was obtained using microanalysis carried out by the University of Otago's Campbell Microanalytical laboratory in New Zealand. Ash in coal samples containing sodium was determined by heating a known weight of the dry coal to 300 °C, followed by heating to 900 °C in a muffle furnace to a constant weight. Ash in coal with little sodium, and acid washed coal samples, was obtained by the same procedure, but heating at 1000-1100 °C in a muffle furnace. Coals and chars were characterized using SEM-EDX, Atomic Absorption Spectrometry (AAS), X-Ray Fluorescence Analysis (XRF) and XRD. Details of each of these techniques have been given previously. Briefly, *SEM-EDX* was performed with the JEOL 840A instrument with a Link Analytical Pentafet X-ray energy dispersive detector, which used the thin window to allow for the detection of carbon and oxygen. *AAS* analysis of inorganic species extracted by water and acid washing from coal samples was performed with a GBC 933 plus Atomic Absorption Spectrometer. The elemental composition of ashes was determined using *XRF;* ash samples were prepared for XRF analysis by pulverizing the ash sample to $<65\mu m$ using an automated mortar and pestle and accurate weights of 0.750g of the ash thoroughly mixed with 4.000gm fusion flux (Norrish 12:22 lithium tetraborate:lithium metaborate). *XPS* was performed using a Kratos Axis Ultra spectrometer (Kratos Analytical UK) with monochromatized Al Kα radiation (1486.6 eV) operating at 150W. Survey and region spectra were collected at 100 eV and 40 eV pass energies respectively. The spectra were charged corrected assuming the C 1s (C-C) peak occurred at 285.0 eV. *XRD* data for ashes were recorded using a Siemans Kristallolex D5000 diffractometer, with the finely ground ashes placed on a zero plate. The SIROQUANT technique developed in Australia by CSIRO was used in the analysis of the phases in these ashes after they had been heated to a series of temperatures. *FTIR* spectra were obtained with a Perkin Elmer 1720X FTIR Spectrometer. CO_2 and CO spectra were recorded in the region 4000-800 cm^{-1}; the concentration for CO_2 was obtained from calibration graphs based on the area of the peaks from 2394.96 cm^{-1} to 2283.3 cm^{-1}, and for CO the area of the peaks from 2226.09 cm^{-1} to 2047.06 cm^{-1}. Calibration of the FTIR cell was performed by measuring these areas for a certified standard mixture of CO and CO_2.

2.4. Semi-Empirical and DFT Computations

Details of computational methods have been given previously [7,8]; 2-D models were initially examined using the ACD/Chemsketch package [14]; molecular mechanics (MM) and SE computations using the PM5 Hamiltonian were carried out with the Fujitsu CAChe *ab initio* 5.04 package;[15] SE computations were carried out with MOPAC2002 [16] at the Australian Partnership for Advanced Computing National Facility, Australian National University, Canberra, Australia (APAC-NF). DFT calculations for selected molecular models (limited to a maximum of 300 atoms) were carried out using the Schrödinger Jaguar package [17] at the Victorian Partnership for Advanced Computing facility, Melbourne, Victoria, Australia (VPAC).

Molecular modelling calculations consisted of single point self-consistent field with the PM5 Hamiltonian (1scf – a self consistent field computation for any specific geometry of a molecular structure), and SE optimization with the PM5 Hamiltonian of the structure to a ground state. SE calculations for difficult structures were performed by varying the minimum allowed ratio for energy change in MOPAC, and also by optimizing to a trust radius of <0.0001. In cases where a self-consistent-field was not achieved the molecular model was rejected. SE optimized the geometry of the molecule to an energy minimum and provided the total energy, heat of formation from its elements in their standard state (ΔH_f), bond lengths, bond angles, bond orders, partial charges, and contributions of σ and π components to bonding with iron clusters. Semi-empirical methods are parameterized with the reference ΔH_f given at 298K for gas-phase systems, and not for liquid or solid phases. ΔH_f for each molecule is calculated as: $\Delta H_f = E_{elect} + E_{nuc} - E_{isol} + E_{atom}$, (where E_{elect} is the electronic energy, E_{nuc} is the nuclear-nuclear repulsion energy, E_{isol} is the energy required to strip all the valence electrons off all the atoms in the system, and E_{atom} is the total heat of atomization of all the atoms in the system (calculated in eV and converted into kcal/molecule in MOPAC). The ΔH_f values for a single molecular system have been used to compare the relative stability of the molecule that had undergone a single change (e.g. abstraction of H commenced with a structure [M ←OH] into the next with [O-M…H]). The simple ideas of single, double, and triple bonds are provided as bond orders; bond orders of less than 0.2 are indicative of "no bond", >0.2 to 1 as single bond, and so on. A comprehensive bond matrix can provide bonds consisting of σ–π–δ components. The default net charges, or partial charge, on each atom is the Coulson charge; Mulliken populations and partial charges can also be provided.

Multi-electron configuration interactions (MECI) computations for coal molecular models could not be carried out for such large molecules, and only single point self-consistent field computations with MECI could be carried out; these were used to indicate the relative differences in the energy of various configurations of coal structures containing transition metal species. MECI results were obtained for small char molecules containing the [Cu_3] cluster. The [charCu_3] molecule required the lowest computer time per cycle; nonetheless, this took >12 months of real time. MECI computations were carried out using following commands in MOPAC: CIS C.I. = 4; ROOT = 1 (1 electron excitations with 4 molecular orbital states giving a total of eight configurations considered for a component of spin = 0.5; number of singly occupied levels =1, with root 1 optimized) [16].

Computations of low-rank coal containing water were performed by adding water molecules into the 3-D space of a model; MOPAC may also compute the impact of water using the Conductor-like Screening Model method (COSMO), and this was carried out for a number of molecular models. For the former method the molecular models containing water molecules was optimized to a ground state with ≤20wt%; models containing a greater number of water molecules were optimized with the additional water molecules located 'outside' of the 3-D structure. Larger amounts of water required additional 3-D space defined by more than two molecular structures of coal, with the excess water situated in the space between them; such large molecular structures could only be optimized using the MOZYME routine in MOPAC, which uses localized molecular orbitals for closed shell systems, and requires lower memory and less computer walltime.

DFT geometry optimization was carried out for small molecules (limited to a maximum size of 300 atoms) which modelled char produced at 800-900 °C, and also of these char models containing transition metal. 1scf-DFT calculations were also carried out using these char models to examine reaction routes for hydrogen and CO formation. The DFT geometry optimization provided the total energy for a given configuration of char molecular model and produced an output file in which each optimization step provided a molecular geometry and its total energy. The data from these output files were used to manually construct a graph of an energy profile to locate the lowest energy configuration of the molecular model. DFT calculations were at B3LYP theory (exact Hartree-Fock, Slater local exchange functional, Becke's 1988 non-local gradient correction, correlation is Vosko-Wilk-Nusair local functional and Lee-Yang-Parr local and non-local functional) and lacvp** or lacvp3** basis sets, which include effective core potentials (ECPs) for transition metals developed at Los Alamos National Laboratory; the basis set is valence-only, containing the highest s and p shells for main group atoms and the highest s, p, and d shells for transition metals including the outermost core orbitals; the 631G basis set developed by Pople and co-worker was used for atoms not described by ECPs. Polarization functions were used on all atoms except for transition metals, and the effective core potentials include one-electron mass-velocity and relativistic corrections.

3. Development of Molecular Structures

3.1. Molecular Model for Low-Rank Coals

Coal has been conceptualized as either a solid macromolecular structure (or component), or a two-component structure in which a mobile component is trapped inside a three-dimensional cross-linked macromolecular structure. The various approaches to constructing coal molecular models have been discussed in [3,7-13].

Our approach has been to develop a molecule based on the average measured properties of low-rank coal. Initially we constructed 2-D fragments, containing six substituted phenyl, one naphthalenyl, and three cyclic alkyls groups, linked with alkyl, ether, carbonyl, and ester groups. Three carboxyl groups were used, an aryl, an alkyl and a tetrahydronapthoic group. The relative distribution of oxygen was based on published data, and our values for Victorian (Australia) and German brown coal samples using our XPS measurements on a dry ash free basis (dafb). The oxygen groups were distributed as carboxyl, phenolic, hydroxyl, methoxyl, carbonyl, and ester. The composition of the 2-D fragments were created to yield the observed properties of brown coal, including $C_{(ar)}/C_{(tot)}$ and $H_{(ar)}/H_{(tot)}$ ratios (ar = aromatic, tot = total) and measured density; they contained either a phenyl group or a sp^3 carbon that linked three portions of this structure.

Table 1 lists the properties of typical 2-D and 3-D molecules and the averaged experimental data for brown coal and lignite [3]. The relatively small amounts of nitrogen and sulphur were varied for a number of molecular models without any noticeable impact on the molecular modelling results. A number of 3-D structures were examined; it was observed that if a relatively large number of small aliphatic ($-CH_2-$), or ether (-O-) links were used to create the 3-D structure, it became very strained and often coal models containing transition metal complexes could not be optimized to a ground state. It was also observed that provided a 3-D structure maximised hydrogen bonding, different configurations for a similar molecule provided similar values for ΔH_f and total energy. The 3-D molecular structures that were developed could be used to model brown coal, lignite and subbituminous coals by modification of various functional groups, particularly oxygen groups; the properties of the coal model were consistent with the measured properties of these coals. All attempts to further modify the original 3-D coal model structure with properties similar to high-rank coal were unsuccessful.

Table 1. Properties of 2-D and 3-D coal models compared with typical measured values for brown coal and lignite [3]

2-D molecular fragment	
Composition	$C_{79}H_{71}NO_{22}S$
MWt	1418.5 au
Analysis	C 66.9%; H 5.1%; O 24.8%; N 1.0%; S 2.3%
Density	1.47 ± 0.06 g/cm^3
$C_{(ar)}/C_{(tot)}$	0.61
$H_{(ar)}/H_{(tot)}$	0.28
Oxygen groups	$O_{(OH)}$ 33%; $O_{(COOH)}$ 19%
3-D molecular model for low-rank coal	
Composition	$C_{258}\,H_{256}\,N_2\,O_{78}\,S$
MWt	4664.9 au
Analysis	C 66.4%; H 5.5%; O 26.8%; N 0.6%; S 0.7%
Density	~1.5 g/cm^3
$C_{(ar)}/C_{(tot)}$	0.6; (O/C) atomic ratio 0.3
$H_{(ar)}/H_{(tot)}$	0.2; (H/C) atomic ratio 1.0
Oxygen groups	$O_{(COOH)}$ 21%; $O_{(Ph-OH)}$ 34%; $O_{(O-CH3), (C-O-C), (C-OH)}$ 25%; $O_{(C=O)}$ 11%; other = 9%
Large 3-D molecular model for low-rank coal	
Composition	$C_{389}\,H_{395}\,N_5\,O_{119}$
MWt	7044.9 au
Analysis	C 66.3%; H 5.7; O 27.0; N 1.0%
Experimental data for brown coal (dafb)	
Analysis	C 67.8%; H 4.9%; O 26.4%; N 0.3-0.6%; S 0.3-0.6%
$C_{(ar)}/C_{(tot)}$	0.57-0.65; (O/C) atomic ratio ~0.3
$H_{(ar)}/H_{(tot)}$	~0.3; (H/C) atomic ratio ~0.9

Table 1. (Continued)

Experimental data for brown coal (dafb)	
Oxygen groups	$O_{(COOH)}$ 17-23%; $O_{(Ph\text{-}OH)}$ 35-38%; $O_{(O\text{-}CH3)}$ ~12%; $O_{(R\text{-}OH)}$~4%; $O_{(RC=O)}$~23%
3-D Molecular model data for lignite	
Analysis	C 73.6%; H 5.8%; O 19.0%; N 0.7%; S 0.7%
$C_{(ar)}/C_{(tot)}$	0.68
$H_{(ar)}/H_{(tot)}$	~0.2_5
Oxygen groups	$O_{(COOH)}$ 18%; $O_{(OH)}$ 43%
Experimental data for Beulah Zap	
Analysis	C 72.9%; H 4.8%; O 20.3%; N 1.1%; S 0.7%
$C_{(ar)}/C_{(tot)}$	0.67
Oxygen group	$O_{(COOH)}$ 22%; $O_{(OH)}$ 41%

A larger structure was also created which contained two smaller organic moieties (which did not significantly alter the elemental composition and functional group distribution for brown coal), held within the large 3-D molecule. This molecular model mimicked the 'two-phase' notion of coal. This large structure required greater computer resources and computations were limited to coal containing K^+, mono-nuclear Fe and Ni complexes. The computations indicated that the larger structure was less strained when octahedral inorganic complexes were added, but with the overall context of these studies, it was not possible to observe significant differences that could justify the extremely large computer resources required for such a structure.

Molecular models of brown coal containing ionic species such as Na^+, K^+, Mg^{2+} and Ca^{2+} and carboxyl anions, were all optimized to a ground state, with oxygen functional groups and water molecules within the coal molecular structure replacing the hydration sphere about the cations. The following factors on the total energy and the ΔH_f value were examined for these SE 3-D optimized structures: (i) the number of fragments, and type of links, used in constructing the structure, (ii) the relative numbers of hydrogen-, σ- and π-bonds in the molecule; (iii) the elemental composition and resulting functional groups distribution, (iv) inclusion of water molecules, and (iv) the addition of inorganic species, especially transition metal complexes.

The coal molecular structure was suitable for SE computations of coal models with transition metal complexes; the distribution and orientation of oxygen functional groups enabled the formation of octahedral and tetrahedral metal structures. The orientation of coal functional groups, particularly carboxyl groups, was critically important for bond formation with multi-nuclear hydroxyl transition metal complexes. The energy changes due to the added inorganic species were usually calculated as the difference between the coal model containing the particular species, and the same coal model with the same inorganic species placed at an 'infinite distance'. This method, however, was difficult to implement for coal models containing transition metal complexes, as the stereochemistry of these complexes is dependent on water molecules coordinated to the transition metal centres. For these models, the relative change in the ΔH_f, and total energy, of the coal model with transition metal complexes containing water molecules coordinated to the metal centres, were compared to that of the same coal molecular model with the appropriate number of water molecules in the coal structure.

Molecular models were examined for coal contained mono-nuclear and di-nuclear aqua-transition metal complexes for divalent Cr, Fe, Co, Ni, and trivalent Cr and Fe; these and other coal models containing multi-nuclear transition metal species were used to develop models for char containing metal oxides and reduced metal oxides/metals. The 3-D character of the coal model is of paramount importance when developing viable structures containing transition metal complexes, especially when these were used to examine pyrolysis chemistry. A brown coal molecule containing bi-nuclear transition metal complexes of Cr, Fe, Co and Ni, constructed to be identical in all respects, with the exception of the transition metals centres; this was done to examine trends in the calculated properties of such molecular structures attributed solely to the transition metal centres of these complexes.

3.2. Char Molecular Models

The transformation of coal models into char models was made based on experimental data obtained from the pyrolysis of aw coal samples, and aw coal samples containing iron and cobalt complexes, over 200-900 °C. The changes made to the coal molecular model followed the pyrolysis chemistry for the loss of carboxyl, carbonyl, ether and hydroxyl groups. Char models containing metal oxides, and also containing metal clusters (Cr, Fe, Co, Ni, Cu) were optimized using SE. Smaller char models, with a maximum of 300 atoms, were constructed for DFT computations; the composition of these structures was similar to chars formed at 800 °C. A few MECI calculations were performed for a selected number of [charCu_3], and molecules for the reaction route to H_2 formation from [charCu_3]; these required extremely lengthy computations (>10,000 hrs of computer walltime).

Molecular models of char were developed by removing various functional groups from the coal model (mainly oxygen functional groups) to mimic the experimentally measured CO/CO_2 ratios and weight loss during pyrolysis. Briefly, this consisted of removing mainly carboxyl groups, and some hydroxyl groups, as CO_2, CO, H_2O to give the experimental CO_2:CO ratios and weight loss; at higher temperatures other functional groups are lost to form H_2 and small amounts of hydrocarbons [3,7,8,11,12]. High CO_2:CO ratios were indicative of temperature regions in which decarboxylation chemistry is predominant (~300 °C). The loss of all carboxyl, carbonyl, ether, and ~40% of phenolic groups from the coal model provided a total weight loss of ~44wt% (similar to the weight loss observed for brown coal at 700-800 C°); high temperature pyrolysis chemistry included formation of H_2 and CO, transformations of metal clusters, and the formation of catalytic sites. Typical data for brown coal to char, and values for the char model, were: weight loss for aw coal of 13.6wt%; weight loss from the char model, 12wt%. Composition of char: C 70.1%, H 4.1%, O 25.4%; composition of char model: C 69.8%, H 5.6%, O 24.3%.

4. RESULTS AND DISCUSSION

The chemistry of metal-mediated pyrolysis, and catalytic steam gasification of low-rank coal, cannot be comprehended without an understanding of the nature of inorganic species added to the coal. If inorganics were added as precipitates, and/or various mixtures of salts, the pyrolysis chemistry will consist of thermal reactions of the various individual inorganic compounds superimposed on the coal pyrolysis chemistry, i.e.; the behaviour of a poorly characterised mixture of organic and inorganic compounds. If the inorganic species chemically bonded to the coal matrix were know, however, the pyrolysis chemistry would be comprehended as arising from the molecular nature of the coal functional groups chemically bonded to the inorganic species.

4.1. Addition of Metal Species to Low-Rank Coals

Many methods have been discussed in the literature for the addition of inorganic species to coal; these include mixing, kneading, pH adjustment often loosely reported as ion-exchange; the salts have included $CaCO_3$, CaO, $Ca(OH)_2$, NaOH, NaCl, KOH, and aqueous solutions of transition metal salts, such as $Fe_2(SO_4)$, $FeCl_3$, $(NH_4)_3Fe(C_2O_4)_3$ $Ni(NO_3)_2$ or $Co(NO_3)_2$. If a metal salt, such as $FeCl_3$, or $NiCl_2$, were mixed with brown coal, and NH_4OH added to adjust the pH to 9-11, the result would include precipitates of $Fe(OH)_3$, or $Ni(OH)_2$, rapidly deposited onto the pores and outer surface of coal particles, and perhaps transition metal/ammonia complexes. Such a procedure would not, in itself, simply result in an exchange of ionic metal species with protons in the coal matrix. Another example is the addition of $CaCO_3$ to brown coal by: (i) physically mixing, (ii) kneading, and (iii) an acid-base reaction. These methods would produce a coal that contained: (i) $CaCO_3$ particles physically mixed with coal particles, (ii) $CaCO_3$ particles within coal pores, and (iii) calcium that has reacted with acidic groups in coal, to form mostly $\{Coal(COO^-)_2(Ca^{2+})\}$. A simple salt, such as NaCl or $FeCl_3$, exists in solution as cations and anions with hydration spheres; when mixed with brown coal, these become ionic species distributed throughout the bulk water held by the coal matrix. Alkaline compounds, such as NaOH and KOH, are strong bases that would react with the weakly acidic oxygen functional groups, such as carboxyl and hydroxyl groups in the coal matrix; consequently these would be present as aqua-cations distributed throughout the bulk water, and carboxylate anions would have formed in the coal matrix. The behaviour of brown coal at elevated pH values, however, is complicated by tendencies to extract humic acids; transition metals form a series of aqua-hydroxyl metal complexes and their distribution is pH dependent, forming various complexes and hydroxide precipitates.

Carboxyl and hydroxyl groups constitute more than half of the oxygen in brown coal. These groups are weakly acidic and under the appropriate conditions, would participate in acid-base chemistry in a mixture of coal and water, of the type (M = metal):

$$\{Coal(COOH)\} + H_2O \leftrightarrows \{Coal(COO^-)\} + H_3O^+ \quad (1)$$

$$\{Coal(COO^-)\} + H_3O^+ + M^{m+} + mOH^- \leftrightarrows \{Coal(COO^-)_m(M^{m+})\} + (m+1)H_2O \quad (2)$$

Conventional ion-exchange resins are well characterised polymers designed to contain active groups (including carboxyl groups) that undergo well-understood ion-exchange processes. Low-rank coals are poorly characterised mixtures of mainly hydrophilic solids, and the term 'ion-exchange' is often used to signify behaviour analogous to that of ion-exchange resins. Most of the chemistry related to the addition of aqua-inorganic species to low-rank coals, however, should include the solution chemistry of the inorganic species and the solid-liquid interface interactions, generally illustrated by equations (1-2).

Main group metal hydroxides (e.g. M = Na^+, K^+, Mg^{2+}, Ca^{2+}) would participate in weak-acid, strong-base chemistry. Transition metal salts, such as $FeCl_3$, would increase the acidity of the mixture by hydrolysis, would exist predominately as solution species, and addition of iron species to the coal would only occur by shifting the equilibrium to the right in equation 2; this requires changing the pH by adding a base. Rapid addition of a base, however, could quickly form a precipitate of the transition metal hydroxide. The rate of addition of the various transition metals to the coal matrix is also dependent on the nature of aqua-complexes formed in solution, and their differing behaviour with change in pH. The addition of transition metal species to the coal matrix, as aqua-complexes chemically associated with oxygen functional groups, requires an understanding of the aqua-chemistry of these species, and of the acid-base behaviour of the solution mixed with low-rank coals. The addition of iron and cobalt aqua-species has been studied systematically [10, 18]; a correlation was observed between the concentration of the transition metal salt mixed with the coal, the pH of the mixture, and the amount of aqua-species added to the coal matrix. The rate at which the transition metal was added to the coal varied considerably; thus, iron could be added relatively quickly, but cobalt species required many hours and higher pH values.

Table 2 lists the amounts of Co(II) and Fe(III) added to the coal, and also the percentage of the salt in the solution, as a function of pH. Almost all of the Fe(III) in the solution was added to the coal matrix at pH 3.5 (adding 10wt% of Fe in coal), whereas, only 57% of the Co(II) was add to the coal at pH 5 (adding 5.7wt% of Co to the coal). The rates of addition also varied considerably; Fe species could be added in a relatively short periods per adjustment of pH, while addition of 92% of the Co required lengthy periods, and complete addition required more than two weeks.

The addition of Ni(II) to brown coal has been discussed in [19]; aw coal and a solution of $NiCl_2$ were mixed and the pH of the solution was adjusted by adding ammonia or hydrochloric acid; the exchange of Ni into the coal was considered complete when the pH remained unchanged for 5 h. The variation of pH with amount of Ni added to the coal was suggested to be due to changes from bidentate-type nickel carboxylate groups (low nickel loading) to bridge-type nickel carboxylate groups (high nickel loading). The pH behaviour of a solution of nickel, however, is complicated, and includes aqua-nickel complexes, $Ni(OH)_2$ precipitate; aqueous ammonia can also form soluble nickel amine species, and an excess of NaOH forms aqua-species such as $[Ni(OH)_4]^{2-}$. It is difficult to conclude from the data provided in [19] that addition of Ni solution species over pH7-9 may be understood by attributing differing bonds to carboxyl groups in coal.

An increase in amounts of various aqua-metal species added to brown coal has been observed with increase in concentration and pH, but the nature of the various species in coal cannot be deduced simply from the pH behaviour of the solution [9]. Detailed studies on the nature of aqua-Fe(III) species present in a mixture of coal and a solution of $Fe(NO_3)_3$ at particular pH values indicate multi-nuclear hydroxyl iron species were added to coal; XPS data, and molecular modelling studies, were also consistent with the addition of aqua multi-nuclear hydroxyl iron complexes, with mono-dentate carboxyl groups forming [O-Fe] bonds, and hydroxyl groups forming coordination bonds [10].

Table 2. Wt% Co(II) and Fe(III), and proportions of solution species, added to coal [3]

pH	wt% Co in coal	% of soln. Co	wt% Fe in coal	% of soln. Fe
2.4	-	-	3	34
3.5	-	-	10	99
5.0	5.7	57	-	-
5.9	7.7	77	-	-
7.0	9.2	92	-	-

Molecular models of brown coal containing a number of various Fe(III) complexes were examined; Figure 1(a) is an example of an iron complex bound to three bidentate carboxyl groups, and Figure 1(b) is an example of a two-iron hydroxyl complex with mono-dentate carboxyl ligands, two bridging hydroxyl groups, two hydroxyl phenolic groups, and water molecules coordinated to the iron centre. The structure in Figure 1(a) is distorted octahedral; the iron centres in Figure 1(b) form a slightly distorted octahedral structure.

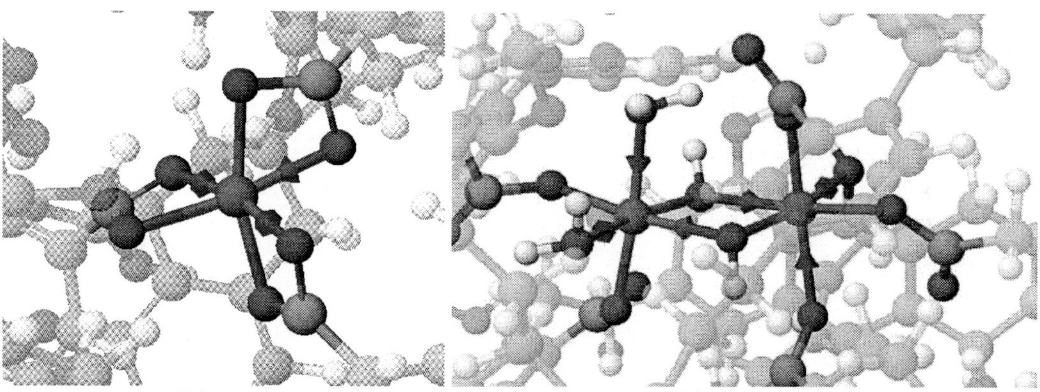

(a) (b)

Figure 1. Fe(III) complexes in coal molecules: (a) Fe(III) with bi-dentate carboxyl groups, (b) $[Fe(OH)_2Fe(2H_2O)]$ with mono-dentate carboxyl groups, bridging hydroxyl, and coordinated H_2O and OH phenol groups.

Molecular models of coal with aqua-metal complexes of Fe, Ni, Co, Cr and Cu were optimised to a ground state; all transition metal complexes were octahedral structures with mono-dentate carboxyl groups, hydroxyl groups, and water molecules forming coordination bonds to metal centres. The total energy of these molecular models were similar, provided the same number of water molecules were present; nickel complexes often formed distorted structures, and instances of hydrogen bonds involving coordinated ligands and water molecules were observed. SE results were also compared to those of molecules optimised using low-level molecular mechanics computations; the trends in the energies of structures optimized using MM calculations often produced contradictory trends to those from SE computations, further illustrating the danger in relying on low-level computational methods in these types of studies [7,8].

The structures that were optimized using SE to a ground state often consisted of a distorted organic molecule and a distorted transition metal octahedral complex; the coal molecular models were stabilized by electronic effects and destabilized by steric factors. The stabilizing factors were due to shifts in delocalized electron density, causing the ionic charge on the metal centre to decrease (donation of electron density from the coal structure was donated to the d-orbitals of the metal centres), and this increased the partial charges on the coordinated ligands. The increase in ionicity of the organic molecule was distributed at a distance from the metal centre because of the extensively delocalized π electron density in phenyl groups. The total energy, and ΔH_f value, of the optimized coal models containing transition metal complexes, resulted from the interplay between steric crowding, 3-D orientation about the metal centre, and electron delocalization. The total energy, ΔH_f (and partial charges) for the same coal model containing different transition mono-nuclear metal centre, $[M(H_2O)_2]^{n+}$ (n=2 and 3), and for bi-nuclear metal centres $[M(OH)_2M]^{2+}$ where M = Cr, Fe, Co, Ni, show relatively small variations, except for coal models containing Ni complexes. Generally, coal molecular models containing transition metal complexes were destabilized by 5% to10% of the ΔH_f value of an equivalent coal model; if the transition metal complex, however, increased the number of water molecules in the coal molecule it was more stable. COSMO computations also confirmed an increase in the stability of the coal models containing transition metal complexes that could be attributed to water molecules. MECI calculations could only be carried out for 1scf calculations and these provided energies of various configurations that may vary by 0 to 3 eV.

The large molecular structure of brown coal containing two smaller organic moieties (to resemble the mobile phase), was optimized to a ground state, and also with the structure containing inorganic the complexes: $[K^+ (^-OOC)]$; $[Fe^{2+}]$; and $[Ni^{2+}]$; Fe and Ni complexes were formed with two mono-dentate carboxyl ligands and hydroxyl coordinated ligands. The coal model containing the potassium anion was optimized with an ionic bond between $[K^+$ $^-OOC)]$ and also six oxygen functional groups and the two water molecules forming a sphere of eight oxygen ligands around the anion. The values of ΔH_f for these molecules were:

{coal($2H_2O$)}	$[C_{389} H_{399} N_5 O_{121}]$ -4,726.14 kcal.
{coal[Fe] $2H_2O$}	$[C_{389} H_{397} N_5 O_{121}$ Fe] -4,638.22 kcal.
{coal[Ni] $2H_2O$}	$[C_{389} H_{397} N_5 O_{121}$ Ni] -4,732.28 kcal.
{coal[K] $2H_2O$}	$[C_{389} H_{398} N_5 O_{121}$ K] -4,855.54 kcal.

The relative differences in ΔH_f for these structures were similar to ones obtained for the smaller 3-D coal molecular model.

4.2. Pyrolysis and Char Formation Chemistry

Major events in the pyrolysis chemistry of low-rank coals include the decomposition of various oxygen functional groups at temperatures between 100-400 °C to yield mainly CO_2, H_2O and CO. The oxygen functional groups are chemically associated with the inorganic species; consequently low-temperature pyrolysis chemistry deals with the thermal reactions of these functional groups bonded to the inorganic. The physical aspects, however, which are relevant to coal pyrolysis include size of coal particles, rates of heating, temperature gradients and mass transfer within coal particles (which depend on the size of the coal sample), and the type of reactor used in performing the experiments.

Five principal stages of pyrolysis have been suggested: the first two at 100-450 °C involving loss of H_2O, and CO_2 *via* decarboxylation, the third at 500-700 °C yields H_2O and some CO, the fourth at 700-900 °C yields tars, CO, hydrocarbons, and hydrogen, and the final stage as high-temperature loss of carbon oxides [1]. Two staged pyrolysis, on the other hand, has been suggested as commencing with decomposition reactions followed by events dominated by heat and mass transfer; a change has been observed at ~400 °C in the Arrhenius plot, based on TG data for brown coal. Cations added to brown coal have been reported not to significantly affect the total volatile matter released during slow pyrolysis but increased the release of CO_2; pyrolysis kinetics indicated Cu^{2+}, Fe^{2+} and Pb^{2+} added to aw coal at pH 5.2 lowered the activation energy compared to that of aw coal [20,21,22,23]

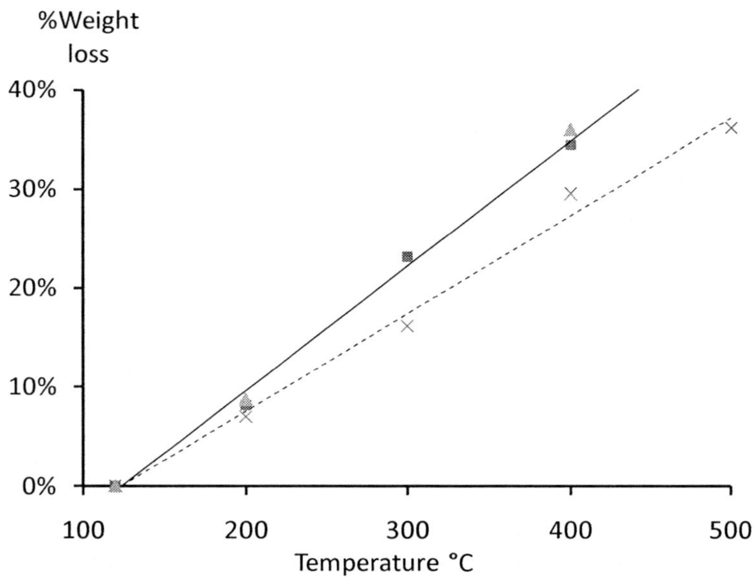

Figure 2. Weight losses for brown coal (▲ 5.7wt% Co(II), ■ 12.4wt% Fe(III); ×aw).

Experimentally measured weight loss of aw brown coal samples, and ones with Fe(III) and Co(II), with increase in temperatures, have been compared with the percentage weight loss of the coal model; the correlation is shown in Figure 2. The data used to obtain values for weight loss from coal molecular models also included CO_2:CO ratios similar to those obtained experimentally at the corresponding weight loss. The agreement between experimental and modelling weight data in Figure 1 is compelling ($R^2 = 99\%$); these results, however, should be viewed in the context of the experimental method used, in which the values were measured at constant temperatures; the CO_2:CO ratios exhibited considerable variations if the temperature changed over any significant time. The values used in Figure 2 were obtained by maintaining the temperature constant for a prolonged period to ensure steady state conditions. The major pyrolysis events over the temperature range shown in the

Figure 2 are loss of CO_2, CO and H_2O; with increasing temperature the amount of CO increases, significant amounts of H_2 are observed, and also small amounts of hydrocarbons.

4.2.1. Low-Temperature Pyrolysis, Loss of CO_2, CO and H_2O

The formation of H_2O, CO_2 and CO has been experimentally related to the decomposition of oxygen functional groups in brown coal; bulk water is lost at 100-110 °C, water that is hydrogen bonded to oxygen functional groups is lost at 100-180 °C, and at higher temperatures water forms *via* pyrolysis reactions, and chemical reactions involving inorganic compounds. A correlation has been reported between CO_2 and H_2O evolved on heating brown coal, and a greater amount of CO_2 was obtained from coal containing Na, K, Mg, Ca and Ba cations. The total yield of gases was related to the loss of carboxyl and hydroxyl groups in the coal, but the formation of CO was not directly related with the loss of any specific oxygen functional group [24,25]. While low-temperature pyrolysis is dominated by decarboxylation chemistry, cross-linking has also been thought to accompany CO_2 and H_2O formation. The decomposition of carboxyl groups into CO_2 and CO continues over a considerable temperature range; different type of carboxyl groups (i.e. alkyl, phenyl) and the various configurations of groups within the 3-D molecular matrix decompose *via* differing reaction routes [3, 11].

The CO_2:CO ratios measured for aw coal over 150-500 °C decreased from an average of 3.1 to an average of 1.4; considerable variations were observed at 500 °C (commencing with an initial ratio of 0.9 to a final ratio of 2.3). These results further illustrate the difficulty in obtaining reproducible data from pyrolysis of low-rank coals. The corresponding ratios for aw coal containing Fe(III) species, however, also increased with an increase in Fe(III) for a given temperature. For example, coal samples containing 1.7wt% Fe, and 12.4wt%Fe, at 200 °C, gave a CO_2;CO ratio of 2.6 and 5.3 respectively; at 400 °C, however, the value of these ratios were reversed, at 3 and 1.7 [11]. The amount of CO_2 for coal containing 12.4wt% Fe(III) was significantly lower at higher temperature; reduced iron species were also observed at these temperatures, consistent with the CO formation mechanisms (discussed below) which included reduction of Fe(III) to Fe(II). A plot of the CO_2:CO ratio and temperature has identified a change in the slope at ~400 °C; this change may be similar to that reported for the change in activation energy for brown coal samples [20-23]. Metal-mediated chemistry is dominated by decarboxylation reactions that include metal carbonate intermediates. Formation of CO_2 from brown coal containing iron species has been modelled *via* an iron carbonate intermediate that ultimately formed a μ-oxo iron bond; the loss of CO_2 and CO may also occur by bond cleavage, reduction of Fe(III) to Fe(II), and the formation of a radical centre. It may be speculated that the observed change in activation energy, and change in CO_2:CO ratios with temperature, may be related to a change in the metal-mediated reaction routes from metal-carbonate complexes yielding CO_2, to reductive metal chemistry yielding CO.

The major route for CO_2 formation at lower temperatures has been modelled as: H transferred from the Fe-OH group, formation of an $R\text{-}CH_3$ group, a CO_2...O-Fe intermediate that forms the carbonato complex, resulting in CO_3 acting as a bridging ligand; the CO_3 of the iron-carbonato complex subsequently decomposes into CO_2 and a μ-oxo iron complex. i.e.:

$$\{coal[RCH_2(COO)_2Fe(OH)_2Fe(COO)_2R]\} \rightarrow \{coal(RCH_3)[(COO)Fe(CO_3)(OH)Fe(COO)_2R]\} \rightarrow \{coal(RCH_3)[(COO)(OH)Fe(O)Fe(COO)_2R]\} + CO_2$$

Further yields of CO_2 would occur *via* the formation of another carbonato complex involving the next carboxyl group bound to iron and the hydroxyl ligand, with formation of the second μ-oxo iron bond.

CO formation increases with increasing temperature; metal-mediated chemistry may occur by (CO) bond cleavage, reduction of Fe(III) to Fe(II), and the formation of a radical. The loss of CO_2 and CO will continue to occur with as the temperature increases, until Fe(0) is formed. Formation of CO and H_2O could also be modelled by the breakdown of a carboxyl group bound to Fe into CO, the formation of a hydroxyl group bound to the iron and a carbon double bond (i.e. [coal($H_2C{=}CH_2$)]). This reaction is more likely for alkyl bonded carboxyl groups at higher temperatures, and can be generally illustrated as:

$$\{Coal[CH_3\text{-}CH_2\text{-}C(O)\text{-}O\text{-}Fe\text{-}O\text{-}]Coal)\} \rightarrow \{Coal[\text{-}CH_2{=}CH_2\ HO\text{-}Fe\text{-}O\text{-}]Coal\} + CO$$

Formation of CO from various ROH coal functional groups may also yield Fe_2O_3, indicating termination of this reaction sequence:

$$\{Coal[(OH)_2Fe(O)(CO_3)Fe\ (OHR)_2]\} \rightarrow Fe_2O_3 + \{Coal[(ROH)_4]\} + CO$$

Reaction between a phenoxyl OH may occur at higher temperatures to form H_2O and a (Ph-O-Fe) bond (these functional groups are assumed to be part of the coal matrix and in close proximity); this could lead to cross-linking:

$$[(Coal)\text{-}Ph\text{-}OH + HO\text{-}Fe\text{-}(O\text{-}coal)] \rightarrow [(Coal)\text{-}Ph\text{-}O\text{-}Fe\text{-}(O\text{-}coal)] + H_2O$$

Elimination of H_2O *via* anhydride formation (also resulting in cross-linking) may occur if carboxyl groups were suitably situated in the 3-D molecular models of brown coal:

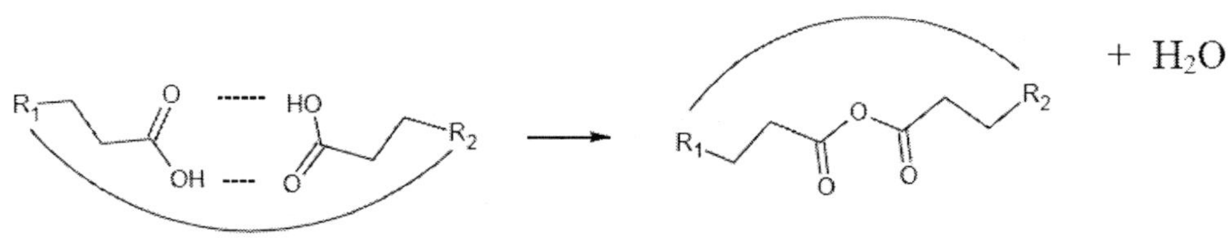

Low-temperature cross-linking is often associated with the evolution of CO_2 from low-rank coals, and at medium temperatures with methane formation. Cross-linking events would compact a structure, which would then undergo less swelling with pyridine [26,27,28]. As the pyrolysis temperature increased, aliphatic structures and the oxygen-containing functional groups would also be removed from chars, while the relative amount of aromatic hydrogen increased [27, 28]. Modelling of cross-linking events may be examined by considering the CO_2:H_2O ratios, the weight loss, and energy changes of the molecular model, with the concurrent cross-linking continuing for a number of these events. It is generally observed that the H_2O:CO_2 ratio from pyrolysis of low-rank coals is >1 and this ratio increases with rank; this cannot result simply from decarboxylation reactions that include the formation of another [C-C] covalent bond, as may be envisaged for cross-linking events. Simple decarboxylation chemistry may be illustrated by equations (A) to (C), Reactions (A) and (C) yield a

[H_2O:CO_2] ratio of 1:1, and (B) would not yield water; overall reactions indicated by (A) to (C) would yield a ratio of $H_2O:CO_2 < 1$.

(A) {2[R-(COOH)]} → {[R-C(O)-R]} + CO_2 + H_2O (ratio $H_2O:CO_2 = 1$)
(B) {[R-(COOH)]} → {R(CH)} + CO_2
(C) {2[R-(COOH)]} → {[R-R]} + CO_2 + H_2O + CO (ratio $H_2O:CO_2 = 1$)

Cross-linking events examined using a brown coal molecular model provided an optimised structure after it had undergone the change typical for such a process and with CO_2 and H_2O placed at an 'infinite distance', i.e.: {R-CH_2-COOH.....OH-CH_2-R′} → {R-CH_2-CH_2-R′} + CO_2 + H_2O. The weight loss for one cross-linking event yielding CO_2 was 1.3wt%; cross-linking *via* anhydride formation yielded one H_2O. It was difficult to locate sufficient groups suitably situated in the molecular model for the number of cross-linking events that would account for a relatively large proportion of the total weight loss. SE results were obtained by manually forcing a number of cross-linking events; these provided an increase in ΔH_f values of +15 kcal for the coal molecule that had undergone four cross-linking events, amounting to a total weight loss of 4.9%. The H_2O:CO_2 ratios for all of these events would be either 1:1, or if anhydride formation were included for a total of five events, would be 1.25:1. These results suggest the energetics for cross-linking events were not unfavourable, but the opportunity for these depends on the stereochemical and conformation aspects of oxygen functional groups in the coal. Anhydride formation was slightly less favoured energetically, but strong hydrogen bonding between the required groups provided the configuration needed for this event. High-rank coals that would undergo plastic/liquid phases may provide additional mobility to overcome stereochemical and conformational barriers to cross-linking events, but this would not be so for low-rank coals. Results from molecular models containing Ni(II) and Fe(III) complexes in a molecule that had undergone the anhydrite reaction/cross-linking indicate this was a very difficult even for the coal molecule containing a simple mono-nuclear structure. The modelling results indicate that while cross-linking events may occur, these are likely to account for a relatively small proportion of the weight loss from low-rank coal pyrolysis, especially for coal samples containing inorganic complexes. Water is also lost from aqua-iron hydroxyl metals complexes when the coal is heated. When coordinated water molecules are driven out, the coal molecule is destabilized and is followed by loss of CO_2 and the formation of the μ-oxy-iron complex. The loss of coordinated water may occur over a significant temperature range; the reactions involving the iron carbonate intermediate yield CO_2 without cross-linking.

4.2.2. Higher-Temperature Pyrolysis, CO and H_2 Formation, and Catalytic Sites

As the temperature increases, reduced iron species are formed *via* reaction routes that yield H_2 and CO; multi-nuclear iron complexes (formed from aqua-hydroxyl iron complexes added to coal) would be distributed in the coal matrix as iron clusters and perhaps colloidal-sized iron species. It is likely that metal clusters would form catalytic sites for steam gasification, but large-sized metal oxide particles would be hindered from forming chemical bonds to char functional groups due to their size relative to the char molecules [12]. The results from all SE molecular modelling show the formation of char molecules containing metal-hydride bonds is energetically favoured; thus intermediate metal hydride species may subsequently decompose to yield H_2. Additionally, [M-CO] complexes have also been shown to form during hydrogen abstraction, and the resulting [M-O-C] group can decompose to form [M-C] and CO; metal-mediated chemistry would therefore yield H_2 and CO. This chemistry, however, is very complicated, and numerous possible reaction routes can be contemplated; the modelling results also indicate that concerted chemistry leading to H_2 and CO formation is likely to be the most energetically favoured. Concerted chemistry would involve intermediates in which metal clusters containing [M....H] metal hydride intermediates would form during the formation of CO. Some of the possible reaction routes leading to the formation of CO and H_2 for [CharFe_3] have been discussed [29]

Experimental data for weight loss and corresponding CO_2:CO ratio were used to develop char models; e.g. for a molecular model of char at 350-450 °C: MF = [$C_{222}H_{182}N_2O_{48}$]; FW = 3645.864; SE ΔH_f -1228.57 kcal, (total energy = -43298.4 eV); elemental composition: C 73.14%, H 5.03%, N 0.77%, O 21.06%; weight loss 25 wt%; CO_2:CO ratio 1: 1.5. The experimental weight loss at this temperature range was 20-25 wt%, and the CO_2:CO ratio 1:1.9. Significant ring condensation was not invoked when forming char molecular structures; nonetheless the proportion of aromatic groups had increased dramatically with increasing weight loss, and also the proportion of two phenyl rings in the molecule. A number of C=C bonds were formed from the elimination of functional groups e.g.: $RCH_2\text{-}CH_2\text{-}COOH \rightarrow RCH{=}CH_2 + CO + H_2O$. The molecular model developed for char formed at 600-700 °C provided a weight loss of 48 wt%, (measured weight loss 45-50 wt% at 600-700 °C). For this char model: MF= [$C_{177}H_{151}NO_{14}$]; FW = 2516.1; elemental composition: C 84.5%, H 6.0%, O 8.9%, N 0.6%; all of the carboxyl groups, over half of the phenolic groups, and most of the carbonyl and ether groups, had been lost. The latter also acted as links for the 3-D macromolecule, and as a result, breaking these links disrupted the 3-D structure; similar char models in which these links were retained also provided an optimized structure.

XRD and XPS experimental data for chars containing iron species identified Fe(III), Fe(II) and Fe(0) phases in char samples, and in char formed at 600-700 °C identified mainly Fe(0). Computations of molecular models of char were therefore carried out for a number of iron oxide species, including Fe_3O_2, Fe_4O_3, and clusters including [Fe_2O], [Fe_3O], [Fe_3], [Fe_4O] and [Fe_4]. The larger sized iron clusters destabilised the char molecule, and [HO→Fe] coordination bonds between the iron clusters and hydroxyl groups in the char molecule also caused changes to that region of the macromolecule. Examples of modelling results are: char model, ΔH_f 1scf-PM5 = 51.5 kcal, SE = -87.4 kcal; [charFe_3O] 1scf-PM5 = 371.6 kcal; [charFe_4O] 1scf-PM5 = 532.6 kcal; [charFe_4] 1scf-PM5 = 573.4 kcal. SE ΔH_f values for char with iron clusters were: [charFe_3O] = 145.8 kcal; [charFe_4O] = 370.8 kcal; [charFe_4] 576.1 kcal. Molecular data for [charFe_3O] were: Fe-O bond orders 0.8 and 0.9, bond lengths 1.88Å and 1.92Å and the third Fe at 2.77Å from the cluster oxygen with a bond order <0.1; Fe-Fe bond orders were 2.7, 1.2, 0.5, and bond lengths 1.74Å, 2.03Å, 2.26Å; one coordinate bond was typical, (Fe←OH bond length of 2.06Å, bond order 0.4), but the second Fe...OH distance was 2.20Å and bond order <0.2, indicating only a weak electrostatic interaction. The data show the char containing iron clusters was less energetically favoured than the char from aw coal.

SE data was also obtained for a number of char samples containing [M_3O_n] clusters, where M=Cr, Fe, Co, Ni and Cu, and n=0-3; the data (while showing the char containing metal clusters was destabilised relative to aw char) generally revealed a trend of increasing stability as the number of oxygen atoms in the cluster (n) increased. The formation of [M-O] bonds between a metal centre and char oxygen groups also increased the stability of the char molecule, reflecting the contribution to ΔH_f by [M-O] bonds. Straight-forward trends in ΔH_f with various metal clusters could not be observed for the SE [CharM$_3$] optimized ground state structures; this was mainly due to the apparent differing tendencies to form [M...H] bonds in these structures. This aspect is dramatically illustrated by the optimized structure of [CharCo$_3$] shown in Figure 3.

The model resulting from 1scf-PM5 computations is typical for [charM$_3$], as shown in Figure 3(a); the SE optimised char model, however, shown in Figure 3(b), illustrates that a ground state could be obtained in which [Co-H] hydrides had formed. This is result is not realistic molecule, but is a dramatic illustration of the tendency to form metal hydrides from these molecules. The chemistry occurs at 700-900 °C for reacting systems, and thus ground state configurations would not be expected to formed at these temperatures; instead, reactive intermediates would exist, and the chemistry would include hydrogen abstraction *via* metal-hydride intermediates that would then break down to yield H_2.

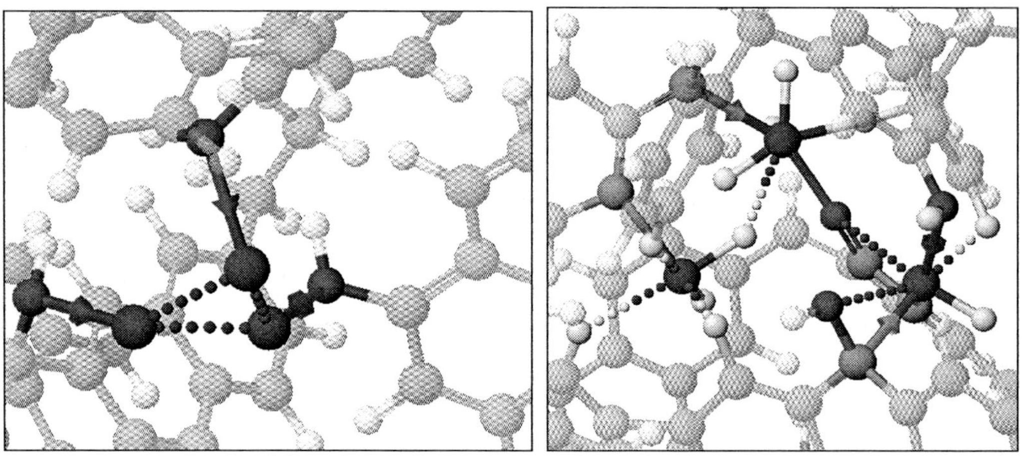

(a) (b)

Figure 3. Structures of [charCo$_3$]: (a) 1scf-PM5 with [Co ←OH] bonds; (b) SE optimized to a ground state structure with [Co-H] bonds (● = Co; ● = O; O = H; ● = C).

1scf-PM5 and DFT computations were carried out for hydrogen formation *via* metal hydride intermediates for the char molecules [charM$_3$], where M = Cr, Fe, Co, Ni and Cu clusters. The results differed considerably; in some cases the formation of metal hydride intermediates were calculated to occur without an energy barrier, but desorption and loss of H_2 required energy. In other cases the reaction route included an energy barrier to [M-H] intermediates, prior H_2 formation. Figure 4 is an example of a reaction route with a significant energy barrier; these are 1scf-DFT computations for H_2 formation from [charCu$_3$], with a mechanistic sequence in which H_2 is formed by H abstraction from two [Cu ←OH] coordination bonds, with oxidative formation of [Cu-O] bonds; the energy barrier for this reaction route was +60 kcal. These results may be contrasted with those for 1scf-PM5 calculations for H_2 formation from [charCr$_3$], in which the reaction leading to the [Cr-H] hydride intermediate was energetically favoured by -140 kcal, but the loss of H_2 required between +20 and +100 kcal (an overall energy of -40kcal for this reaction route to H_2

formation). Results for H_2 formation from [charNi$_3$], however, commenced with an energy barrier of +10 kcal, and loss of H_2 was energetically favoured by -25kcal.

Accurate values of the energy of ground states of various molecules containing transition metals are obtained using MECI calculations; these could not be carried out for most structures, but results were obtained for CharCu$_3$. These SE-MECI computations were performed for a number of the models shown in Figure 4; [Char(Cu$_3$)] models [Cu ←OH] with two coordination bonds, and also the various [CharCu$_3$(H)] models used in computations for H_2 formation. The hydride structures listed as CharCu$_3$(H1) to CharCu$_3$(H5) in Figure 4 were all optimised to a ground state that was energetically favoured by -20 to -29kcal compared to the starting molecule CharCu$_3$. The lowest energy configuration for CharCu$_3$ was -1.2947eV, compared with values of between -1.5614 eV to -1.3558 eV. These results show a favourable change in energy for hydrides, but relatively small variation in the ΔH_f values for the various copper hydride structures. The molecules were all optimised to a ground state, and the orientation of the two [Cu..H] bonds in each of the various structures has been reflected the variations in ΔH_f. This data further re-enforces the favourable energy for the formation of stable metal hydrides, as indicated by the extreme results for [CharCo$_3$].

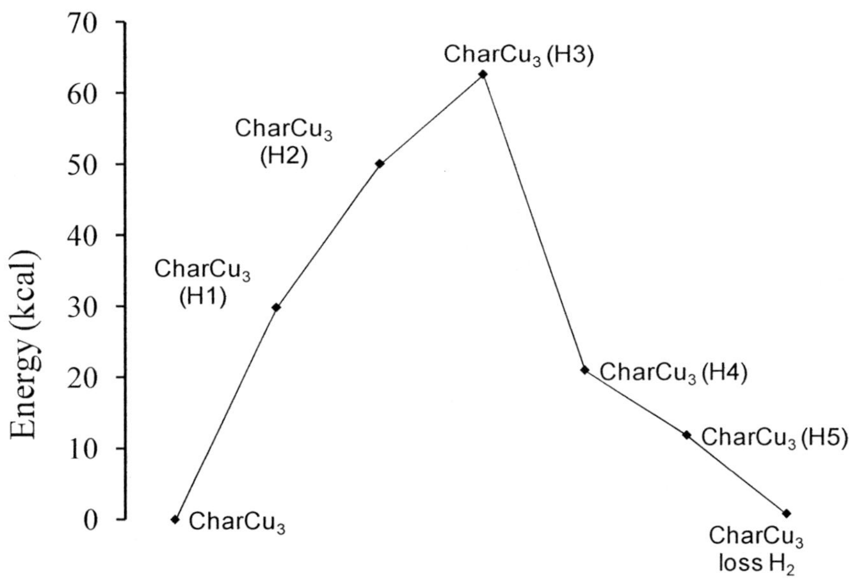

Figure 4. 1scf-DFT energy changes during H_2 formation from [CharCu$_3$].

The MECI results all identify the lowest lying configuration as a doublet, with the absolute values of the lowest level at between -1.2947eV and -1.5614eV; the lower energy configurations (compared to normal ground states) can be obtained for complexes with single occupied levels. Results from calculations which removed MECI computations all provided ΔH_f values for a ground state that were of a higher energy than the MECI results (e.g. ΔH_f for [CharCu$_3$]: with MECI = -80.14kcal; without = -18.37 kcal). It is obvious these results show the formation of energetically favoured metal hydrides.

The results all show differences in the computed energies for reaction routes that yield H_2 during pyrolysis of chars containing metal species. These differences are due to the differences arising from: (a) energy required to abstract H from a [M ←OH] coordination bond, (b) the stabilities of [M-H] bonds, (c) the steric constraints encountered when two [M..H] H atoms migrate to form a hydrogen molecule [M...(H_2)], and (d) the energy required to desorb the H_2 molecule from the metal centre. The reaction for H_2 formation from [charFe$_3$], shown in Figure 5 below, is an example of an energetically favoured route. 1scf-DFT calculations for [charFe$_3$] were for: (i) a structure with a coordinated [HO→Fe] bond, {Char[(HO)→(Fe)(Fe)$_2$]}, (ii) formation of a hydride by cleavage of the [HO→Fe] group, to form the [C-O-Fe] bond in {Char[(C-O-Fe)Fe$_2$(H)]}, (iii) additional H abstraction to form the di-hydride, and (iv) formation and desorption of H_2:

{Char(OH)C-OH)(Fe$_3$)} → {Char(OH)(C-O-)(Fe$_3$)(H)} → {Char(C-O-)$_2$(Fe$_3$)(2H)} → {Char(C-O-)$_2$(Fe$_3$)} + H_2

The relative variations in observed for structures optimised to a ground state, and also variations in energies obtained for the various intermediates, introduces some uncertainty into the SE and DFT computations of reaction schemes. The complexity of pyrolysis chemistry, and our relatively poor understanding of coal and char molecular structures, will inevitably render any computational effort uncertain. Consequently, these computations are shown to be consistent with reproducible experimental data; the results also show that it is imperative that computations for such studies are sufficiently detailed, and computational methods rigorous, to enable a reasonable comparison between computational results and experimental data.

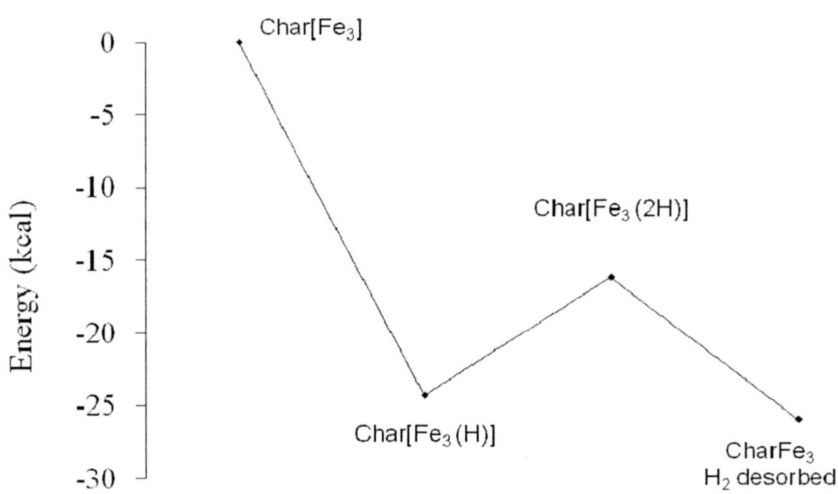

Figure 5. 1scf-DFT energies of H2 formation for [CharFe3] (Reprinted with permission from Domazetis et al., Energy Fuels, 2007, 21, 2531-2542 [160] Copyright 2007 American Chemical Society).

DFT geometry optimization of [CharM$_3$] (M = Fe, Ni, Co, Cu) provided similar structures to those shown in Figure 6; Figure 6(a) is the optimised structure for [CharDCu$_3$], and is similar to Figure 3(a). Figure 6(b) also shows that loss of H_2 yields a structure with two [Cu-O] bonds (and H_2 at a distance).

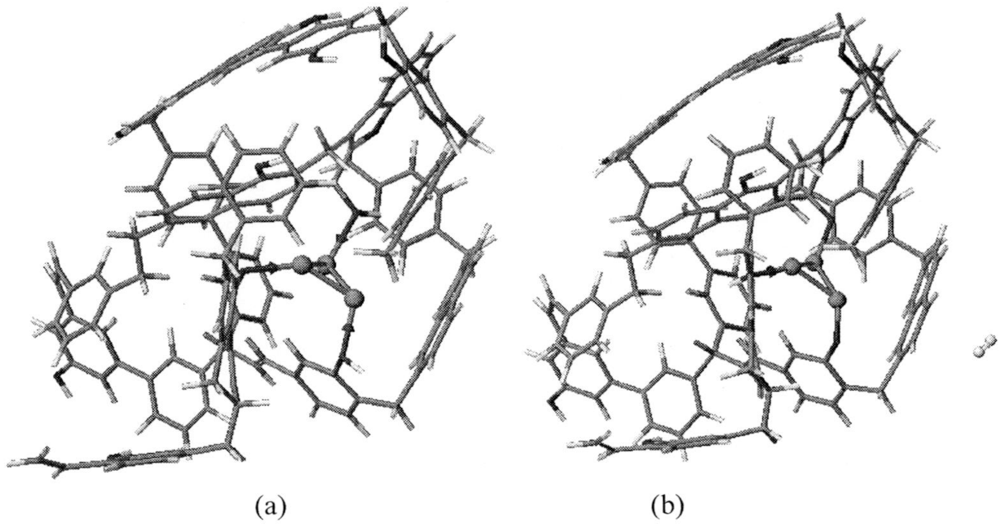

Figure 6. DFT optimised structures of (a) [CharCu$_3$], (b) [CharCu$_3$] after loss of H_2 (Cu = green, C = grey, oxygen = red, hydrogen = white).

The change in energy (in kcal) with each step in the DFT geometry optimisation of [char] and [charM$_3$] is shown in Figure 7. The difference in the minima of the energy profile for the various metal clusters varies by just under 20 kcal. This is consistent with the metal clusters forming similar coordination bonds [M ←OH] and the molecular structure experiencing similar steric crowding from the metal clusters.

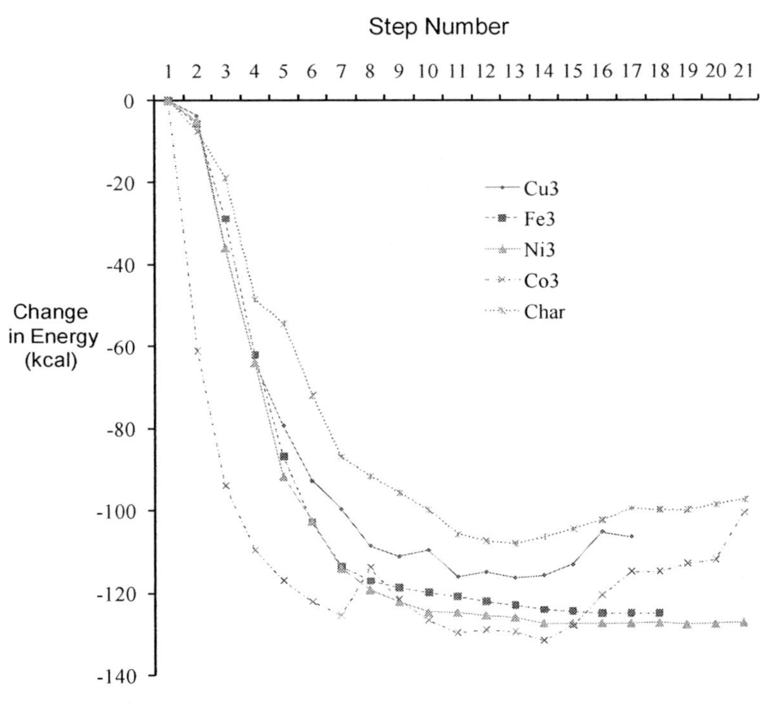

Figure 7. DFT geometry optimisation of Char, and CharM$_3$ (M=Cu, Fe, Ni, Co).

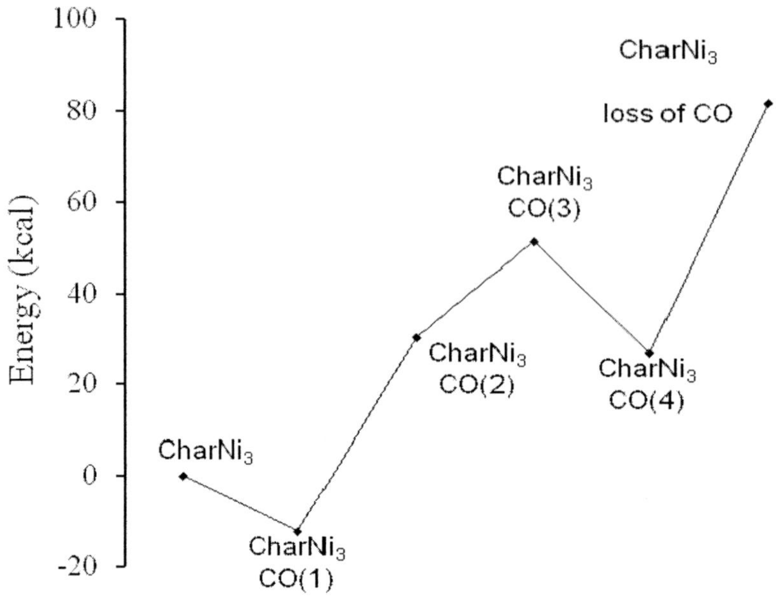

Figure 8. 1scf-PM5 modelling of CO formation from [charNi$_3$] after H_2 formation.

Reaction routes discussed for the breakdown of carboxyl groups into CO_2 and CO have included abstraction of hydrogen by iron centres. Hydrogen abstraction from [Fe←OH] groups may lead to the oxidative formation of [-C-O-Fe] bonds, iron-hydrides, and these would ultimately yield H_2; the breakdown of the newly formed [-C-O-Fe] bonds would yield CO. Relative ΔH_f data indicated a number of structures considered for these reaction routes may be energetically favoured, such as a structure containing two [Fe-O-C] bonds (yielding H_2), and a structure which formed a mono-Fe-hydride and the [Fe-C] bond from the [Fe-C(H)] group. Variable overall changes in ΔH_f were obtained for the various reaction sequences. DFT geometry optimisation was also been carried out for [charM$_3$] models formed after H abstraction to form [M-O-C] groups in [char(HO→M$_3$)], and also for the structures formed during CO formation, and also of the structures after CO formation. All molecular models provided DFT results similar to the energy profiles in Figure 7.

Since the ΔH_f values for formation and loss of CO varied, molecular modelling was carried out for a number of reaction routes for [charM$_3$]; an example is shown in Figure 8 for CO formation from [charNi$_3$] after yield of H_2. The reaction route for CO formation from [charNi$_3$] in Figure 8, includes an energy barrier of ~45kcal for desorption of CO from the nickel cluster; this barrier is almost the same as the energy barrier obtained for the reaction route leading to CO formation, and is consistent with the stability (and thus the tendency to form) gaseous nickel carbonyl.

A number of reaction routes were examined for the [charFe$_3$] model. The results, while displaying a range of energies, were all dependent on the nature of the molecular complex used as the 'starting' intermediate. The simplest modelling approach was to commence with a [charFe$_3$] model formed after the loss of H_2; this model has two [Fe-O-C-C] groups, and each of these may eliminate CO and form a new [Fe-C] group; this reaction scheme gave a maximum energy barrier of between 100 and 185 kcal and is not considered to be the major route for CO formation. An alternative reaction route which commenced with an intermediate [char Fe_3..2(H)] species provided a relative difference for CO formation of -170 kcal, but the loss of CO from the Fe_3 clusters was endothermic, yielding an overall energy difference in ΔH_f for CO formation of -100 kcal. Additional calculations in which the molecule commenced with an [Fe (Fe-O) Fe-C] group that yielded CO, indicated another favourable reaction route, which was favoured by -80kcal; 1scf-DFT calculations in which the reaction commenced with [charFe_3] and H_2 formation, leading to CO formation, provided an overall reaction route without a significant energy barrier.

Figure 9 shows some of the numerous structures considered when modelling the formation of CO from [charFe_3]; structure 9(a) is an Fe_3(hydride) intermediate, with CO displaced from a phenyl group; structure 9(b) shows an Fe_3 cluster with a CO molecule situated parallel to the iron cluster metal centres, and structure 9(c) CO situated end-on prior to desorption.

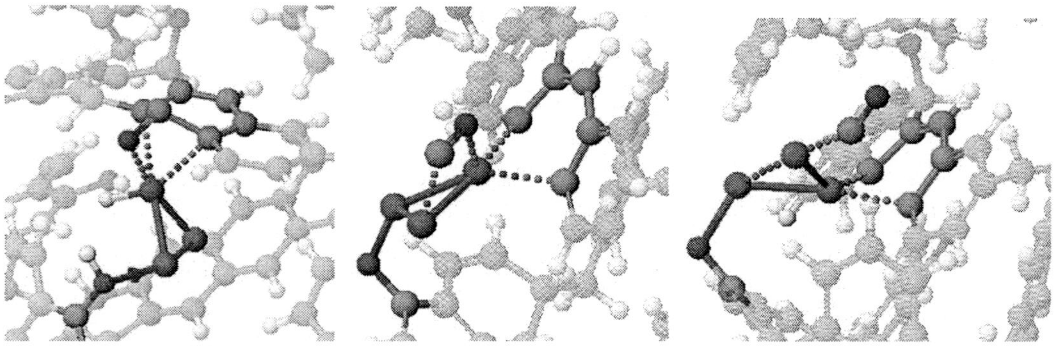

(a) (b) (c)

Figure 9. Three structures considered in modelling schemes for CO formation (a) hydride-CO intermediate, (b) CO parallel to Fe_3 cluster, (b) CO 'end-on' to Fe_3 cluster.

The difference in the ΔH_f values for structures 9(b) and 9(c) was calculated at -23 kcal. Values for structure 9(a) were complications by tendencies for the iron centres to form π complexes with phenyl rings; the elimination of CO from [Fe_2(Fe-O-C-(Ph)] group was generally more likely in a concerted reaction in which [Fe_3...H] intermediates formed. Additional computations were carried out to examine the orientation of [Fe...(CO)] intermediates, and these also provided a range of relative energy differences [7, 8, 12, 29]. The overall result for such chemistry is critically dependant on the nature of the starting molecule, because the energy changes with each step in are relative to this value. Since the chemistry is continuous and occurs at high temperatures, selecting the right 'starting' molecule is extremely difficult; it is impossible to decide on a 'start' for concerted chemistry.

The results for modelling CO formation will generally reflect the tendency to dissociation of CO from the [M...CO] complex. Generally, the CO adsorption on the 3d metal surfaces of Sc to Fe indicates a tendency toward CO dissociation, whereas for Co to Cu, it tends to remain adsorbed on the surfaces [30]; the modelling results may also be indicative of the behaviour of metal-carbonyls. Metal carbonyl formation has been discussed in catalytic mechanisms for the Fischer-Tropsch synthesis, with Co, Ni, and Ru carbonyl formation observed to be high at elevated pressure and low temperature [31] and nickel carbonyl formation also observed in nickel catalysts during methanation of CO, and iron carbonyl species thought to be involved in gasification and methanation [32].

The modelling results show that a number of reaction routes can yield H_2 and CO and that specific reactions (i.e. hydrogen abstraction and elimination of [Fe-O-C-C] groups) lead to the formation of these gases. The data also show that concerted chemistry, in which the reaction sequence commenced with hydride formation, with intermediates leading to CO formation prior to H_2 loss, was the most favoured. Concerted reaction schemes, however, were extremely difficult to model, and examinations of various possible reaction sequences that may be implicated in concerted chemistry provided a range of ΔH_f values for the loss of CO. All such data indicated energetically favoured reaction routes exist for loss of CO and H abstraction from char functional groups containing iron clusters; H_2 and CO may form *via* metal-mediated chemistry by cleavage of [O-H] and [C-H] groups in char; hydride/ CO/iron structures and formation of H_2 and CO may continue until either all OH and CH groups were eliminated or the iron moiety became situated at a relatively large distance from such groups.

A comprehensive reaction scheme that included all of the possible iron complexes from the pyrolysis of brown coal containing various iron hydroxyl complexes would require a very large number of reaction routes. Nonetheless, even such large schemes would not include the formation of other products, such as methane, nor of radical chemistry. It is also likely that some iron clusters may be more reactive than others and large sized iron species would form from large iron hydroxide particles in the coal.

4.2.3. Catalytic Steam Gasification

The formation of H_2 and CO during high-temperature pyrolysis is directly relevant to catalytic steam gasification of coals; this chemistry leads to the formation of the catalytic sites; H_2 and CO formation from gasification reactions include H abstraction and elimination of CO from the [M-O-C] groups formed after insertion of [O] into the [M-C] group.

Additionally, when the char substrate is consumed during catalytic steam gasification, the concentration of transition metal species increases, and this may cause agglomeration of metal oxides into larger sized particles; it is likely these would be involved in post-gasification chemistry. A number of modelling results were obtained for various [charM_3] models (M_3 = Cr, Co, Fe, Ni). The most complicated reaction route that has been modelled was for CO formation from [charCr_3...H_2O], shown in Figure 10. This reaction sequence consisted of structures that commenced with a H_2O molecule adsorbed on the Cr centre, and ultimately led to the formation of CO while a second H_2O molecule was coordinated to the Cr centre, in effect 'displacing' the CO. This reaction route provided a favourable means for desorption and loss of CO and would appear to avoid loss of gaseous metal carbonyl compounds. The pyrolysis chemistry of [charFe_3] ultimately forms [Fe-C] groups; molecular modelling results show H_2O molecules can form coordinated bonds with the iron centre of these groups [12].

Reaction routes were modelled for H_2 formation from the [char(Fe_2)(C-Fe...OH_2)] that had formed when a water molecule was chemisorbed on the [Fe-C] group. Various reaction routes may also be considered for Fe catalytic steam gasification.

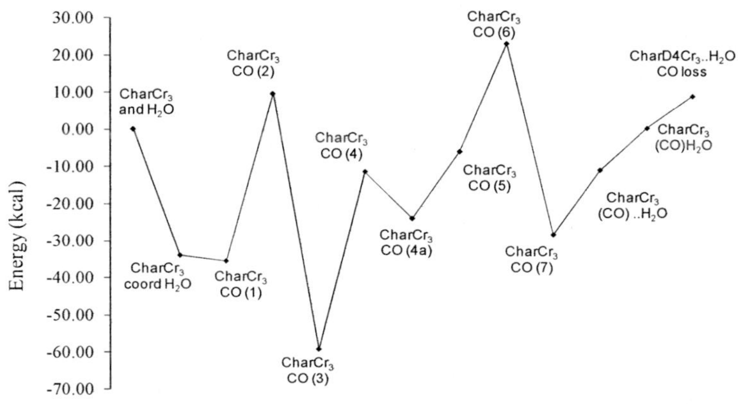

Figure 10. 1scf-PM5 modelling of reaction route to formation of CO from [charCr$_3$H$_2$O].

Figure 11 is an example of hydrogen formation from the reaction of H_2O adsorbed on the [Fe-C]; this reaction route was modelled to commence at the end of H_2 formation during pyrolysis, so that the end result was the yield of 2 molecules of H_2, the first was H_2 from the char molecule resulting from pyrolysis chemistry (after CO loss and formation of [Fe-C] site for coordination of H_2O), and the second H_2 from H_2O. The reaction route commenced with the molecular model [charFe$_3$] with H_2O at a distance; this is followed by an iron di-hydride (formed during pyrolysis), which desorbed H_2, followed by adsorption of a H_2O molecule to the newly formed [Fe-C] site. The H_2 from pyrolysis was placed at a distance and would not participate in any further chemistry; the second H_2 was formed by H abstraction from H_2O, and the formation of an [Fe-O-C-C] site. This reaction route would subsequently form CO by elimination from the [Fe-O-C-C], to yield CO and a new [Fe-C] site; the catalytic cycle would then continue by the coordination of another H_2O molecule to the newly formed [Fe-C] site. In this case, the reaction sequence ended before the adsorption of another H_2O molecule.

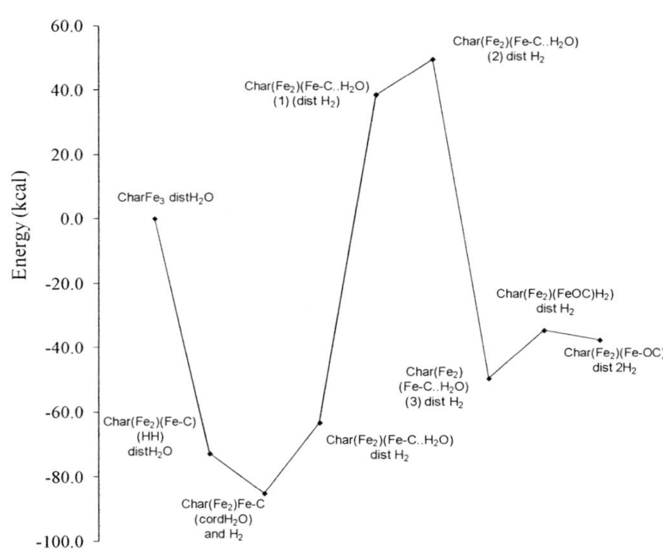

Figure 11. 1scf-PM5 modelling of H_2 formation from [char(Fe$_2$(Fe-C)] and H_2O.

Further modelling of catalytic steam gasification requires molecular dynamics calculations, in which water molecules impact the surface of char, with the number of catalytic sites [M-C] related to the proportion of transition metal distributed throughout the char as metal clusters. The impact of large metal oxide particles formed within the char molecule are not considered to act as sites for reaction between [C] and steam; these would lead to the formation of larger pores within char and in this way increase the surface area available for reactions with H_2O [12].

Experimental data show that brown coal containing aqua-iron complexes, when heated to 900 °C in an atmosphere of steam, loses a greater weight compared to an aw coal sample with a considerably greater yield of hydrogen (e.g. in 15 min, char with iron measured conversion of 76wt%, which was 20wt% higher than aw coal, and at in 60 min conversion for char containing iron species was almost 100wt%) [8,11,12]. The experimental data is consistent with molecular modelling that shows low-energy barriers to reactions involving water molecules and char models containing [M-C] groups. The composition of the resulting gas, however, includes a greater yield of H_2 than would be obtained from the reaction between steam and char. A satisfactory mass balance was obtained by assuming: (a) steady flow rate of helium, (b) constant concentration in the gas stream, and (c) 20 wt% of char was carbon reacting with steam (i.e. $C + H_2O \rightarrow H_2 + CO$) over a period of 15 min. The theoretical yield of H_2 and CO gave a H_2:CO ratio of 1.4:1, considerably lower than the experimental value of: 4.3 (H_2) : 1 (CO). The observed yield of the gases could be reproduced theoretically by assuming additional chemistry yielded H_2 and CO_2 by consuming a portion of the CO. If it is assumed that 58% of the total CO from gasification had then reacted with H_2O to yield additional H_2 and CO_2, the calculated and observed yields become similar; i.e. calculated yield H_2 30.9 mol%, CO_2 10.3 mol% and CO 6.8 mol%, (CH_4 0 mol%) remainder He carrier gas; measured yield H_2 29 mol%, CO_2 10.7mol%, CO 6.7 mol% (CH_4 2 mol%) and He carrier gas; without He the gas composition would be: H, 59.8%; CO_2, 22.0%; CO, 14.0%; CH_4, 4%.

Catalytic gasification of brown coal containing Fe, Co and Ni, with steam and H_2 has been discussed in [33]; the reactivity of brown coal with iron at 650 °C was lower than that without catalyst, and conversions at 120 min were 88 wt% and 60 wt% with and without iron, respectively. These results may be rationalised on the basis that the active site [Fe-C] is not formed at the low temperature, and is formed as the char is heated to 600-800 °C; thus at low temperatures, few catalytic sites have formed for iron systems. An additional factor is likely to be due to the method used in adding the metal complexes to brown coal. All of the coal samples were mixed with inorganic salts by kneading, and it is likely that poorly characterized mixtures of salts and aqua-species were formed by this method. Greater reactivity was reported in [33] for brown coal with added Ni and Co, and also when Fe/K and Fe/K were added to brown coal. Gasification of brown coal in steam, treated by immersion of the coal in a solution of $Fe(NO_3)_3$ followed by drying at 50 °C, also gave lower weight losses at 650 °C [34].

Pyrolysis and steam gasification have been discussed for brown coal containing nickel species added by impregnating the coal with aqueous solution of hexa-ammine nickel(II) carbonate and ammonia to give a loading of nickel on coal of 12-14 wt% [35]. The gasification was carried out using a laboratory atmospheric fluid bed reactor; increased yields of H_2 were observed with ~23% of the steam consumed; this was thought to be due to catalytic reaction with steam and also cracking of tar and light hydrocarbons. Although at the lower temperature the coal conversion was low, the final composition of the gas contained a high proportion of H_2. This was reported to be due to a number of reactions, including reaction of CO with H_2O to produce H_2 and CO_2. A satisfactory mass balance was obtained for these experiments, and the calculated dry gas composition at 0.1 MPa and 600 °C was: H_2, 67.4; CO, 5.3; CO_2, 26.9; CH_4, 0.4 mol%. This is fairly close to the observed composition of product gas: H_2, 63; CO, 10; CO_2, 24; CH_4, 3 mol%.

The experimental data for brown coal containing Fe, and also Ni, is consistent with significant catalysis of the reaction between char and steam. A specific catalytic activity (g-char min^{-1} per g-Fe), as the difference in weight loss minus that of aw coal per gram of iron added to the coal, was reported at 0.5 to 0.1 for different amounts of iron and char conversion in [34]. These values may be compared to the apparent turn-over number at 800-900 °C of 0.8 to 0.2 mole C/mole Fe, per min, (or 3.7 to 0.9 g-char min^{-1} per g-Fe) for brown coal containing aqua-iron species [12].

Catalytic steam gasification offers a method for increased hydrogen production using low cost coal and relatively inexpensive catalysts particularly iron compounds.

CONCLUSIONS

High level computational molecular modelling can optimise 3-D molecular models of low-rank coals containing a number of inorganic complexes. The coal molecular model containing iron complexes is transformed into char consistent with experimental data for metal-mediated pyrolysis chemistry; mechanisms for CO_2 formation occur *via* an iron carbonate intermediate, and CO *via* reductive mechanisms that yield iron oxides and iron clusters. Mechanisms for the formation of H_2 and CO, examined using SE and DFT molecular modelling of char containing transition metal clusters, show that a number of reaction routes can yield H_2 and CO. Concerted reaction mechanisms involving transition metal hydride/carbonyl intermediates appear to be energetically favoured; these mechanisms, however, are extremely difficult to model. The reactions examined show the [Fe-C] group would be the catalytic site for steam gasification. Molecular modelling of catalytic steam gasification is consistent with experimental data, in which a significant amount of H_2 and CO is formed from the reaction of steam with char. Experimental data also shows a greater yield of hydrogen occurs due to post-gasification reactions.

ACKNOWLEDGMENT

Contributions by staff at La Trobe University Departments of Chemistry, Earth Sciences, Physics and the Centre for Materials and Surface Science, are gratefully acknowledged. Computational molecular modelling was performed with support from e-grants and computer facilities from VPAC, and allocations under the Merit Allocations Scheme at APAC-NF. Some of the results were obtained under the La Trobe University industry collaboration research program with Clean Coal Technology Pty Ltd. Rheinbraun GBT supplied samples of German brown coal. The unexpected, sudden death of Dr. B.D. James has meant that this research will no longer have the benefit of his insightful input. We mourn his passing.

REFERENCES

[1] Suuberg, E.M.; Peters, W.A.; Howard, J.B., *Ind. Eng. Chem. Process Des. Dev.* 1978, *17,* 37-46.

[2] Solomon, P.R.; Serio, M.A.; Suuberg, E.M., *Prog. Energy Combust. Sci.* 1992, *18,* 133-220.

[3] Domazetis, G.; James, B.D.; Liesegang, J., "Pyrolysis of low-rank coal and metal-mediated chemistry preceding catalytic steam gasification." In *Pyrolysis: Types, Processes, and Industrial Sources and Products*. F Columbus (Ed), Nova Science Publishers, Inc, NY, 2009. In press.

[4] (a) Shinn, J.H., *Fuel.* 1984, *63,* 1187-1196. (b) Nakamura, K.; Takanohashi, T.; Iino, M.; Kumagai, H.; Sato, M.; Yokoyama, S.; Sanada, Y., *Energy Fuels.* 1995, *9,* 1003-1010. (c) Takanohashi, T.; Kawashima, H., *Energy Fuels.* 2002, *16,* 379-387. (d) Mathews, J.P.; Hatcher, P.G.; Scaroni, A.W., *Energy Fuels.* 2001, *15,* 863-873. (e) Derbyshire, F.; Marzec, A.; Schulten, H-R.; Wilson, M.A,.; Davis, A.; Tekely, P.; Delpuech, J-J.; Jurkiewicz, A.; Bronnimann, C.E.; Wind, R.A.; Maciel, G.E.; Narayan, R.; Bartle, K.; Snape, C., *Fuel.* 1989, *68,* 1091-1106. (f) Kidena, K.; Murata, S.; Nomura, M., *Fuel Processing Technol.* 2008, *89,* 424-433.

[5] (a) Patrakov, Yu.F.; Kamyanov, V.F.; Fedyaeva, O.N., *Fuel.* 2005, *84,* 189-199. (b) Faulon, J-L.; Carlson, G.A.; Hatcher, P.G., *Org. Geochem.* 1994, *21,* 1169-1179. (c) Hatcher, P.G.; Lerch III, H.E.; Koytra, R.K.; Verheyen, T.V., *Fuel.* 1988, *67,* 1069-1075. (d) Hatcher, P.G.; Clifford, D.J., *Org. Geochem.* 1997, *27,* 251-274. (e) Hatcher, P.G., *Org. Geochem.* 1990, *16,* 959-968. (f) Drobniak, A.; Mastalerz, M., *Int. J. Coal Geology.* 2006, *66,* 157-178.

[6] Haenel, M.W., *Fuel.* 1992, 71, 1211-1223.

[7] Domazetis, G.; James, B.D.; Liesegang, J., *Energy Fuel.* 2008, *22,* 3994-4005.

[8] Domazetis, G.; James, B.D.; Liesegang, J., *J. Mol. Model.* 2008, *14,* 581-597.

[9] Domazetis, G.; Liesegang, L.; James, B.D., *Fuel Process. Technol.* 2005, *86,* 463-486.

[10] Domazetis, G., Raoarun, M., James, B.D., Liesegang, J., "Studies of Mono- and Polynuclear Iron Hydroxy Complexes in Brown Coal," *Energy Fuels* 2005, *19,* 1047-1055.

[11] Domazetis, G., Raoarun, M., James, B.D., "Low-Temperature Pyrolysis of Brown Coal and Brown Coal Containing Iron Hydroxyl Complexes," *Energy Fuels,* 2006, *20,* 1997-2007.

[12] Domazetis G., Raoarun, M., James, B.D., Liesegang, J., "Molecular modelling and experimental studies on steam gasification of low-rank coals catalysed by iron species," *Applied Catalysis A: General,* 2008, *340,*105-118.

[13] Domazetis, G., Raoarun, M., James, B.D., Liesegang, J., Pigram, P.J., Brack, N., Glaisher, R., "Analytical and Characterization Studies of Organic and Inorganic Species in Brown Coal," *Energy Fuel,* 2006, *20,* 1556-1564.

[14] ACD/ChemSketch Freeware, version 10.00, 2006, *Advanced Chemistry Development,* Inc., Toronto, ON, Canada, www.acdlabs.com.

[15] CAChe ab initio version 5.04, Fujitsu Ltd, 2000-202; Oxford Molecular Ltd, 1989-2000.

[16] MOPAC 2002 Version 2.5.2, Stewart JJP. Fujitsu Ltd, Tokyo, Japan.

[17] Jaguar, version 6.5, Schrödinger, LLC, New York, NY, 2005.

[18] Del Papa, L., B.Sc. Honours Thesis, 2008, La Trobe University.

[19] Shirai, M.; Arai, M.; Murakami, K., Aggregation and Role of Ion-Exchanged Nickel Species during Pyrolysis of Brown Coal: Influence of Extent of Metal Loading, *Energy Fuels.* 2000, *14,* 1038-1042.

[20] Hill, J.O.; Ma, S.; Heng, S., *J. Thermal Anal.* 1989, *35,* 2009-2024.

[21] Ma, S.; Huang, G.; Hill, J.O., *Thermochimica Acta.* 1991, *184,* 233-241.

[22] Ma, S.; Hill, J.O.; Heng, S., *J. Thermal. Anal.* 1991, 37, 1161-117.

[23] Ma, S.; Hill, J.O.; Heng, S., *J. Thermal. Anal.* 1989, *35,* 977-988.

[24] (a) Schafer, H.N.S., *Fuel.* 1980, *59,* 295-301. (b) Schafer, H.N.S., *Fuel.* 1980, *59,* 302-304.

[25] (a) Schafer, H.N.S., *Fuel.* 1979, *58,* 667-672. (b) Schafer, H.N.S., *Fuel.* 1979, *58,* 673-679.

[26] Solomon, P.R.; Serio, M.A.; Despande, G.V.; Kroo, E., *Energy Fuels.* 1990, *4,* 42-54.

[27] Ibarra, J.V.; Moliner, R.; Gavikn, M.P., *Fuel.* 1991, *70,* 408-413.

[28] Ibarra, J.V.; Moliner, R.; Bonet, AS.J., *Fuel.* 1994, *73,* 918-924.

[29] Domazetis G.; Raoarun, M.; James, B.D., *Energy Fuels.* 2007, *21,* 2531-2542.

[30] Jiang, L.; Xua, Q., Theoretical study of the interaction of carbon monoxide with 3d metal dimmers, J. Chem. Phys. 2008, *128,* 124317-12426.

[31] Schulz, H., Short history and present trends of Fischer–Tropsch synthesis, App. Catalysis A: General, 1999, *186,* 3-12.

[32] Bartholomew, C.H., Mechanisms of catalyst deactivation, App. Catalysis A: General, 2001, *212,* 17–60.

[33] Ohtsuka, Y.; Tamai, Y.; Tomita, A., Iron-Catalyzed Gasification of Brown Coal at Low Temperatures, Energy Fuels, 1987, *1*, 32-36.

[34] Yamashita, H.; Yoshida, S.; Tomita, A., Local Structures of Metals Dispersed on Coal. 2. Ultrafine FeOOH as Active Iron Species for Steam Gasification of Brown Coal, Energy Fuels, 1991, *5*, 52-57.

[35] Tomita, A.; Watanabe, Y.; Takarada, T.; Ohtsuka, Y.; Tamai, Y., Nickel-catalysed gasification of brown coal in a fluidized bed reactor at atmospheric pressure, Fuel, 1985, *64,* 795-800.

In: New Trends in Chemical Physics Research
Editor: James N. Lapherty
ISBN: 978-1-61668-535-5

Chapter 10

DOUBLE RARE-EARTH - BORON SUBMONOLAYER TO MULTILAYER ADSORBED FILMS AS A PROTOTYPE OF RARE-EARTH HEXABORIDE SURFACE

T. T. Magkoev*[1], I. V. Tvauri[1], A. M. Turiev[1], N. I. Tsidaeva[1], D. F. Remar[1] and K. Fukutani[2]

[1]Department of Physics, University of North Ossetia, Kesaev 121-83, Vladikavkaz 362020, Russian Federation

[2]Institute of Industrial Science, The University of Tokyo, Meguro-ku, Tokyo 153-8 505, Japan

ABSTRACT

Submonolayer to multilayer coadsorption of lanthanum with boron and gadolinium with boron on the surface of Mo(110) has been studied by means of Auger electron spectroscopy (AES), electron energy loss spectroscopy (EELS) and work function (ϕ) measurements. The equilibrium state of double adsorbate systems achieved either by adsorption of rare-earth metal (REM) on boron precovered Mo(110) surface held at room temperature or after moderate annealing of the system with opposite order of adsorption (B on REM films) is the layer which is the inhomogeneous mixture of boron and REM atoms with preferential concentration of boron in the surface area of the mixed film.

The work function of such films even at REM to boron concentration ratio much higher than 1/6 are very close to the values of corresponding bulk LaB_6 and GdB_6, favoring assumption of surface rearrangement as the dominant reason of high electron emission efficiency of hexaborides.

Almost total similarity of the results for La-B and Gd-B systems can be viewed as the consequence of weak participation of Gd f-electrons in determining the thermionic properties of corresponding double layers.

* Corresponding author: Tel.: +7 918 8224 595; fax: +7 928 4804 412. E-mail address: t_magkoev@mail.ru

Keywords: Rare-earth metals; Boron; Physical vapor deposition; Adsorption; Work function measurements; Auger electron spectroscopy; Electron energy loss spectroscopy.

1. INTRODUCTION

Rare-earth metal hexaborides are materials of long standing technological relevance mainly as low temperature thermionic emitters [1,2]. Among all the hexaborides LaB_6 is one which is most widely used for emitter fabrication due to unique combination of low work function (2.7 eV in average) and stability [3,4]. Unique emission properties of LaB_6 are seen from comparison of its emission current density (65 A/cm^2 at T=1873 K) with that of tungsten (1,75 A/cm^2 at T=2700 K) [4]. More recently LaB_6 receives additional interest because of potential application in electron emitting tip devices, nanoscopy and nanolitigraphy [5-7]. Application importance motivated an extensive research of LaB_6, as well as other hexaborides, which has been done for last several decades. Most of the studies were focused on elucidation of the origin of low work function. So far such an extensive study yields however no clear evidence whether the low work function is an intrinsic property of the material or is due to REM^+ - B^- dipole layer which forms on the surface.

Nishitani et al [8] on the basis of X-ray photoelectron spectroscopy (XPS), ion scattering spectroscopy (ISS), low energy electron diffraction (LEED) and work function measurements of LaB_6 single crystal planes found that the work function anisotropy for (100), (110), (111) planes (2.3, 2.5, 3.3 eV, respectively) correlates with the surface structure. The origin of the unusually low work functions of the LaB_6 (100) and (110) clean surfaces is interpreted by electric dipole moments produced by positive charges of surface lanthanum ions. Swanson et al [9] studied the crystallographic dependence of the work function of LaB_6 by thermionic and retarding methods. Work function increases in the sequence (346)-(100)-(110)-(111) and ranges from 2.41 eV to 2.9 eV. The sequence of the work function increase is the same as the increase in surface B/La ratio as determined by Auger electron spectroscopy, thus pointing at the importance of the surface dipole layer. Later, Gesley and Swanson [10] studying the orientation dependence of field electron emission of LaB6 emitter found that the (310) surface has the lowest work function – 2.5 eV - of any single crystal plane of LaB_6. They also found that the relative orientation dependence of the work function is $\phi(310) < \phi(210) < \phi(100) < \phi(110) < \phi(111) < \phi(211)$. The authors concluded that this ordering is in accordance with the relative dipole contribution to the work function. Such conclusion is also in lines with the results of Ociepa and Mroz [11] and Ociepa [12], who measured the work function coverage dependence of LaB_6 evaporated films on Ta and observed a steep decrease of the work function at the initial stage of film growth. This fact as well as the obtained minimum values of ϕ (2.8, 2.7, 1.7 eV) which are close to those of bulk LaB_6 are attributed to dipole layer which forms on the surface. Chambers and Swanson [13] analyzing La ($M_5N_{45}N_{45}$) and B (KL_1L_1) Auger angular intensity distribution as a function of polar and azimuthal angles for $LaB_6(001)$ concluded that outermost layer is enriched with La, thus producing a positive La-B dipole moment.

On the other hand, Goldstein and Sczostak [14] on the basis of AES, LEED and work function measurements have shown that $LaB_6(001)$ shows no La enrichment even when heated as high as 1700°C. The LEED pattern is (1 x 1) indicating no surface reconstruction.

Generally similar result was observed later by Ozcomert and Trenary [15] who used scanning tunneling microscopy (STM) to obtain atomically resolved image of LaB_6. Berrada et al [16] on the basis of XPS, AES and work function measurements concluded that the emission mechanism in hexaborides (LaB_6 and CeB_6) is not due to surface dipole layer but is an intrinsic property of the material. Similarly, Mogren and Reifenberger [17] studying energy distributions of electrons emitted from a LaB_6(001) field emission tip found the importance of the bulk states for the electron emission mechanism. Theoretical analysis of the temperature dependence of the work function of LaB_6 carried out by Bulyga and Solonovich [18] points at the decisive role of the bulk electronic structure of material.

As seen from this brief overview, almost all of the studies on LaB_6 and other hexaborides focus on the bulk material. At the same time in order to distinguish between the two ongoing viewpoints on low work function of LaB_6, as well as other hexaborides, it would be informative to study some model systems which emphasize the role of the "pure" REM-B dipole layer. In relation to this the aim of the present work is to investigate the properties of double submonolayer to multilayer films which are obtained by the controlled growth of La-B and Gd-B films on metallic substrate (Mo(110)). Apart from the application importance such investigation is also interesting from the fundamental standpoint. These model systems are prototypes of the systems obtained by coadsorption of electronegative (B) and electropositive (La, Gd)) atoms, and thus allow gaining additional information on the chemisorption process in such complex systems, as well as in simple systems obtained by adsorption of a single component. Motivation of the choice of La and Gd with electronic configurations La [Xe]$5d^1 6s^2$ and Gd [Xe]$4f^7 5d^1 6s^2$ is to determine the effect of f-electrons on the properties of the systems under consideration. Adsorption of some rare earth metals on a number of substrates has been quite extensively studied previously [19-26], whereas almost no data are reported on their cooperative adsorption with boron. Also extremely lacking are the works studying adsorption of boron alone on metal substrates [27,28].

2. Experimental

Measurements have been carried out in ultra-high vacuum system (base pressure: $2x10^{-10}$ Torr) consisting of stainless steel chamber equipped with retarding four-greed hemispherical electron analyzer with coaxial electron gun for Auger electron spectroscopy and electron energy loss spectroscopy, low-energy electron gun, operating at energies lower than 10 eV for work function measurements (Anderson method), sample holder allowing heating to temperatures as high as 2800 K, quartz microbalance and adsorbate sources. The chamber was also equipped with the leak valve for controlled oxygen dose and residual gas mass-spectrometer. Energy resolution of the electron spectrometer derived from the elastic peak width was lower than 1%. The sensitivity of AES measured via the controlled dose of oxygen was 0.03 monolayers. The Auger spectra were measured in dN/dE whereas EELS in N(E) form with constant $\Delta E/E$ and 7V modulation. EELS in the configuration used was mainly sensitive to collective and interband transitions. The Mo(110) sample of a rectangular shape (10 x 5 x 0.3 mm^3) was mounted on the sample holder with the WRe 0,3 mm diameter wires spot welded at the four corners of the back of the sample. Direct current through these wires allowed heating of the sample to about 1400 K. Higher temperatures were achieved by electron bombardment of the sample by the tungsten filament placed in a tantalum screen behind the sample. The Ta screen was used to avoid electron bombardment of sample holder parts during high temperature annealing when high voltage is applied. The temperature was measured by W-WRe thermocouple spot welded to the back upper side of the sample as well as by optical pyrometer through the UHV window. The latter allowed to control temperature distribution along the sample at higher temperature and achieve uniformity by different combination of connections of sample holder wires. The uniformity of temperature distribution is a critical factor in measuring the adsorption energy via the well-known isotherm method [29] which was used and in this case for La-Mo(110) and Gd(110) systems.

Lanthanum and gadolinium were deposited on Mo(110) by thermal evaporation through direct heating of the 0.2 mm thick and 1 mm wide M-shape tantalum foil to which the corresponding bulk materials (ca. 0.5 mm^3) of 99.97% (La) and 99.98% (Gd) purity were spot-welded in ethanol environment. The latter was necessary to avoid oxidation by the ambient oxygen during welding. The foils with welded La and Gd were placed in tantalum screens with a nozzle on an upper side. Constructed in this way sources of rare-earth metals had high stability of the produced flux. Boron evaporation needs higher temperature which was achieved by electron bombardment. This was done from the 0.2 mm diameter grounded tungsten spiral (F) wrapped around the peace of bulk boron fixed in a positively biased (+2 kV) tantalum rod and placed in tantalum screen (1) to avoid hitting of the spiral by evaporated boron species (Figure 1). In this case it is also critical that the plane of the spiral is lower of the plane of the upper side of the screen (1), otherwise the boron species can hit the spiral. Initially, the constructed boron source did not contain this screen (1), so that the boron species were also deposited on the spiral. In this case the emission current was unstable, and moreover the lifetime of the spiral was becoming extremely low. It was concluded from this that boron strongly interacts with heated tungsten leading to dramatic transformation of its stability. Finally, the construction of the source shown in Figure 1 was used for boron evaporation. It exhibited remarkable stability of produced flux and long lifetime. All Ta foils and W filaments used for construction of the sources were preliminary outgased in a separate chamber at a pressure of ca 10^{-8} Torr.

After placing into UHV chamber the sources were carefully outgased for prolonged time until no contaminations were detected by AES. Deposition flux of the adsorbates was measured by quartz microbalance and verified by AES via the monitoring the Mo signal attenuation and the growth mode. The frequency of the quartz microbalance critically depends on the temperature, therefore it was placed in a tantalum screen for thermo-stabilization. For correct determination of the flux, and hence the adsorbate coverage, the similarity of deposition geometries for the sample and the quartz was achieved by their strictly coaxial and coplanar placement on a manipulator. Combination of quartz microbalancing (assuming unity sticking probabilities on the sample and the quartz) and AES allowed measurement of the surface concentration with the accuracy of ca. 0.15 of one monolayer coverage. It was assumed that the values of surface concentrations (n) of 1.1 x 10^{15} cm^{-2} for La, 1.3 x 10^{15} cm^{-2} for Gd and 2.7 x 10^{15} cm^{-2} for B correspond to unity coverage (θ=1).

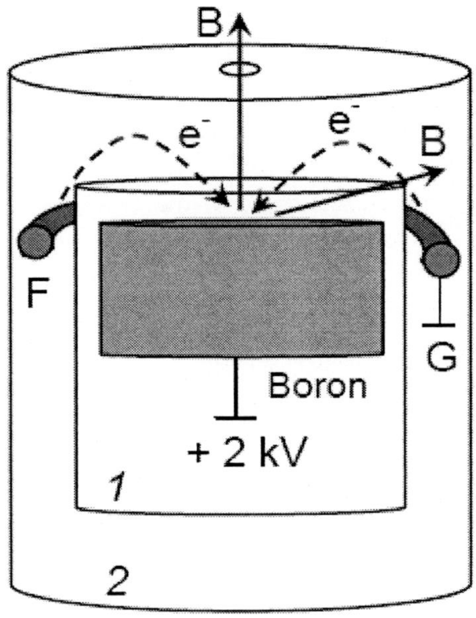

Figure 1.

The Mo(110) crystal was cleaned by the usual procedure of heating at 1300 K in and oxygen ambient (10^{-6} Torr) to remove carbon and sulfur contaminations and subsequent high-temperature (2700 K) flashing cycles in UHV. The surface was considered clean if no contaminations can be monitored by AES at the most sensitive scale. After each set of adsorbate deposition and measurement the surface was cleaned by cycles of high temperature flashing and the cleanliness was checked by AES. Sometimes, after a certain set of measurements, even after high temperature flashing the carbon C(KLL) signal could be detected by AES. In this case annealing of the sample in oxygen ambient to remove carbon contamination was applied.

3. Results and Discussion

3.1. Single Adsorbate Systems

3.1.1. La-Mo(110) and Gd-Mo(110)

Before studying cooperative adsorption of La and B and Gd and B on Mo(110) it is informative to investigate initially adsorption of each of these elements alone. The coverage dependence of work function and the Mo MVV (221 eV) Auger signal peak-to-peak intensity upon adsorption of La and Gd on Mo(110) are shown in Figure 2. Plotting of La and Gd Auger uptake curves is rather ambiguous because of their intersection with Mo Auger lines at energies lower than 170 eV. Linear decrease of Mo MVV Auger intensity (Figure 2, inlay) is an evidence of two-dimensional growth of the film at least in the submonolayer range. This growth mode is favored by the thermodynamic argument that the effective free surface energies of La and Gd are generally lower than that of the Mo. Moreover, as can be seen from the below result on adsorption energy measurement the bonding energy of both La and Gd with Mo is significantly higher than the energy of La-La and Gd-Gd interaction. Auger uptakes curves in the range from 1 to 2 and 2 to 3 monolayers due to higher point scattering can not unambiguously distinguish between subsequent monolayer growth of the film or formation of three-dimensional islands. However, previous detail LEED and TDS studies of Kolaczkiewicz and Bauer [19] indicating layer-by-layer growth mode for parent system: Gd/W(110) may serve as an indirect evidence of the same growth mode and in the present case.

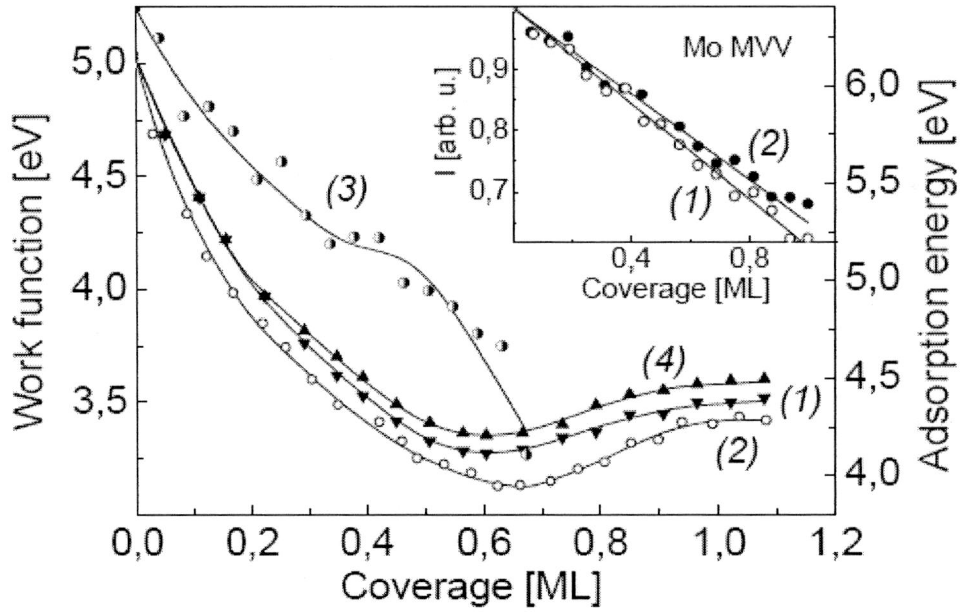

Figure 2.

The Anderson method is sensitive only to the work function change which was determined by measuring the linear part shift of the monitored retarding curve. In order to get the absolute values of ϕ, the work function of clean Mo(110) was assumed to be 5.0 eV [30]. Special care to ensure the adsorbate cleanliness was taken to measure the work function plots, because it was found that even hardly detectable by AES oxygen contamination strongly affects the plot, especially increasing the initial slope of the curve and total work function change. As seen in Figure 2, the work function versus coverage plots for both La and Gd are qualitatively similar (curves 1 and 2). Moreover the shape of these plots resemble that occurring upon adsorption of alkali and alkali-earth atoms on refractory metal surface: steep work function decrease at low coverage followed by the work function minima region associated with metallization of the two-dimensional film and subsequent work function growth and stabilization at a coverage of one monolayer roughly corresponding to the work function of bulk material [31]. For such systems widely explored point of view is that the work function change is due to charge transfer between adatom and the substrate and hence formation of the surface dipole layer changing the work function. This idea initially proposed by Gurney [32] was later confirmed by numerous theoretical calculations for a variety of adsorbate systems [33]. Key factor determining whether the work function increases or decreases upon adsorption is the relation between electronegativities of adsorbate and substrate atoms. The general trend is that the larger the electronegativity difference the larger is the work function change, at least at the very early stage of adsorption. Pauling's electronegativities for La, Gd and Mo are 1.10, 1.20, and 2.16, respectively, thus generally determining the work function decrease upon La and Gd adsorption on Mo(110). Initial dipole moments (μ) determined from the work function versus coverage plots using Helmholtz equation $\Delta\phi=4\pi n\mu e$ at $n \rightarrow 0$ are 2.0 and 1.8 D for La and Gd, respectively. The proximity of these values as well as the general similarity of $\phi(\theta)$ plots (Figure 2, curves 1 and 2)) indicate that the same electronic mechanism is responsible for work function change for both La and Gd. Taking into account the electronic configuration of La [Xe]$5d^1 6s^2$ and Gd [Xe]$4f^7 5d^1 6s^2$ one can assume that the adsorption properties for both La and Gd are determined mainly by 5d6s electrons, whereas 4f electrons are weakly involved in the formation of chemisorption bond. The latter is also evidenced by the fact that work function versus coverage plots for La and Gd are generally of the same shape as for alkali and alkali-earth adsorbate metals containing no f-electrons. Also, taking into account that for transition metal atoms adsorption properties including the work function versus coverage plots [34,35] are strongly dominated by d-electrons, one can expect even more pronounced influence of more complicated f-bands on such properties distinguishing behavior of Gd from La and from alkali and alkali-earth metals. Additionally, general similarity of work function versus coverage plots measured by Shakirova and Serova [22] along the rare-earth metal series by field emission microscope indicates at indecisive role of 4f-level filling on the work function.

Although no pronounced effect of f-electrons on the work function change is observed, one can not exclude their involvement in the adsorption mechanism. Unlike the similarity of the values of initial dipole moments for La (2.0 D) and Gd (1.8 D) the adsorption energies (E_a) measured by isotherm method via the monitoring the adsorbate thermodesorption flux in equilibrium are considerably different: 4.9 eV (La) and 6.4 eV (Gd) at coverage close to zero. Corresponding coverage dependences for La and Gd are qualitatively similar, except of the initial and final values of adsorption energy. The plot for Gd/Mo(110) is shown in Figure 2 (curve 3). This E_a versus coverage plot and its correlation with the work function plot is similar to that exhibited by alkali- and alkali-earth atoms on refractory metal surface [31]. The initial adsorption energy decrease is associated with the electrostatic dipole repulsion which increases with coverage, followed by the subsequent energy stabilization associated with phase transition due to metallization of the film (correlation with the work function minima) and subsequent drop corresponding to the formation of the bulk like adsorbate when the desorption energy is close to sublimation energy of the corresponding material (La-3.9 eV, Gd – 3,6 eV). Considerably higher value of E_a for Gd compared with La is an evidence that apart from the 5d6s-electrons 4f-electrons may also be responsible for such adsorption energy enhancement. This, however, might not be the pure adsorbate-substrate charge transfer effect, since initial dipole moments of Gd (1.8 D) and La (2.0 D) are close to each other. Strong compensation of the repulsive dipole-dipole interactions by attractive, presumably indirect electronic interactions which was found for rare-earth metals on W(110) [19] may account for higher value of E_a for Gd.

Relation of the adsorption energy at initial coverage when it is dominated mainly by the adsorbate-substrate interaction and at approximately unitary coverage when the E_a is close to sublimation energy, provides support for the above assumption of the two-dimensional growth mode of the films for both La and Gd derived from the Auger uptake curves (Figure 2, inlay). Higher value of adatom-substrate interaction energy compared to that of La-La and Gd-Gd interaction favors formation of two-dimensional layer rather than the three-dimensional islands. Thermal annealing of La and Gd films reveals no significant change of the work function. Change of the work function versus coverage plot for Gd deposited on Mo(110) held at 1000 K is shown in Figure 2 (curve 4). Similar trend also holds for La. Approximately up to 1/4 of monolayer the curve almost totally reproduces that occurring at room substrate temperature, and afterwards only a rather slight increase of the work function occurs. This means that the films formed at room temperature are close to their equilibrium state, and annealing mainly leads to their structural ordering resulting in a slight work function increase [19]. At low adsorbate coverage the similarity of work function plots for low and high substrate temperature suggests that at this coverage the film is in the form of two-dimensional lattice gas. In this low coverage region the coverage increase leads to the adatom spacing distance decrease and thus to the adsorption energy gradual drop due to enhanced interdipolar interaction as seen from the corresponding E_a versus coverage plot (Figure 2, curve 3).

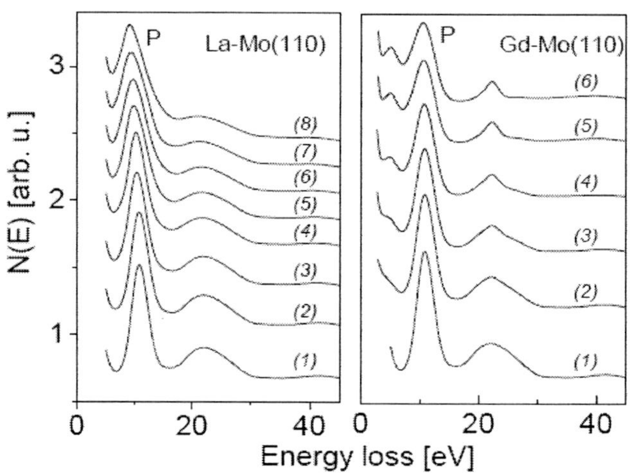

Figure 3.

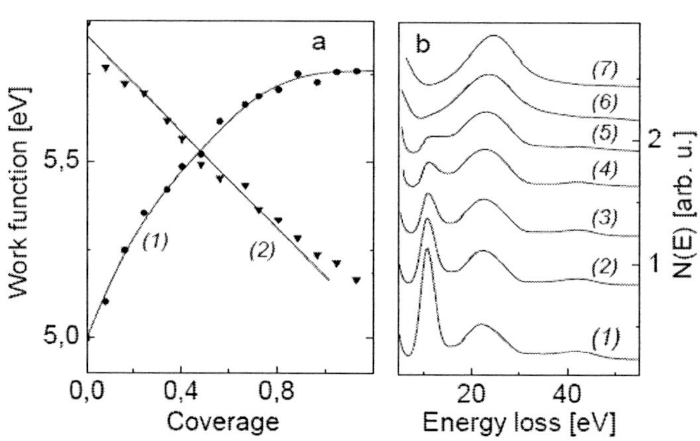

Figure 4.

Annealing of La and Gd one monolayer films at 1000 K does not result in a noticeable change of the Auger spectra. This also provides support for the above assumption of general equilibration of the films formed at room substrate temperature. However, in the case of Gd, annealing of multilayer film results in the transformation of the film as followed from the dramatic change of the Auger spectra. Annealing results in the substantial growth of Mo and a slight decrease of Gd Auger signal intensity. Taking into account that in this case there is almost no change of the work function (Figure 2, curve 4), such inconsistent behavior of Mo and Gd Auger intensities may be explained assuming formation of three-dimensional islands on top of the complete Gd layer on Mo(110). In case of the formation of such islands directly on Mo(110) (not Gd flat film) one would expect much more pronounced increase of the work function. Similar behavior of work function versus coverage plots for substrate temperatures of 300 and 1000 K (Figure 2, curves 1 and 4)) supports the assumption of flat Gd film on which surface the three-dimensional islands form. In the case of annealing of several monolayer thick (2, 3 and 5) La films to the temperatures as high as 1200 K no dramatic change of the Auger spectra and the work function was observed, suggesting morphological stability of the film formed at 300 K. The fact that unlike La, the Gd multilayer film tends to form three-dimensional islands upon annealing provides additional support of the above assumption of strong attractive interactions for Gd, compensating electrostatic repulsion.

Electron energy loss spectra upon adsorption of La and Gd on Mo(110) are shown in Figure 3. The spectrum of bare Mo(110) (curves 1) consists of three losses at 10.5, 21 and 47 eV which are attributed to surface and bulk plasmons and interband 4s4p transitions, respectively [36,37]. The observed relative variation of the intensities of the first two lines with the primary electron energy supports this interpretation.

As seen in Figure 3 in both cases there is a gradual change of the EEL spectra form from that characteristic for Mo(110) to those exhibited by bulk La and Gd [38,39]. No noticeable change of the surface and bulk plasmons bands of Mo at the early stage of film growth suggests that there is no change of the collective-like nature of Mo valence electrons. The effect produced by La and Gd adsorption does not noticeably influence both the tangential (bulk plasmon) and longitudinal (surface plasmon) [40] collective modes of the Mo(110) electron density. This is an evidence that adsorption of La and Gd does not dramatically change the electronic state of Mo(110) compared to that occurring upon chemical compound formation, in which case usually plasmon modes change drastically [41]. At intermediate coverage the features of bulk La and Gd appear in the corresponding EEL spectra and at a coverage exceeding 2 monolayers the spectra correspond to those of bulk adsorbate materials (Figure 3, curves 6 and 8).

The spectrum of Gd consists of three pronounced features at loss energies of 5, 10.5 and 22 eV attributed to f-f intraband transitions, plasmon loss and 5p one electron excitation, respectively, whereas the La spectrum is dominated by one line at 9 eV which is associated with the metal plasmon mode [38,39].

Annealing of both La-Mo(110) and Gd-Mo(110) systems at submonolayer and multilayer coverage to temperatures up to 1200 K does not noticeably change the EEL spectra, indicating of no chemical interaction in the adsorbate system or interdiffusion of its components. The latter is also confirmed by AES results showing no Auger energy shift and corresponding intensity variation of both substrate and adsorbates upon annealing.

3.1.2. B-Mo(110)

Behavior of boron on Mo(110) is significantly different. Unlike La and Gd, boron adsorption results in the work function increase (Figure 4a, curve 1). Work function versus coverage continuously increases until saturation at 5.8 eV occurring at adsorbate surface concentration of ca. 2.7 x 10^{15} cm^{-2}. Comparison of Mo and B atom size shows that the above surface concentration of boron corresponds to coverage of one monolayer. No noticeable work function change was observed above this coverage, which is an indirect evidence of formation of complete layer totally covering the Mo(110) surface. The latter is confirmed by the way the Mo MVV (221 eV) Auger signal attenuates with boron coverage increase (Figure 4a, curve 2): To good approximation linear decrease indicates at the formation of two-dimensional boron film, at least, in the submonolayer region. Auger uptake curve of KLL boron line (179 eV) can not be unambiguously drawn due to its strong intersection with Mo MNV signal (188 eV). For boron coverage corresponding to second and third monolayers the conclusion about the growth mode based on the form of Mo Auger signal intensity decrease is not straightforward due to enhanced data point scattering at higher adsorbate coverage. However taking into account that at boron coverage corresponding to ca. 5 monatomic layers the Auger spectrum of B-Mo(110) system in the energy range of 50 to 300 eV consists of only one B KLL line favors assumption of layer-by-layer growth mode up to this coverage.

Apart from the Mo Auger intensity decrease adsorption of boron at substrate held at room temperature does not result in the change of the Auger peak shape and energy. However, the EEL spectra undergo dramatic transformation upon boron adsorption. Already at the very early stage of film growth the Mo(110) surface plasmon mode drastically attenuates and subsequently loosing intensity disappears at boron coverage of ca. 0.8 monolayers (Figure 4b). At the same time the Mo(110) bulk plasmon mode is altered less, at least at low boron coverage. At coverage exceeding one monolayer the EEL spectrum consists of one broadened band centered at ca. 24 eV which is attributed to the bulk plasmon of boron [42]. Unlike the surface plasmon area the transformation of the spectra in the bulk plasmon region to good approximation can be viewed as superposition of Mo and B plasmon lines with the opposite change of intensity. Assuming the three-fold hollow of Mo(110) plane as an adsorption site of B, in lines with the case of oxygen-Mo(110) system [43], and taking into account about as twice as smaller size of boron atom (2.3 Å) compared to Mo (4.2 Å), the boron film is almost coplanar with the Mo(110) topmost layer. This may affect the lateral integrity of the outermost Mo(110) layer violating the longitudinal collective motion of electron density responsible for surface plasmon mode [40]. A rather small initial dipole moment of B (0.2 D) derived from the work function versus coverage plot (Figure 4a, curve 1) is an additional evidence of negligible shift of boron outwards of the Mo(110) plane. Moreover taking into account the relation between Mo and boron electronegativities (2.16 and 2.04, respectively) one should expect not the work function increase as observed, but its decrease. To account for the observed increase of the work function it can be assumed that boron atoms are shifted even inwards of the Mo(110) plane to form the effective dipolar layer with the negative Mo^{x-} end shifted towards the vacuum.

Thus significant change of the electronic state of Mo(110) upon boron adsorption manifested in the dramatic change of the surface plasmon mode relates mainly to the topmost substrate layer. This can be viewed as a precursor of stronger interaction and possible chemical bond formation which occurs upon annealing of B-Mo(110) system. Even a slight thermal treatment causes dramatic change of the properties of the system which differs from both La and Gd on Mo(110). Transformation of the Auger spectra as well as the dependence of the work function, and Mo and B Auger intensities on the annealing temperature are shown in Figure 5.

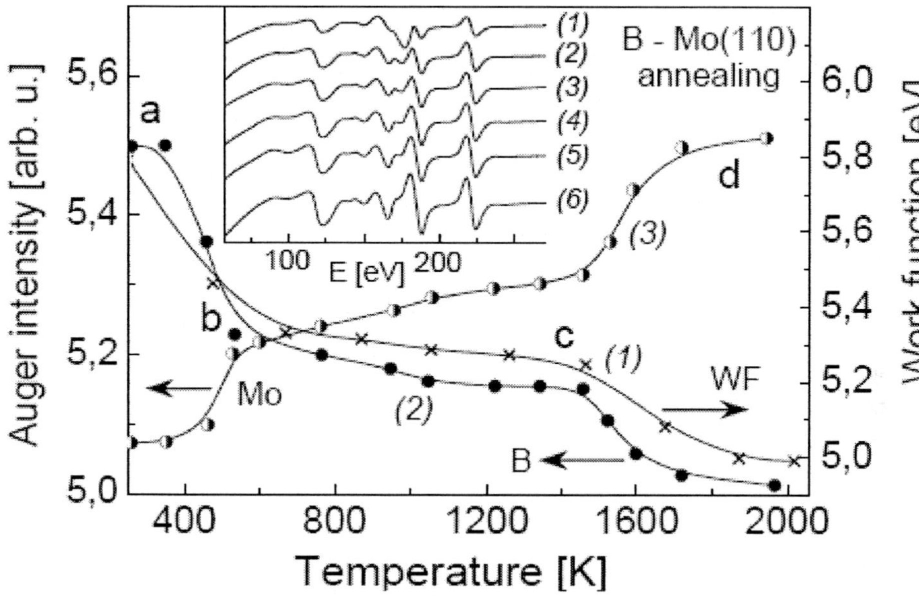

Figure 5.

It is seen that at a temperature of about 450 K all the measured characteristics of the system notably change: Work function decrease is accompanied with the B Auger intensity abrupt drop and corresponding Mo Auger signal increase. Subsequent growth of annealing temperature enhances this tendency (part ab) until at the temperature of ca. 700 K and up to ca. 1500 K there is only moderate change of the properties (part bc), again followed by dramatic change (part cd). In the whole annealing temperature interval the change of both boron and Mo Auger line positions were not detected (see inlay). This means that the observed temperature dependent variation of the work function and the Auger intensities are mainly due to the diffusion of B into Mo(110) without compound formation. The latter however can not be totally ruled out taking into account the stabilization part (bc) which may be an evidence of formation of molybdenum borides limiting the B-Mo(110) interdiffusion process. The observed strong tendency of boron diffusion into the Mo(110) bulk at elevated temperature supports the above assumption of boron layer inward shift upon adsorption at room temperature as a precursor of boron diffusion into the bulk. Behavior of the B-Mo(110) at higher annealing temperatures (1500-2000 K) as can be seen in Figure 5 is dominated by the boron concentration decrease and its total disappearance at ca. 2000 K, as judged from AES and the work function approaching that of bare Mo(110). The EEL spectrum at the above temperature is also close to that of clean substrate after high temperature flashing. Therefore assuming molybdenum borides formation in the surface and subsurface area upon annealing to moderate temperatures (part bc) one should expect their high temperature dissociation and subsequent desorption of boron from the substrate or its diffusion into the bulk, not accessible with the techniques used, almost totally restoring the initial state of Mo(110) surface (part cd).

Additional support of this interpretation of behavior of B-Mo(110) system upon annealing is provided from measurements of boron Auger intensity decay rate at fixed temperature. The corresponding plots for four different annealing temperatures of the similarly prepared four B-Mo(110) systems are shown in Figure 6. Assuming a temperature activated process the activation energy for the corresponding process can be determined from these dependences redrawn in a semilogarithmic plot of the relaxation time (τ) after which the B KLL intensity have dropped to certain level versus the reciprocal temperature (inlay). The corresponding activation energy of the process occurring in the temperature range of 450 to 600 K (plot ab) is 0.4 eV, whereas the same value for the process in the 800-1400 K annealing interval (plot bc) is by the order of magnitude higher: 3.8 eV. Such a dramatic difference is a result of different processes determining the boron concentration variation in the corresponding temperature range. The initial part of the curves (ab) (Figure 5) may be attributed mainly to the diffusion of boron atoms into the bulk with no chemical reaction whereas in the higher temperature regime (part bc) the boron concentration decay is limited by the bond braking of Mo-B surface compound formed, totally disappearing from the surface area upon further temperature rise (part cd).

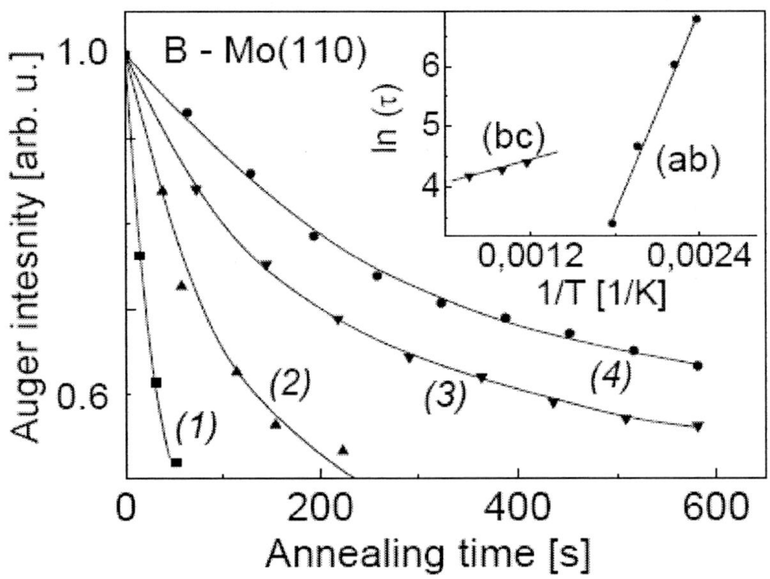

Figure 6.

3.2. Double Adsorbate Systems

3.2.1. La-B-Mo(110) and Gd-B-Mo(110), B-La-Mo(110) and B-Gd-Mo(110)

Cooperative adsorption of La and Gd with boron was studied for different orders of adsorption of the corresponding components. This order was find to be a crucial point determining the properties of the systems formed at the substrate held at room temperature. Even small amount of preadsorbed boron dramatically changes the work function versus coverage plots for subsequent adsorption of both La and Gd compared to that on bare Mo(110). The corresponding dependences for different initial boron coverage combined in one plot are shown in Figure 7. There are several points differentiating the work function versus coverage plot from that of both La and Gd on bare Mo(110): Minima region of $\phi(\theta)$ curve is no longer available at all initial boron concentrations studied; the saturation work function is considerably lower; more La and Gd adsorbate concentrations are needed to stabilize the work function; initial slopes of La and Gd work function plots determining the initial dipole moments are notably higher and increase with the preadsorbed boron concentration. It is noteworthy that the stabilization work function is close to the values of bulk lanthanum and gadolinium hexaborides: 2.7 and 2.9 eV, respectively.

The AES and EELS results may help to account for the above differences in the work function plots for La(Gd)-B-Mo(110) and La(Gd)-Mo(110). Auger spectra upon boron adsorption on Mo(110) held at room temperature followed by deposition of La and Gd are shown in Figure 8. The common feature for both elements is inconsistent attenuation of Mo and B Auger intensities upon rare-earth metal deposition: the Mo Auger signal drops dramatically whereas that of boron decreases insignificantly (curves 1-4). Such inconsistent Auger intensity attenuation can not be reconciled assuming La and Gd three-dimensional island formation on top of the preadsorbed boron layer. In the latter case one should expect correlation of Mo and B Auger amplitude decrease. Therefore it is reasonable to attribute the above changes in the Auger spectra to the outward diffusion of boron from the B-M(110) interface into the layer of rare-earth metals. The possibility of such boron layer redistribution is presumably undermined by the fact that the boron film formed on Mo(110) at the substrate held at room temperature is not in equilibrium state as follows from the low measured value of diffusion activation energy: 0.4 eV. This value evidences of weak bonding of as-deposited boron with Mo(110), which can easily be compensated by much higher energy of adsorption of both La (4.9 eV) and Gd (6.4 eV) serving as a driving force for B-Mo(110) bond breaking and outward diffusion of the formed boron atoms. Additionally, the intermixing can be governed by the thermodynamic argument: The free surface energy of the mixture may be lower than the corresponding values for respective pure components. The latter, for instance, is the case for oxygen and nitrogen adsorption and incorporation on metal surfaces and the formation of oxides, nitrides and oxynitrides with the free surface energy significantly lower of the values of the corresponding pure metals [44,45]. The fact that the attitude to intermix might be an inherent property of La and Gd and B layers is also evidenced from AES results for thick boron and lanthanum and gadolinium films. Curves 5 in Figure 8 correspond to 5 monolayers of preadsorbed boron, whereas curves 6 correspond to about 20 monolayers of La and Gd. It is clearly seen that even at this high coverage of La and Gd the boron Auger signal intensity is rather big. Beginning from the coverage of about 8 monolayers and up to 20 monolayers the La/B and Gd/B Auger intensity ratio remains almost constant indicating at the diffusion of a certain part of boron atoms to the film/vacuum interface area. It may turn out that here is a definite favorable ratio of La (Gd) and boron atom concentrations in one or several outer layers maintaining of which is a driving force for boron outside diffusion. This ratio, however, as evidenced from the amounts of deposited La (Gd) and boron is far from 1/6 occurring in bulk hexaborides.

The work function, nevertheless, in both cases corresponds to the saturation values for the La-B and Gd-B systems: 2.7 and 2.9 eV, respectively (Figure 7) which are close to the values of the corresponding bulk hexaborides.

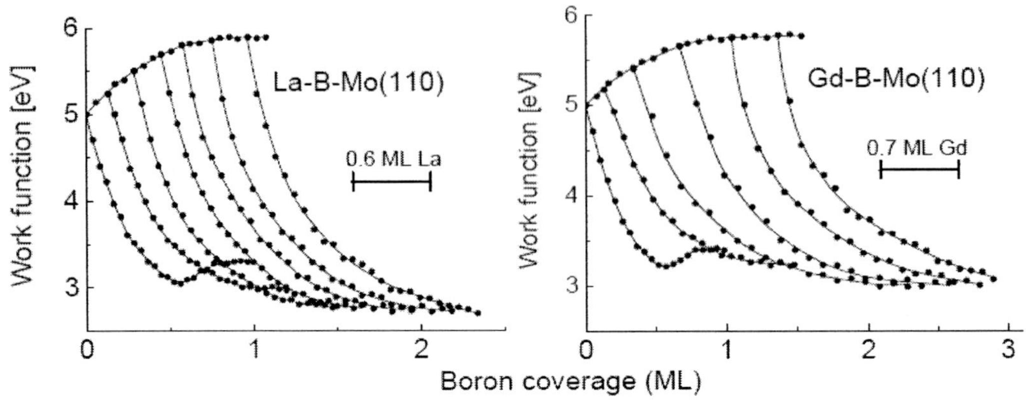

Figure 7.

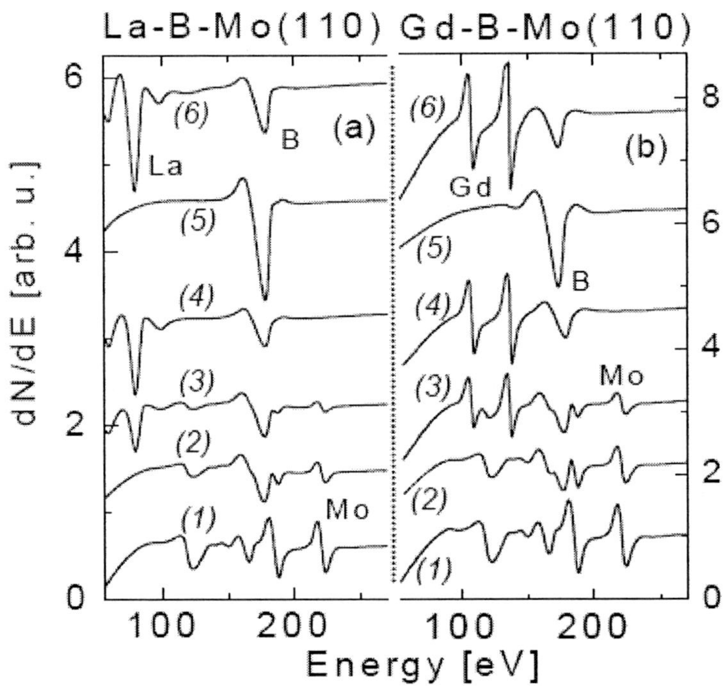

Figure 8.

Closer inspection of Auger spectra reflecting the interdiffusion process (Figure 8) reveals no energy shift of both La and Gd and B and Mo Auger lines. It means that intermixing occurs via the separate atoms with no chemical bond formation, which otherwise would change the Auger energy and be a limiting process to hamper diffusion of boron throughout the rare-earth metal film at room temperature.

The same evidence is provided by the EEL spectra upon La and Gd adsorption on the boron film. The corresponding spectra for Gd-B-Mo(110) are shown in Figure 9a. As it is shown above, the B-Mo(110) EEL spectrum consists of one broad line centered at 24 eV (Figure 9a, curve 1).

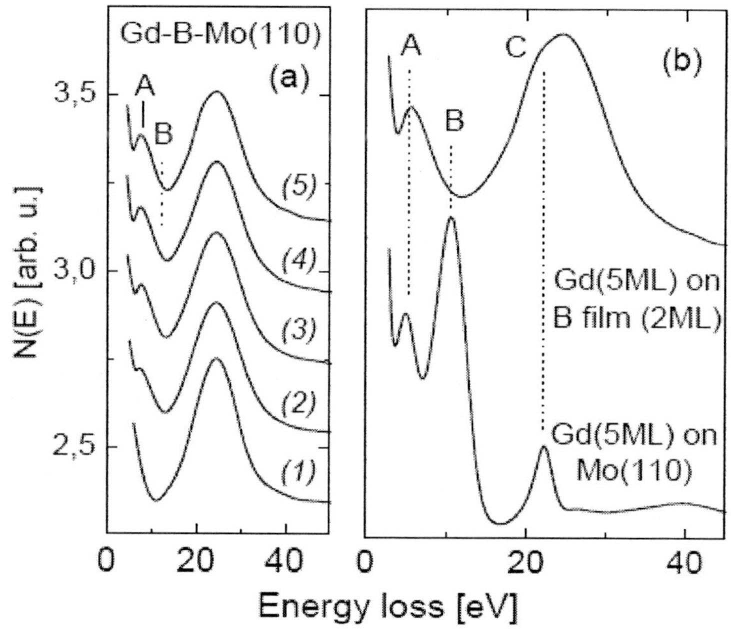

Figure 9.

This line is almost no affected by the subsequent Gd (as well as La) deposition. In the letter case the only apparent feature is appearance of a new loss (A) at energy of about 5 eV. With account of its intensity and energy position it is reasonable to assign this to the loss occurring in Gd film deposited onto Mo(110) (Figure 3). As has been noted above this loss is due to one-electron excitation in bulk gadolinium [35,36]. Moreover, it is very sensitive to whether or not the Gd atom forms chemical bonds [35]. Therefore its presence in the EEL spectrum of Gd-B mixed system suggests that Gd is in the atomic form. Even no similar loss feature exists for La, in view of the above considerations, the latter can be expected and for La too. In the EEL spectra of both La and Gd deposited on boron film the plasmon line characteristic for bulk metals (located at position denoted as "B" in Figure 9) can not be detected even at their coverage as high as twenty monolayers, as seen from comparison of the spectra on bare Mo(110) and that precoverd by boron (Figure 9b). At the same time appearance of plasmon losses even in two-dimensional metal films and/or islands is a common feature for almost all metals studied [40]. This means that the collectivization of the La and Gd valence electrons in the mixed system is disturbed in both lateral and normal directions thus supporting assumption of the presence of boron not only in the subsurface area but also in the topmost layer of the film. The latter, therefore, can be viewed as consisting of La and Gd film with incorporated boron atoms to prevent La-La and Gd-Gd electron lateral coupling to form collective vibration.

According to rather high boron Auger signal in La-B and Gd-B double systems (Figure 8, curves 6), one can not exclude the possibility that boron resides not only in the topmost La-B (Gd-B) mixed layer but also diffuses towards the vacuum to form the pure boron overlayer. The latter, however, seems unlikely if take into account the following results of work function measurements of the systems obtained by the reversed order of La (Gd) and boron adsorption, namely the work function increase when boron is on top of La or Gd layer. The corresponding plots are shown in Figure 10.

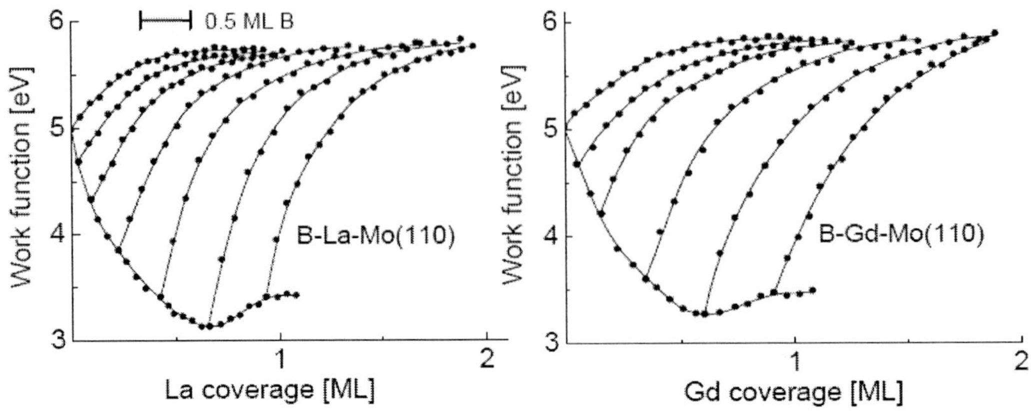

Figure 10.

The work function was measured upon gradual increase of boron coverage on the Mo(110) precoverd by different amounts of lanthanum and gadolinium. At all initial metal coverage the work function gradually increases until saturation at the same value of ϕ and coverage as on bare Mo(110). The Auger intensities of La (Gd) and Mo gradually decrease until at a boron coverage of about 5 monolayers only one B KLL peak is seen in the spectrum. This means that unlike the previous adsorption order there is no noticeable component intermixing in such systems at room substrate temperature. To good approximation the observed linear decrease of La and Gd Auger intensities up to boron coverage of one monolayer evidences of the formation of two-dimensional boron film, totally covering the La and Gd layers. The latter possibly determines the fact that the work function stabilizes at monolayer boron coverage. This is also supported by the fact that the saturation work function value for boron on Mo(110) with and without La and Gd layers is the same: In the case of formation of three-dimensional islands one had to expect rather different values of ϕ at unitary coverage because of the different work functions of bare Mo(110) and that precoverd by La or Gd, supporting the boron islands.

The B-REM double layers formed in this order of deposition are not stable. Even slight annealing causes dramatic change of the work function and the Auger intensity. As an example, the corresponding plot for B-La-Mo(110) system is shown in Figure 11. For Gd the plots are qualitatively similar.

Even slight annealing (up to 400 K) results in a noticeable decrease of the work function and boron and La(Gd) Auger intensity change. Further increase of annealing temperature facilitates this trend until at about 600 K there is dramatic change of these values followed by stabilization in the temperature range of ca. 800-1300 K. At higher temperature the boron Auger peak almost disappears and that of La basically remains unchanged, and the work function approaches the value characteristic for La-Mo(110) system. At annealing temperature of about 1600 K and higher the work function slightly increases accompanied by some La Auger intensity decrease.

The plots shown in Figure 11 can be explained by strong intermixing of boron and the rare-earth metal films upon annealing. An abrupt change of the plots occurring at about 600 K is likely due to enhanced diffusion of boron atoms from the surface to subsurface area of the REM film.

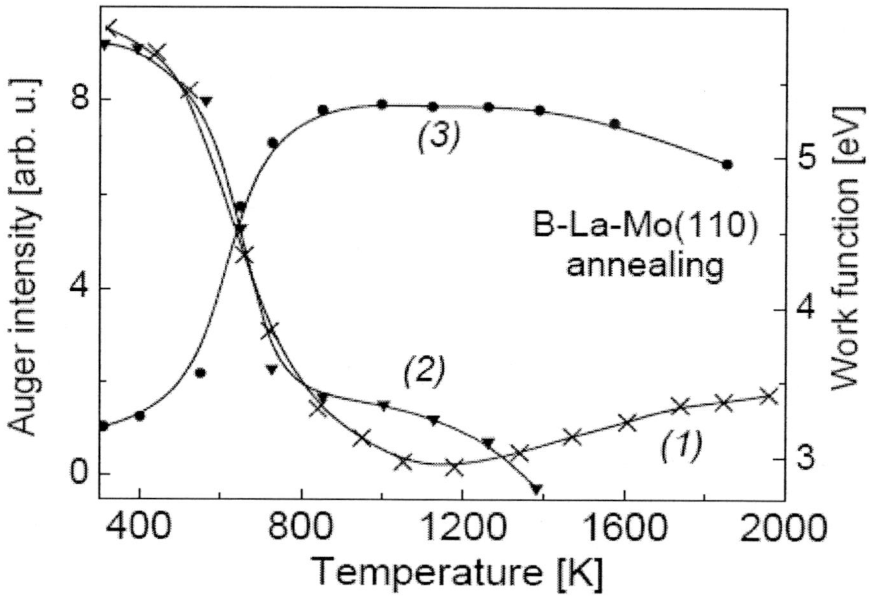

Figure 11.

Subsequent stabilization area at 800-1300 K is presumably due to chemical bond formation in the REM-B system hindering the diffusion process. However this bond might not be strong since rather moderate temperature rise to ca. 1500 K results in boron total disappearance from the film. The latter is probably due to REM-B bond breaking, and in view of the above behavior of B-Mo(110) system, to subsequent diffusion of B into the Mo(110) lattice. It is interesting to note that rather small amount of boron which remains in the surface area after annealing to about 1000 K results in work function value of 2.9 eV (Figure 11) which is close to that of bulk hexaborides.

The same situation is seen from the work function plots shown in Figure 7. As can be seen, only 1/3 of boron monolayer is needed to achieve the work function close to that of hexaborides. Even as high as twenty monolayer thick La or Gd films deposited on this preadsorbed boron layer does not change the value of the work function. According to Auger results, the boron submonolayer film peak intensity upon La and Gd layer thickness growth from 1 to 20 monolayers changes rather insignificantly. This is an evidence that the systems tends to maintain a certain ratio of REM to boron concentration in the surface region. In spite of the fact that this ratio is far from that of bulk hexaborides (1/6) the work functions in both cases are nevertheless similar. Such similarity supports the viewpoint that the low work function of hexaborides is a result of surface layer rearrangement, rather than the intrinsic property. This surface rearrangement, however, can not be viewed as only formation of metallic overlayer due to La or Gd enrichment. Basically this is also favorable situation with the regard of achieving of high thermionic efficiency (as seen in Figure 11 the resulting work function when boron is almost totally gone is rather low: ca. 3 eV). However notably better situation is achieved when there is a mixture of metal and boron atoms. As it is mentioned above when one monolayer of La or Gd film is deposited on boron precoverd Mo(110) at low initial coverage (not exceeding ca 1/5 of monolayers) the B(KLL) Auger intensity does not change, which indicates that boron is not beneath of the metal layer but is coplanar with it.

Outward shift should also be excluded since in this case one have to expect work function increase as for B-La(Gd) system (Figure 10). The further increase of the metal thickness probably maintains this situation because there is no change of boron 1/5 monolayer Auger intensity. Therefore it seems reasonable to expect that low work function of the double La(Gd)-B-Mo(110) is due not to the situation when electropositive La(Gd) atoms are located above the electronegative boron atoms, but to their coplanar mixture.

The fact that there is tendency to maintain a certain La (Gd) to boron concentration ratio is also apparent from the case when La or Gd is deposited on thick boron layer. For boron coverage of about 5 monolayers, with the subsequent increase of La or Gd coverage the boron Auger signal initially decreases and then remains at a certain level even for thick La and Gd films (Figure 8). The latter means that when the preferred La(Gd)/B concentration ratio is reached at the surface, the remaining boron atoms are distributed over the thick La (Gd) film. The fact that much lower relative amount of boron compared to that in hexaborides is necessary to achieve about the same value of the work function is in lines with the maintenance of high thermionic efficiency of real hexaboride cathodes even after prolonged time of operation when boron for the most part is gone from the sample. For instance, Mrozckowski [4] studying the electron emission characteristics of sputtered lanthanum hexaboride found that coated tungsten filaments run over 1000 hours without any reduction of the emission current.

At the same time, according to AES and XPS results there is considerable decrease of the amount of boron during operation which significantly changes the initial 1/6 La to B concentration ratio.

Additional support of the above assumption that two-dimensional mixture of rare-earth metal and boron reduces the work function is provided by the comparison of the work function for different order of adsorption of the same amounts of REM and B (Figure 12). Curve 1 corresponds to boron adsorption on the La preadsorbed film, while curve 2 corresponds to opposite adsorption order. It is seen that at initial La and B coverage not exceeding ca. 0.15 monolayers the curves are close to each other whereas at higher coverage there is substantial deviation. The latter is expected if take into account the corresponding work function plots (Figures 7 and 10), indicating of significantly different state of the double layers obtained at different adsorption orders. However, similarity of the plots at coverage lower than 0.15 monolayers suggests that the state of the layer does not depend on the adsorption order.

Taking into account a rather small surface concentration one can expect that boron and metal atoms form two-dimensional layer on the Mo(110) surface. Since the coverage is rather small, about the same layer can be formed regardless of the deposition order. At higher coverage, when some additional atoms can be adsorbed above this two-dimensional layer, the work function changes depending on the adsorption order. It is worthy of noting that initially when formation of two-dimensional layer is expected, the adsorption of boron on La precoverd surface results in work function decrease, even though electronegativity of B is higher than that of La so that an opposite trend should be expected.

Further concentration growth above ca. 0.15 monolayers leads to work function increase which is reasonable to assign to boron atom adsorption on top of the La atom thus forming a negative dipole layer. Unlike the case of boron adsorption on La-Mo(110), for the opposite adsorption order there is no change of the work function direction at a coverage higher than ca. 0.15 (Figure 12, curve 2).

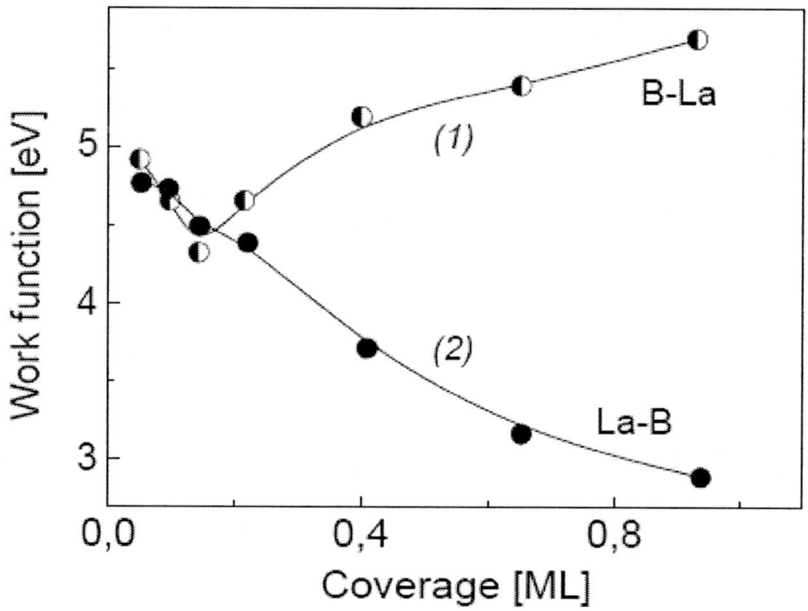

Figure 12.

Moreover there is no also change of the shape of the curve in this area meaning that the work function decrease mechanism at lower and higher coverage is basically the same. In lines with the above Auger results showing diffusion of boron from B-M(110) interface to the La film this means that such work function plot smooth variation is an evidence of coplanar coexistence of La and B atoms in all submonolayer to monolayer coverage. Even if no adsorption of La on top of boron atom is expected in the submonolayer range the positive dipole layer reducing the work function may also be formed in this case of coplanar placement taking into account much larger atomic diameter of La (5.48 Å) compared to that of boron (2.32 Å).

As seen in Figures 7 and 10 the initial slopes of all the work function plots for double systems, regardless of the adsorption order, increase with the increase of the surface concentration of the preadsorbed component meaning corresponding increase of the effective dipole moment values. These values, derived from Helmholtz equation, for a certain component versus the work function of the substrate changed by preadsorption of the other component are shown in Figure 13. These plots have common character: To good approximation all of them are linear and if extrapolated to zero dipole moments give the work function value close to that exhibited by the corresponding bulk component. Such similarity suggests that the dipole formation mechanism for single La, Gd and B atoms is generally the same and is mainly determined by participation of free-like s- or sp-electrons. Assuming the validity of the above mentioned Gurney's charge transfer model of work function change, the corresponding adatom-substrate levels schematically are shown in Figure 14. Variation of the substrate work function should result in the change of the filling of the atomic level and thus in the change of its dipole moment. In case of rather smooth dependence of density of states on the energy as for sp-levels one can expect almost linear change of atomic charge with the filling of the atomic level due to substrate work function change, and thus the similar linear change of the dipole moment (Figure 14, a).

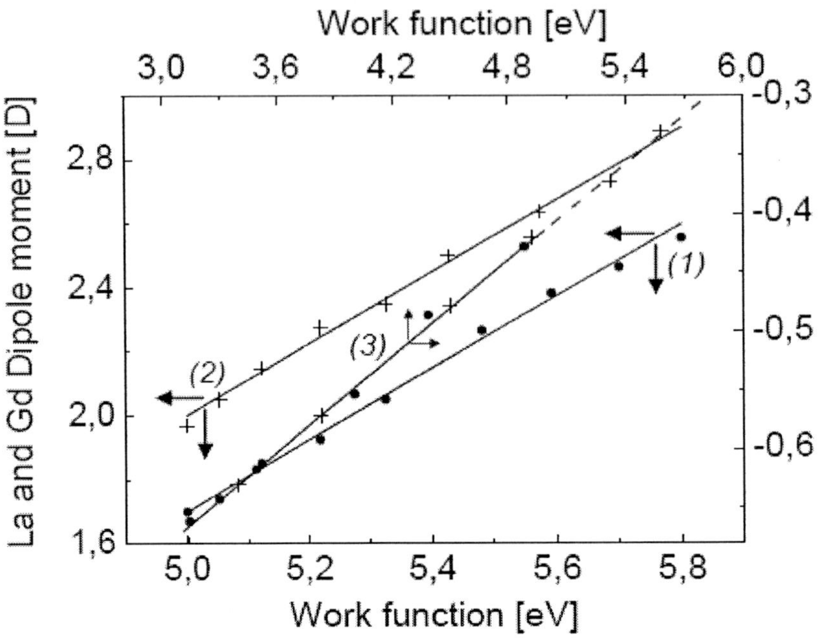

Figure 13.

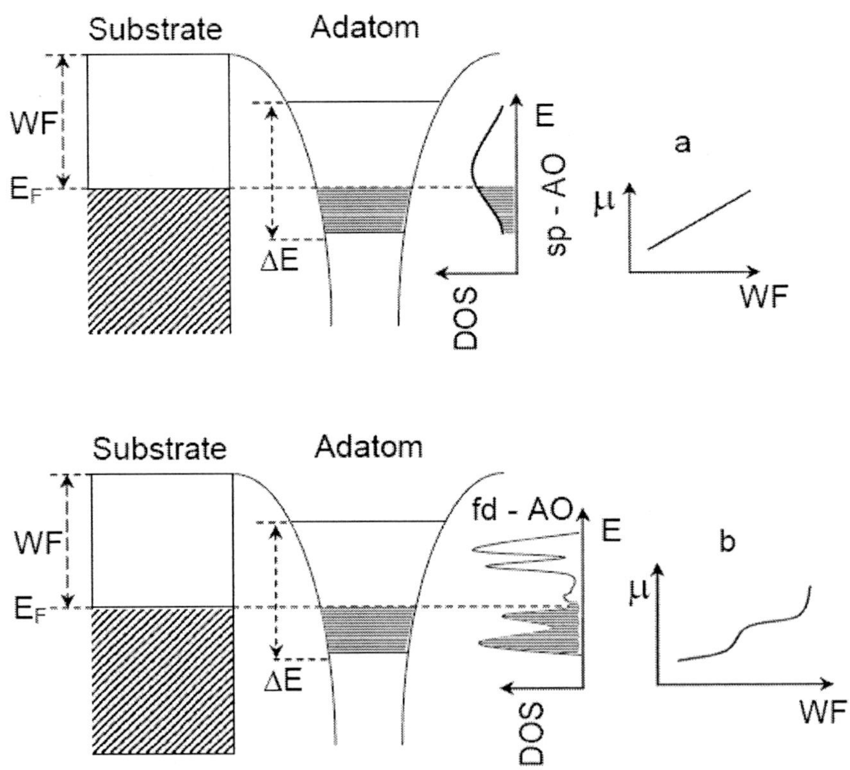

Figure 14.

In the case of much more complicated DOS for fd-levels the linear behavior can hardly be expected (Figure 14, b). The latter is additional evidence against considerable influence of f-electrons on the work function change. At the same time, however, in some cases they might provide considerable input into the formation of chemisorption bond as evidenced, for instance, from noticeably higher adsorption energy for Gd than La on Mo(110). Nevertheless, for the coadsorption with boron no considerable differences between La and Gd were found, except slight difference of saturation work function values (Figure 7). This indicates that f-electrons of Gd play indecisive role in the formation of low work function, morphology and stability of the systems formed upon cooperative adsorption with boron.

CONCLUSION

Comparative study of adsorption of La, Gd and B, separately, on Mo(110) surface, on one hand, and cooperative adsorption of La with boron and Gd with boron, on the other hand, has been carried out by means of Auger electron spectroscopy, electron energy loss spectroscopy and work function measurements. Even a small fraction of monolayer of preadsorbed boron dramatically changes the work function versus coverage plots compared to those occurring upon adsorption of La and Gd on their own: The shape of the work coverage versus coverage plots changes and the lowest work function value achieved is considerably lower of that of pure La and Gd films. Differences are less pronounced for opposite adsorption order: The work function versus boron coverage gradually increases until the value exhibited by the boron film on bare Mo(110). The former, according to Auger and EELS results, is due to boron outward diffusion from B-Mo(110) interface into the lanthanum and gadolinium layers occurring even at room substrate temperature, whereas in the latter case no intermixing of boron deposited on metal preadsorbed layers was observed. However, even slight thermal treatment of the latter system dramatically changes its properties as a consequence of rare-earth metal and boron layers interdiffusion. Generally such treatment of B on REM films results in the system obtained by the opposite order of adsorption, i.e. REM on B film. Therefore intermixed REM-B layer is preferable state of coadsorbed REM and boron films, the topmost layer of which presumably consists of no single component but of both REM and boron atoms. The work functions of such mixed systems are very close to the values exhibited by corresponding hexaborides. Maintenance of such low value of work function in the wide range of REM to boron concentration ratio considerably differing from that of 1/6, points at the preferential role of the surface rearrangement in determining the high thermionic efficiency of bulk hexaborides, rather than to their intrinsic property. Almost total similarity of behavior of the double systems with La and Gd indicates at indecisive influence of Gd f-electrons on determining the work function, morphology and stability of the systems.

ACKNOWLEDGMENTS

The work was carried out under Russian fund for basic research (RFBR) and Japan Society for the promotion of science (JSPS) cooperation program (grant # 09-02-92109-ЯФ_a) and was supported by the Federal Agency of Science and Innovations of Russian

Federation under contract # 02.552.11.7035 and the Russian Federal Agency of Education under contract # 2.1.1/3938.

REFERENCES

[1] Lafferty, J.M. *J. Appl. Phys*. 1951, vol. 22, pp 299-303.
[2] Davis, P.R.; Schwind, G.A. *Appl. Surf. Sci*. 1986, vol. 25, pp 355-264.
[3] Davis, P.R.; Gesley, M.A.; Schwind, G.A.; Swanson, L.W.; Hutta, J.*J. Surf. Sci*. 1989, vol. 57, pp 381-390.
[4] Mrozckowski, S.J. *J. Vac. Sci. Technol*. 1991, vol. A9, pp 586-571.
[5] Marchenko, A.A.; Cherepanov, V.V.; Taraschenko, D.T.; Kazantseva, Z.I.; Naumovets, *A.G. Surf. Sci*. 1998, vol. 416, pp 460-467.
[6] Nakamoto, M.; Fukuda, K. *Appl. Surf. Sci*. 2002, vol. 202, pp 289-296.
[7] Late, D.J.; More, M.A.; Misra, P.H.; Singh, B.H.; Kukrej, L.M.; Joag, D.S. *Ultramicroscopy* 2007, vol. 107, pp 825-830.
[8] Nishitani, R.; Aono, M.; Tanaka, T.; Oshima, C.; Kawai, S.; Iwasaki, H.; Nakamura, S. *Surf. Sci*. 1980, vol. 93, pp 535-544.
[9] Sawnson, I.W.; Gesley, M.A.; Davis, P.R. *Surf. Sci*. 1981, vol. 107, pp 263-269.
[10] Gesley, M.A.; Swanson, I.W. *Surf. Sci*. 1984, vol. 146, pp 583-592.
[11] Ociepa, J.G.; Mroz, S. *Vacuum* 1979, vol. 29, pp 241-245.
[12] Ociepa, J.G. *Thin Sol. Films* 1984, vol. 120, pp 123-128..
[13] Chambers, S.A.; Swanson, I.W. *Surf. Sci*. 1983, vol. 131, pp 385-392.
[14] Goldstein, B.; Sczostak, D.J. *Surf. Sci*. 1978, vol. 74, pp 461-468.
[15] Ozcomert, J.S.; Trenary, M. *Surf. Sci*. 1992, vol. 265, pp L227-L331.
[16] Berrada, A.A.; Mercurio, J.P.; Etourneau, J.; Alexandre, F.; Theeten, J.B.; Duc, T.M. *Surf. Sci*. 1978, vol. 72, pp 177-183.
[17] Mogren, S.; Reifenberger, R. *Surf. Sci*. 1991. vol. 254, pp 169-177.
[18] Bulyga, A.V.; Solonovich, V.K. *Surf. Sci*. 1989, vol. 223, pp 578-576.
[19] Kolaczkiewicz, J.; Bauer, E. *Surf. Sci*. 1986, vol. 175, pp 487-496.
[20] Stenborg, A.; Bauer, E. *Phys. Rev*. B. 1987. vol. 36, pp 5840-5846.
[21] Gonchar, F.M.; Medvedev, V.K.; Smereka, T.P.; Babkin, G.V. *Fiz. Tverd. Tela* 1990, vol. 32, pp 1872-1876.
[22] Shakirova, S.A.; Serova, E.V. *Surf. Sci*. 1999, vol. 422, pp 24-29.
[23] Ramstad, A.; Raaen, S.; Barrett, N. *Surf. Sci*. 2000, vol. 448, pp 179-186.
[24] Asha, A.M.; Nix, R.M. *Surf. Sci*. 1995, vol. 322, pp 41-48.
[25] Ashman, R.C.; Foerst, C.J.; Schwarz, K.; Bloechl, P.E. *Phys. Rev*. B. 2004, vol. 70, pp 155330-155337.
[26] Netzer, F.P. J. *Phys.: Condens. Matter* 1995, vol. 7, pp 991-997.
[27] Tucker, C.W. *Surf. Sci*. 1966, vol. 5, pp 179-188.
[28] Miysaki, A. *Jap. J. Appl. Phys*. 1978, vol. 17, pp 719-727.
[29] Yates, J.T. *Experimental Innovations in Surface Science : a Guide to Practical Laboratory Methods and Instruments*; American Institute of Physics: New York, NY, 1998; pp 1-814.

[30] Fomenko, V.S. *Emission Properties of Materials*; Naukova Dumka Publishers: Kiev, 1981, pp 1-542..
[31] Bonzel, P.; Bradshaw, A.M.; Ertl, G. *Physics and Chemistry of Alkali Metal Adsorption*; (Elsevier: Amsterdam, 1989; pp. 1-369..
[32] Gurney, R.W. *Phys. Rev*. 1935, vol. 47, pp 479-483.
[33] Scheffler, M.; Stampfl, C. In: *Handbook of Surface Science. Electronic Structure*; Eds: Horn, K.; Scheffler, M. Elsevier: Amsterdam, 2000; pp 418-527.
[34] Plummer, E.W.; Rhodin, T.N. *J. Chem. Phys*. 1968, vol. 49, pp 3479-3485.
[35] Magkoev, T.T.; Christmann, K.; Lecante, P.; Moutinho, A.M.C. *J. Phys.: Condensed. Matter* 2002, vol. 14, pp 273-279.
[36] Weaver, J.H.; Lynch, D.W.; Olson, C.G. *Phys. Rev*. B. 1974, vol. 10, pp 501-507.
[37] Ballu, Y.; Lecant, J.; Rousseau, H. *Phys. Rev*. B. 1976, vol. 14, pp 3201-3210.
[38] Netzer, F.P.; Strasser, G.; Rosina, G.; Matthew, J.A.D. *Surf. Sci*. 1985, vol. 152/153, pp 757-563.
[39] Valle, F.D.; Modesti, S. *Phys. Rev*. B. 1989, vol.40, pp 933-938.
[40] A. Liebsch, *Electronic Excitations at Metal Surfaces*, (Plenum Press, New York, London, 1997).
[41] Raether, H. *Excitation of Plasmons and Interband Transitions by Electrons*; Springer-Verlag: Berlin, New York, 1980; pp 1-184..
[42] Rogers, J.W.; Knotek, M.L. *Appl. Surf. Sci*. 1982, vol. 13, pp 352-359.
[43] Ertl, G.; Kuppers, J. *Low Energy Electrons and Surface Chemistry*; Verlag-Chemie: Weinheim, 1974; pp 1-217.
[44] Campbell, C/T/ *Surf. Sci. Rep*. 1977, vol. 27, pp 1-142.
[45] Franchy, R. *Surf. Sci. Rep*. 2000. vol. 38, pp 195-269.

In: New Trends in Chemical Physics Research
Editor: James N. Lapherty
ISBN: 978-1-61668-535-5

Chapter 11

OPTIMIZATION OF POLYMER SOLAR CELLS: AN APPROACH TO INCREASE THE OPEN CIRCUIT VOLTAGE

P. Morvillo [*] ***and E. Bobeico***

ENEA, C.R. Portici,

ENEA, C.R. Portici, P.le E. Fermi 1, Portici (NA), Italy

ABSTRACT

A promising approach to decrease the cost of photovoltaic devices, mainly made with inorganic materials, is the fabrication of solar cells based on organic semiconductors, like semiconducting polymers. During the last years new polymers have been prepared and tested as donor materials in solar cells with the aim to improve the light absorption and the efficiency of the corresponding device. Actually, power conversion efficiencies surpassing 5% have been demonstrated for polymer-fullerene solar cells, but to use these devices in commercial applications this value need to be increased further; in addition other aspects, like the large area processing and the lifetime need also to be improved. The power conversion efficiency for a solar cell is directly proportional to the short circuit current (I_{sc}) and to the open circuit voltage (V_{oc}). The maximum V_{oc} for polymer solar cells is related to the difference between the highest occupied molecular orbital (HOMO) of the electron donor and the lowest unoccupied molecular orbital (LUMO) of the electron acceptor. In addition, even the relative position of donor LUMO and acceptor LUMO is fundamental for an efficient charge transfer from the donor polymer to the acceptor fullerene.

In this paper we will briefly introduce the basic working principles of polymer solar cells (materials, device architecture, device physics, efficiency limits). Next we focused our attention on the properties of the most used electron acceptors, namely the fullerenes, and we implemented a theoretical approach to design fullerene derivatives to be used as electron acceptor in polymer solar cells. We calculated the LUMO levels of fullerene

[*] Contact Author: Pasquale Morvillo. E-mail: pasquale.morvillo@enea.it, phone: +390817723260, fax: +390817723345. Address: ENEA, C.R. Portici, P.le E. Fermi 1, 80055 Portici (NA), Italy

derivatives successfully used as electron acceptors and we correlated the obtained values with the V_{oc} of the corresponding device. Furthermore, we investigated the possibility to tune the LUMO level varying the side group attached to the fullerene cage.

1. Introduction

Photovoltaics is the process of energy harvesting from sunlight and converting it into electricity using solar cells [1]. The use of this technology is increasing and today it is considered as an alternative to conventional fossil fuel electricity generation. Although the development of solar power engineering is very intensive, actually the cost of this kind of energy is still higher than that obtained from conventional sources limiting large scale applications. Today, commercial photovoltaic cells are mainly based on inorganic materials, such as silicon (Si), gallium-arsenide (GaAs), cadmium-telluride (CdTe), and cooper-indium-selenide (CIS) [1-4]. The production of these cells still requires many high temperature (up to 900°C) and high vacuum processes with numerous photolithographic steps leading to relatively high manufacturing costs.

A promising approach to decrease the cost of photovoltaic devices is the fabrication of solar cells based on organic semiconductors, like conjugated polymers [5] in combination with fullerenes [6-12]. Usually these materials can have very high optical absorption coefficients (compared to silicon) which offer the possibility for the production of very thin solar cells. In addition they are soluble in organic solvents, making the device realisation relatively easy (the active layer can be deposited using spin casting technique or ink jet printing [13], or screen printing [14], for large area processing). Moreover, because these thin films are flexible, they can also be deposited on flexible plastic substrate making the whole device flexible and with the possibility to use roll-to-roll methods.

The most efficient device architecture for polymer solar cells is based on the bulk heterojunction (BHJ) concept [15-17] in which the active layer, sandwiched between two electrodes with different work functions, consists of a blend of an electron donating material, e.g. a p-type conjugated polymer, and an electron accepting material (n-type), such as fullerene derivatives. Actually power conversion efficiencies exceeding 5% for single junctions [18,19] and 6.5% for tandem cells (obtained stacking two series-connected cells with different absorption spectra [20]) have been reached. Although the device efficiency need to be increased further to use polymer solar cells in commercial applications, other aspects, like the large area processing [21-23] and the lifetime of such devices [24-27] are also crucial. Recently, improvements in all these fields have been achieved, but not on the same materials and it is still a challenge to combine all these desirable properties.

During the last years new polymers have been prepared and tested as donor materials in solar cells [11, 12] with the aim to improve the light absorption and the efficiency of the corresponding device. Alignment of the energy levels, in particular the highest occupied molecular orbital (HOMO) and the lowest unoccupied molecular orbital (LUMO), between the donor polymer and the acceptor fullerene are of great importance for an efficient charge transfer. Therefore, for each donor polymer it is necessary to find a proper electron acceptor.

In this paper we will briefly introduce the basic working principles of polymer solar cells (materials, device architecture, device physics, efficiency limits). Next we will focus our attention on the properties of the most used electron acceptors, namely the fullerenes, and, finally, we will show how it is possible to design fullerene derivatives for polymer solar cells using a theoretical approach.

2. WORKING PRINCIPLES OF POLYMER-FULLERENE SOLAR CELLS

The conversion of light into electricity in the polymer-fullerene solar cells involves five different steps in the commonly accepted mechanism:

1) incident photon absorption by the active layer, resulting in the exciton formation;
2) exciton diffusion to the interface region of donor-acceptor material (i.e. polymer-fullerene interface);
3) exciton dissociation with the formation of free charges;
4) charge carrier transport under an electric field to the anode (holes) and cathode (electrons);
5) charge collection by the electrodes to supply an external current [28].

To allow an efficient collection of the sunlight, the organic material should have an absorption spectrum similar to the solar emission spectrum and should be sufficiently thick. Instead to inorganic photovoltaic devices, in polymer solar cells, excitons rather than free charge carriers are created upon light absorption. The dissociation of the exciton takes place at the donor-acceptor interface. Since the diffusion length of the exciton is around 10 nm, for an efficient charge separation, donor and acceptor should be mixed giving a nanometer-scale phase separation (the so called bulk heterojunction). Furthermore, the photogenerated charges must be transported to the collecting electrodes: the hole are transported by the p-type semiconductor (polymer) to the anode and the electrons by the n-type material (fullerene) to the cathode. For this reason, continuous pathways should exist for donor and acceptor from the interface to the respective electrodes.

A typical structure of a polymer-fullerene bulk heterojunction solar cell consists of a photoactive thin film sandwiched between two different electrodes. One of these electrodes must be transparent to allow the sunlight to reach the photoactive film: usually it is made by indium-tin-oxide (ITO) coated with poly(3,4-ethylenedioxythiophene):poly(styrene sulfanate) (PEDOT:PSS).

A schematic representation of this kind of the device structure is shown in Figure 1.

Some commonly used donors (semiconducting polymers) and acceptors (fullerene derivatives) materials are shown in Figure 2.

Actually the record of the power conversion efficiency for a single junction device has been demonstrated using regioregular poly(3-hexylthiophene) (P3HT) (electron donor) blended with [6,6]-phenyl-C61-butyric acid methyl ester (PCBM) (electron acceptor) [18, 19]. The power conversion efficiency (η) of a photovoltaic cell is directly proportional to short circuit current density (J_{sc}), the open circuit voltage (V_{oc}) and the fill factor (FF) (eq. 1):

$$\eta = \frac{J_{sc} * V_{oc} * FF}{P_{inc}} \quad (1)$$

where P_{inc} is the incident power irradiation.

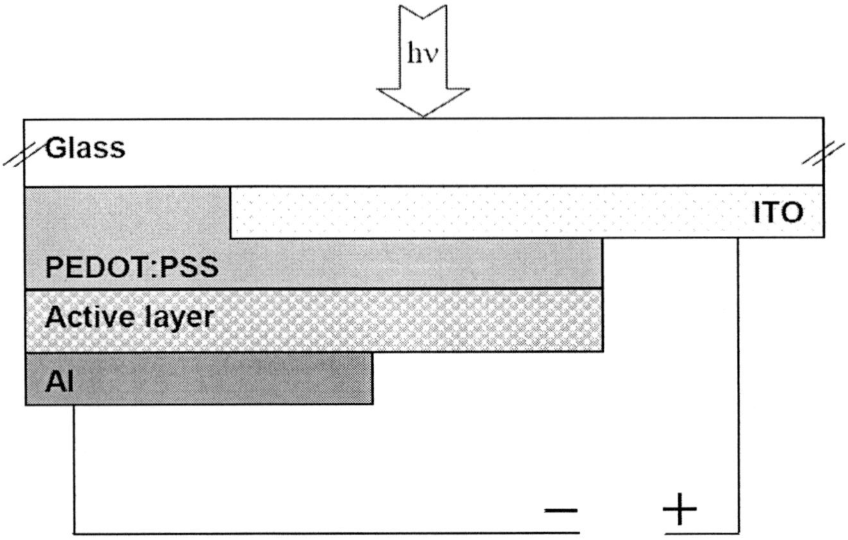

Figure 1. Schematic device structure of a polymer-fullerene bulk heterojunction solar cells. The active layer is sandwiched between two contacts: an indium-tin-oxide electrode coated with a hole transport layer PEDOT:PSS and an aluminium top electrode.

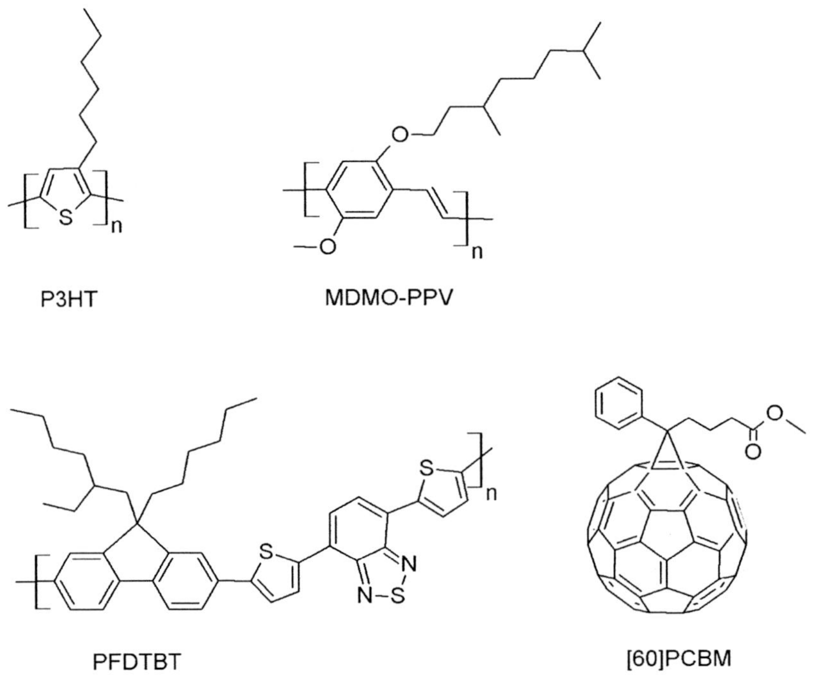

Figure 2. Some donor and acceptor materials commonly used in polymer–fullerene solar cells: P3HT, poly(3-hexylthiophene); MDMO-PPV, poly[2-methoxy-5-(3,7-dimethyloctyloxy)]-1,4-phenylenevinylene; PFDTBT poly[2,7-[9-(2'-ethylhexyl)-9-hexylfluorene]-alt-5,5-(4',7'-di-2-thienyl-2',11',3'-benzothiadiazole)]; [60]PCBM, [6,6]-phenyl-C61-butyric acid methyl ester.

In a polymer-fullerene solar cell, if the absorption of the acceptor can be neglected (this is almost true for C_{60} derivatives), the J_{sc} depends on the energy gap of the photoactive donor polymer. If we consider a single passage of the light inside the active layer (without taking in account reflection from the back contact, double passage, interference and scattering of the light) and ideal conditions (each photon incident on the cell of energy greater than the band gap gives rise to one electron flowing in the external circuit), the maximum J_{sc} can be calculated considering the photon flux for the Air Mass 1.5 spectrum and integrating this distribution from low wavelengths up to value corresponding to the energy gap of the donor polymer [11]. If we consider the absorption of the active layer obtained from optical simulations on realized devices [29-32], the maximum achievable J_{sc} is very close to the highest value reached till now and only few improvements are possible. A more accurate evaluation of the maximum J_{sc}, should be done considering the internal quantum efficiency [33], which includes optical (reflection and multi passage of light) and electrical losses (carrier transport and lifetime).

The V_{oc}, for a device with ohmic contacts (where the Fermi levels of the electrodes are pinned to the energy levels of the relevant molecular orbitals [34]), is related to the energy difference between the highest occupied molecular orbital (HOMO) of the electron donor and the lowest unoccupied molecular orbital (LUMO) of the electron acceptor [34-37]. Instead, for non-ohmic contacts, the experimental V_{oc} is in agreement with the work function difference of the electrodes as expected from the MIM model [38].

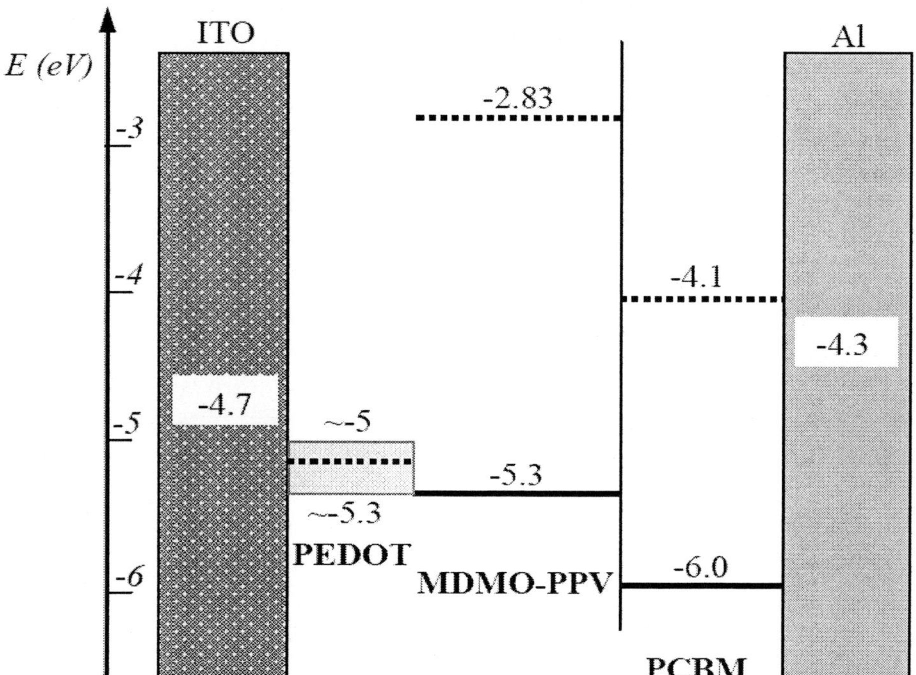

Figure 3. Energy level diagram for a MDMO-PPV:[60]PCBM bulk heterojunction solar cell.

In addition, even the relative position of donor LUMO and acceptor LUMO is fundamental for an efficient charge transfer from the donor polymer to the acceptor fullerene.

Usually, the LUMO of the donor polymer needs to be 0.3-0.5eV higher than the LUMO of the acceptor molecule [39, 40]. We have to remark that the V_{oc} depends not only from the position of the frontier orbitals of the donor and acceptor, but also from other factors like the injection barriers, the ideality factor, the recombination, the mobility and the blend morphology.

It has been estimated that for poly[2-methoxy-5-(3,7-dimethyloctyloxy)]-1,4-phenylenevinylene (MDMO-PPV) based devices, the V_{oc} can be increased from the actual value of 0.850 V (using [60]PCBM as electron acceptor) to 1.3 V (with a new acceptor having LUMO level 0.5eV below the donor LUMO level), raising the efficiency by a 50% (Figure 3) [36].

Also in the case of P3HT based device, there is still room to increase the V_{oc}, because the energy difference between the LUMO levels of donor and acceptor [60](PCBM) is around 1.1eV.

In other words, to maximize the V_{oc} in a BHJ solar cells, it's necessary to find the appropriate acceptor for each used donor polymer. On the other hand, a lot of time is required to investigate experimentally new structures for acceptors with the aim to optimize the LUMO for a given donor polymer and maximize the V_{oc}. So it can be useful to implement a theoretical approach to study the influence of the acceptor LUMO level on the V_{oc} of the device.

3. Basic Properties of Fullerenes

Since the discovery [41] and isolation of macroscopic quantity of C_{60} [42], the fullerenes have received an increasing interest in material science due its remarkable electrochemical and photophysical properties [43]. C_{60} is a good electron acceptor, and its triply-degenerate low-lying LUMO allows for the reversible addition of up to six electrons [44]. The ultrafast photoinduced electron transfer from a π-conjugated polymer to C_{60} [45] has established fullerene derivatives as the most important n-type materials for polymer solar cells.

The fullerenes are the third allotrope of carbon (after diamond and graphite) [46]. They are discrete molecules, with a spherical structure, consisting of pentagons and hexagons. Based on the Euler's theorem, the fullerenes, independent of their size, have to contain exactly 12 pentagons. Depending on the number of hexagons, fullerenes of different size are possible and depending on the distribution of the pentagons on the surface, different isomers are also possible (for molecules having the same number of carbons). The smallest possible fullerene (containing only pentagons) is C_{20} but theoretical investigation have shown that only fullerenes fulfilling the isolated pentagon rule (IPR) (i.e. molecules where pentagons are completely surrounded by hexagons) are stable.

The first fullerene that satisfy the IPR rule (and also the most abundant studied) is C_{60}. This molecule consists of 12 pentagons and 20 hexagons and has the I_h symmetry. After C_{60}, the next possible fullerenes that follow the IPR are C_{70}, C_{72}, C_{74}, C_{76}, C_{78}, C_{82}, C_{84}. Fullerenes containing more than 60 carbon atoms are called higher fullerenes. While fullerenes up to C_{76} have only one IPR isomer, there is a dramatic increase in the number of possible fullerene isomers expanding cage size. The number of IPR isomers are 5 for C_{78}, 7 for C_{80}, 9 for C_{82} and 24 for C_{84}. The fullerene molecule contains bonds located at 6-6 ring junction or at 5-6 ring junction. In C_{60}, which contains only one type of 6-6 bond, primary addition occurs preferentially at this place, which have higher double bond character. In higher fullerenes, not all the 6-6 bonds are equivalent and the possible number of isomers deriving from functionalization can be very high. In practice, however, additions to higher fullerenes often show a remarkable regioselectivity [47].

4. Fullerenes as Electron Acceptor in Polymer Solar Cells

[60]PCBM

The application of fullerenes as electron acceptors in polymer solar cells started with the discovery of the ultrafast photoinduced electron transfer between the soluble poly(para-phenylenevinylene) derivative and the buckminsterfullerene C_{60} [45]. By the way, the poor solubility of C_{60} hampers its application in devices realized mainly by wet techniques. The solution to this problem was offered by a highly soluble C_{60} derivative ([60]PCBM) (Figure 4) which is also the most commonly used fullerene derivative in polymer solar cells [48].

Although the chemical functionalization of C_{60} can improve the solubility and the processability in the commonly used organic solvents, it can also change the electronic properties of the molecule. For this reason, it is important to understand how the LUMO level changes upon the fullerene modification. In fact, the position of the LUMO level (optimized for the LUMO level of the donor) isn't the only requirement for a fullerene derivative to be used as electron acceptor in polymer solar cells. It is also important to avoid concomitant intramolecular photoinduced electron or hole transfer from an electronrich addend to the fullerene cage. This process was observed in a 4-dialkylaminophenyl-substituted fulleropyrrolidine [49].

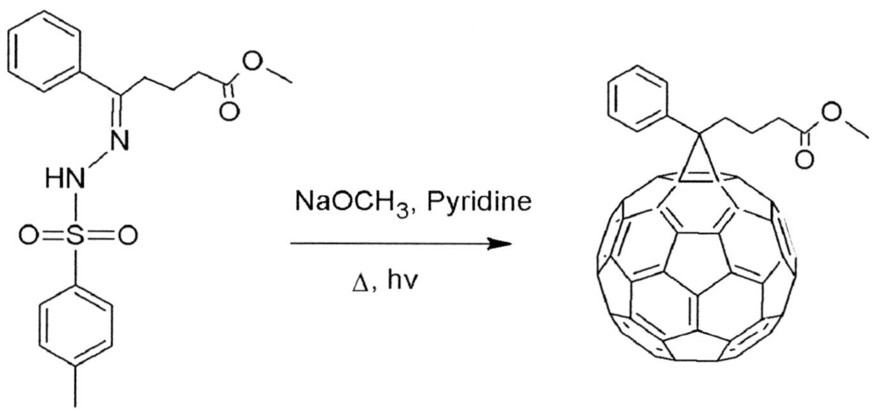

Figure 4. Synthesis of [60]PCBM.

[70]PCBM

Since higher fullerene derivatives have a stronger and broader absorption compared to [60]PCBM, it has been proposed that they can improve the current output of the corresponding polymer-fullerene devices. [70]PCBM (obtained as a mixture of isomers, because C_{70} has different types of bonds where an addend can arrange) has already been used with success as electron acceptor in BHJ solar cells (Figure 5). It has been demonstrated that devices based on a blend film of MDMO-PPV:[70]PCBM show better power conversion efficiency compared to solar cell realized using a blend of MDMO-PPV:[60]PCBM [50]. The increase was due to the much stronger optical absorption of [70]PCBM in visible wavelengths compared to [60]PCBM. This difference derives from the reduced symmetry of C_{70} relative to C_{60}: the lowest-energy transitions became allowed resulting in a dramatic increase in the light absorption. The first reduction potential (V vs Fc/Fc+) of [70]PCBM is a little bit lower than the corresponding C_{60} analogous (-1.089V against -1.078V) for this reason the LUMO level is almost the same and, consequently, also the open circuit voltage of the device.

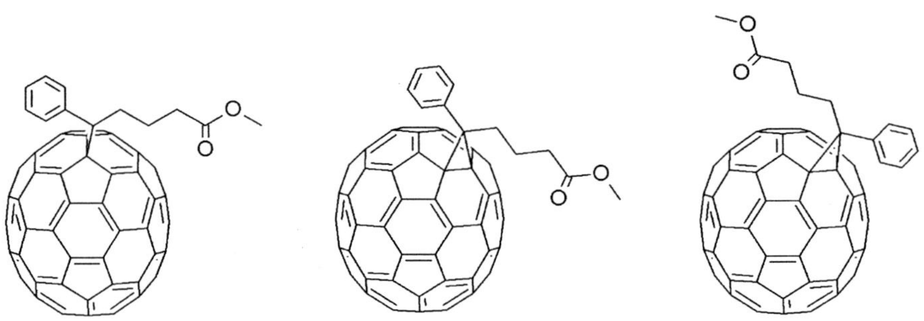

Figure 5. Chemical structure of the three [70]PCBM isomers.

[84]PCBM

Since C_{84} has a better absorption than C_{70}, [84]PCBM was prepared with the aim to further improve the current output of the photovoltaic device. This higher fullerene was obtained as a mixture containing three major isomers and it was tested in a bulk heterojunction with MDMO-PPV as donor polymer. Since the LUMO level of [84]PCBM is lower compared to the ones of [60]PCBM and [70]PCBM, there is a loss in the V_{oc} [51]. Nevertheless, [84]PCBM can be a promising acceptor for donor polymers with a LUMO level lower compared to the one of MDMO-PPV.

Other Fullerene Derivatives

[60]PCBM is the best performing fullerene used in combination with P3HT. Other derivatives have been prepared with the aim to improve the performance of the devices by placing substituents on the phenyl ring of [60]PCBM [36], by replacing the phenyl ring with the thiophene ring [52] (to improve the miscibility with polythiophenes polymers) or by changing the addend attached to the fullerene cage [53-55].

Polymers with lower gap (compared with P3HT and MDMO-PPV) can have a shift in one or both HOMO and LUMO level. If the LUMO level is lowered, an acceptor with a proper LUMO is required in order to have an efficient exciton dissociation. For these polymers, acceptors with a lower LUMO level (compared to [60]PCBM), like 3'-(3,5-Bis-trifluoromethylphenyl)-1'-(4-nitrophenyl) pyrazolino[60]fullerene ([60]BTPF) [56] and 3'-(3,5-Bis-trifluoromethylphenyl)-1'-(4-nitrophenyl) pyrazolino[70]fullerene ([70]BTPF) [57, 58] have been proposed and successfully used in bulk heterojunction solar cells.

5. Design of Fullerene Derivatives for Polymer Solar Cells

Recently we have shown the possibility to tune the LUMO level of the acceptor to increase the V_{oc} of a BHJ solar cells using a quantum chemical approach [37]. The theoretical LUMO levels (calculated for isolated molecules in gas phase) correlate well with the first reduction potential of these species (experimentally determined with cyclic voltammetry) and with the open circuit voltage of the corresponding photovoltaic device based on the blend of MDMO-PPV with the acceptor molecules.

Theoretical Approach

The LUMO levels of the investigated compounds were calculated using the density functional theory (DFT). This methodology has proven to be an important tool in quantum chemistry to describe the electronic structure of molecules at a reduced computational cost. The LUMO levels obtained with the above procedure, differ from the experimental data for several reasons: the values are calculated for isolated molecules in gas phase, while the experimental LUMO are related to molecules in solution or in solid phase. In addition, the physical meaning of the orbitals (Kohn-Sham) computed with DFT are under debate. By the way, it has been shown that LUMO levels calculated with DFT method usually exhibit satisfactory linear relationship with experimental or calculated electron affinity.

All calculations were carried out using the PC GAMESS/Firefly QC package [59], which is partially based on the GAMESS (US) [60] source code. The geometric structures of the investigated fullerene derivatives were preliminary optimized under no constraints using the PM3 semiempirical method [61]. The harmonic vibrational frequencies calculated at PM3 theory level were employed to confirm that the PM3-optimized geometries correspond to minima on the potential energy surfaces. Then the PM3 optimized geometries were subjected to further geometry optimization and energy calculation using DFT method with 3-21G* basis set [62, 63] and Becke's three-parameter functional (which defines the exchange functional as the linear combination of Hartree-Fock, local and gradient-corrected exchange terms) in combination with the correlation functionals of Lee, Yang and Parr (B3LYP) [64, 65]. Since these supramolecules contain over 60 atoms, a smaller basis set is chosen to save computation time. Indeed, this method has proved to be very adapted to study macromolecular systems with fullerene units [66]. Afterwards, we investigated the localization of the frontier orbitals using a single point calculation at the same level of theory.

The prediction of the geometries of fullerenes using DFT B3LYP/3-21G* is fairly good although this agreement is accredited to fortuitous error cancellation from exchange-correlation functional and basis set.

Geometric and Electronic Structure of [60]PCBM

The geometries of C_{60} and [60]PCBM were optimized using the DFT (B3LYP/3-21G*) method. The obtained structures (reported in Figure 6) are consistent with the literature and the prediction of bond lengths and angles is in excellent agreement with the experimental values. The influence of the addend on the geometry of the C_{60} is restricted to the neighbouring pentagons and hexagons. The lengths for the bonds of the other rings, are very close to the values reported for the C_{60} molecule (1.458 Å for the bond fusing five- and six-member rings and 1.401Å for the bond connecting six-member rings).

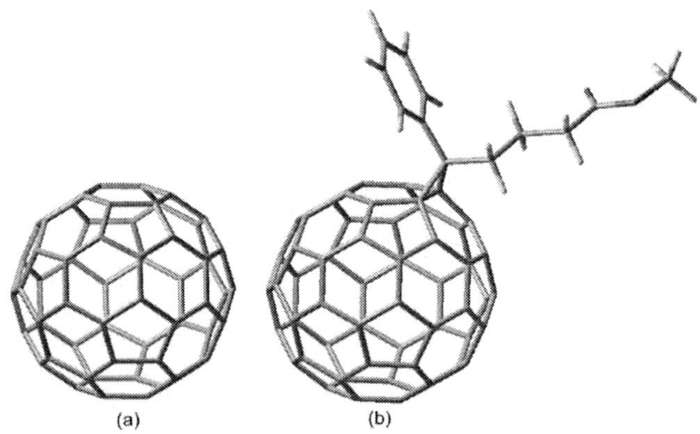

Figure 6. C_{60} (a) and [60]PCBM (b) structure optimized using DFT B3LYP/3-21G* method.

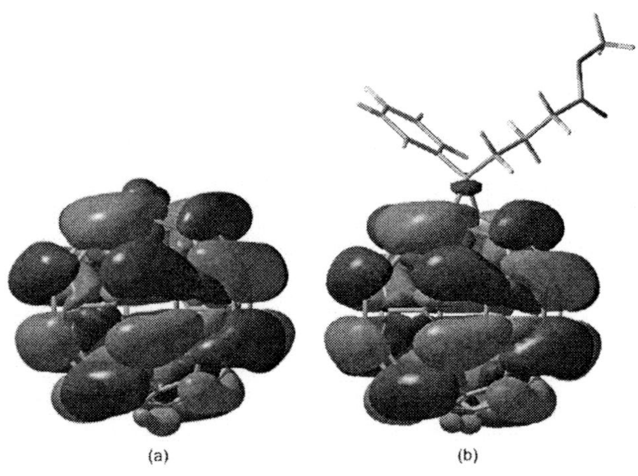

Figure 7. Isosurface of the LUMO level of a) C_{60}, b) [60]PCBM obtained using DFT B3LYP/3-21G* method.

In Figure 7, the Kohn-Sham LUMO levels of C_{60} and PCBM are shown. As we can see, also in [60]PCBM, it is fully located on the fullerene cage (i.e. there is no intramolecular photoinduced electron transfer to the addend). By the way the value of the LUMO level of [60]PCBM is significantly raised compared to C_{60}, minimizing the energy losses in the electron transfer from the LUMO level of the donor polymer and increasing the gap between the HOMO of the donor and the LUMO of the acceptor (resulting in an increase of the open circuit voltage of the corresponding device).

Tuning the LUMO Level of Fullerene Using Substituents on the Phenyl Ring of [60]PCBM

Recently, a new class of [60]PCBM derivatives have been proposed with the aim to influence the reduction potential of these species by placing substituents on the phenyl ring [36]. Electron-donating (methoxy groups) substituents have been successfully used to raise the LUMO levels of these species. The studied molecules were tested as electron acceptors in blend with MDMO-PPV in the bulk heterojunction solar cells and it was possible to increase the V_{oc} of the corresponding device.

We calculated the LUMO level using the DFT (B3LYP/3-21G*) method for the fullerene derivatives reported in Figure 8. In Figure 9, we report the first reduction potential (V vs Fc/Fc+), obtained from cyclic voltammetry data [36], versus the calculated LUMO level. The correlation is very nice although the LUMO levels were calculated for isolated molecules in gas phase, while the electrochemical measurements were made in solution. This correlation can be useful to foresee the first reduction potential of unknown fullerene derivatives once the LUMO level has been calculated. This relationship was also proven for fullerene derivatives that differ not only by a side chain [34]. We have shown that calculated LUMO, LUMO+1 and LUMO+2 levels of [60]PCBM, N-3-(2-ethylhexyloxy)benzyl azafulleroid, C_{60} and N-3-(2-ethylhexyloxy)-benzyl ketolactam correlate well with first, second and third reduction potential of these species in solution [67].

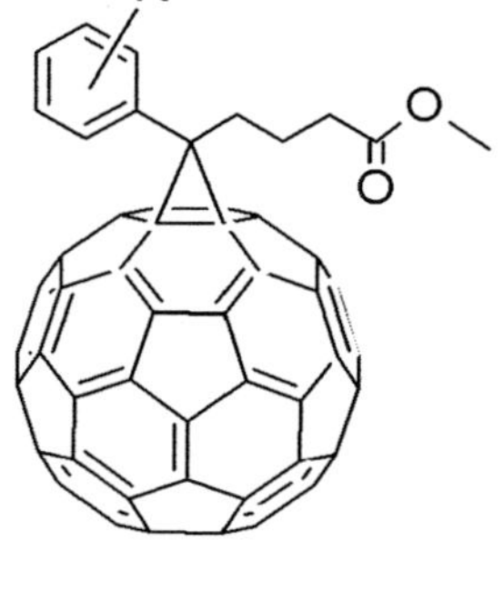

R = 2-OCH_3; 4-OCH_3; 2,5-OCH_3; 3,4-OCH_3;
3,4-methylenedioxy; 2,3,4-OCH_3

Figure 8. Chemical structure of the investigated fullerene derivatives.

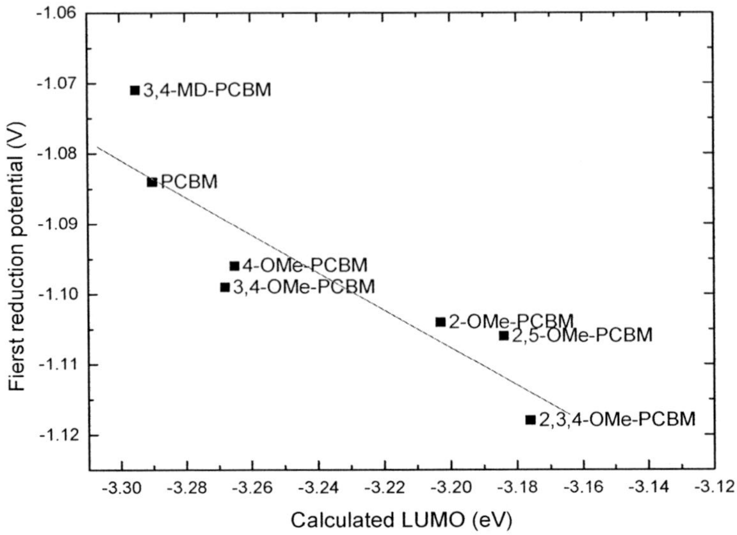

Figure 9. First reduction potentials of fullerene derivatives [36] versus LUMO levels calculated with DFT B3LYP/3-21G* level of theory.

The effect of methoxy group on the LUMO level of [60]PCBM decreases going from the ortho to the para position. So it is important that the electron donor substituent is closer to the fullerene cage. Furthermore, a significant influence on the LUMO level is observed with increasing the number of alkoxy substituents.

In Figure 10 and 11, we report the V_{oc} of the photovoltaic devices realized using these acceptors and MDMO-PPV, as donor polymer, as a function of the calculated LUMO levels.

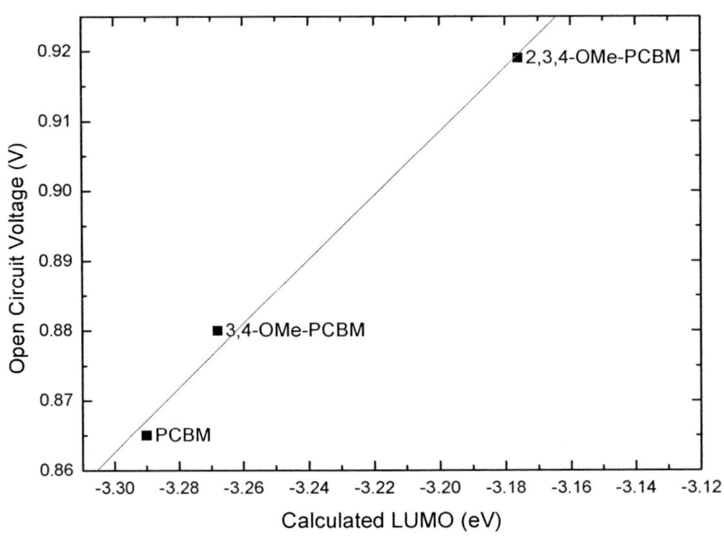

Figure 10. Open circuit voltage of MDMO-PPV:fullerene devices [36] versus theoretical LUMO levels (B3LYP/3-21G*) of acceptors for blend film dissolved in chlorobenzene.

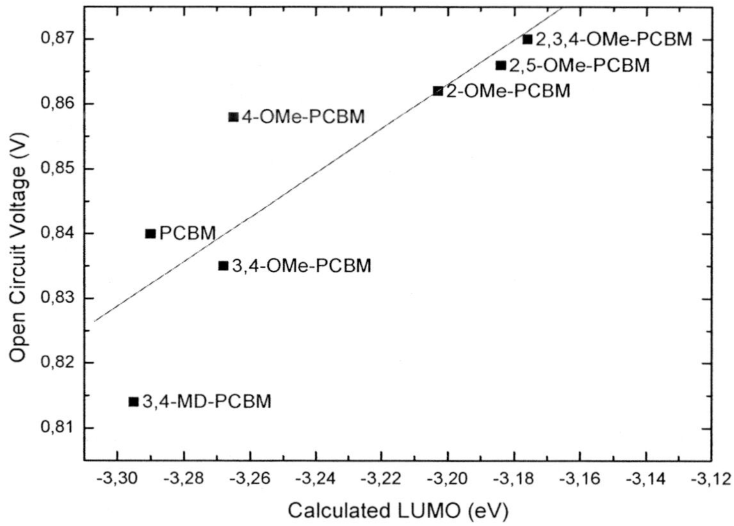

Figure 11. Open circuit voltage of MDMO-PPV:fullerene devices [36] versus theoretical LUMO levels (B3LYP/3-21G*) of acceptors for blend film dissolved in o-dichlorobenzene.

Also in this case, we can see a good linear correlation, although the V_{oc} was not obtained from fully optimized devices. Not all the acceptors were soluble in chlorobenzene; for this reason, two set of data are shown, related to devices based on blend dissolved in chlorobenzene or o-dichlorobenzene. Since the V_{oc} of polymer–fullerene solar cells is affected by the morphology of the device, we cannot directly compare these two set of data.

Tuning the LUMO Level of Fullerene Changing the Atom Attached to the Cage

All the above mentioned fullerene derivatives contains a side group attached to the C_{60} cage through a carbon atom in a three-membered ring. Another approach to tune the LUMO level of fullerenes is to change the atom attached to the cage. It has been demonstrated that the position of the LUMO level shows a remarkable linear relationship with the electronegativity of the atom attached to the C_{60} cage [68]. The LUMO level can be significantly raised when the attached atom is silicon [69].

Such fullerene derivatives have been already prepared in the past [68, 70, 71]: they are stable and show a reversible reduction reaction, which is a necessary requirement to use a molecule as electron acceptor in polymer solar cells. We calculated the LUMO level for a fullerene silirane derivative analogue of [60]PCBM (Figure 12). The calculated value, -3.12 eV, is 0.18 eV less of the LUMO level of [60]PCBM.

As found for [60]PCBM, the HOMO and LUMO are fully located on the fullerene cage, indicating that, also in this case, there is no intramolecular photoinduced electron transfer. We estimated that the value of the LUMO can be still sufficient for an efficient charge transfer from donor polymers like P3HT or MDMO-PPV. From the linear correlation shown in Figure 11 between V_{oc} and LUMO levels, we can expect an increase of V_{oc} up to 1V for devices realized with this new fullerene silirane derivative. Also for P3HT based device, we can estimate a V_{oc} greater than 1V.

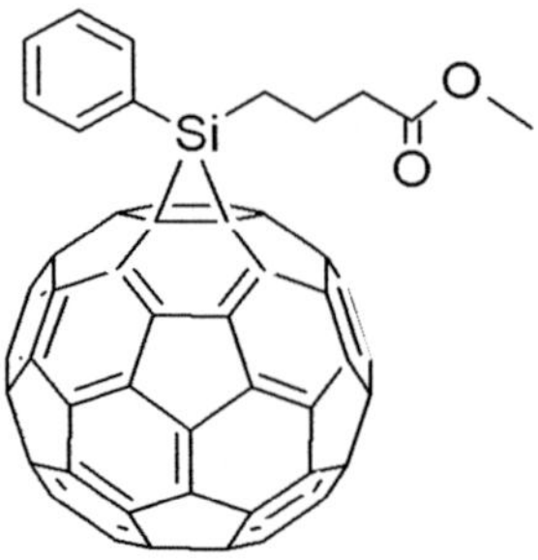

Figure 12. Chemical structure of the fullerene silirane derivative analogue of [60]PCBM.

Conclusion

In this paper we have briefly reviewed the basic working principles of polymer solar cells (materials, device architecture, device physics, efficiency limits) and the properties of the most used acceptors, the fullerenes. Since the efficiency of a solar cells is directly proportional to the open circuit voltage that, for these devices, is related to the difference between the HOMO of the donor and the LUMO of the acceptor, we have proposed a theoretical approach to design fullerene derivatives with a proper LUMO level. We calculated the LUMO level for fullerene derivatives successfully used as electron acceptor for polymer solar cells. The calculated value, correlate well with the first reduction potential of these species and with the V_{oc} of the corresponding polymer-fullerene device. We have shown that this approach can be useful to design new fullerene derivatives in order to maximize the open circuit voltage and improve the performance of such solar cells.

References

[1] Green, M. A. SOLAR CELLS: Operating Principles. Technology and System Applications, Publisher: Prentice Hall, Inc., *Englewood Cliffs*, N.J., 1982 (reprinted 1986, 1992); pp. 103-200.

[2] Goetzberger, A.; Hebling, C.; Schock, W.-W. Photovoltaic materials, history, status and outlook, *Mater. Sci. Eng.* R 2003, 40, 1-46.

[3] Contreras, M. A.; Egaas, B.; Ramanathan, K.; Hiltner, J.; Swartzlander, A.; Hasoon, F.; Noufi, R. Progress towards 20% efficiency in Cu(In,Ga)Se polycrystalline thin-film solar cell, *Prog. Photovolt.: Res. Appl.* 1999, 7, 311–316.

[4] Green, M.A.; Emery, K.; Hishikawa, Y.; Warta, W. Solar Cell Efficiency Tables (Version 31), *Prog. Photovolt*: *Res. Appl.* 2008, 16, pp. 61–67.

[5] Skotheim, T. A.; Reynolds, J. R. (Eds.), *Handbook of conducting polymers*, CRC, Boca Raton, 2006, vols. 1–2.

[6] Brabec, C. J.; Dyakonov, V.; Parisi, J.; Sariciftci, N. S. (Eds.), *Organic photovoltaics: concepts and realization,* Publisher: Springer, Berlin, 2003, vol. 60.
[7] Günes, S.; Neugebauer, H.; Sariciftci, N. S. Conjugated polymer-based organic solar cells, *Chem. Rev*. 2007, 107, 1324-1338.
[8] Spanggaard, H.; Krebs, F.C. A brief history of the development of organic and polymeric photovoltaics, *Sol. Energy Mater. Sol. Cells* 2004, 83, 125-146.
[9] Hoppe, H.; Sariciftci, N. S. Organic solar cells: An overview, *J. Mater. Res*. 2004, 19, 1924-1945.
[10] Brabec, C. J.; Sariciftci, N. S.; Hummelen, J.C. *Plastic Solar Cells, Adv. Funct. Mater*. 2001, 11, 15-26.
[11] Bundgaard, E.; Krebs, F.C. Low band gap polymers for organic photovoltaics, *Sol. Energy Mat. Sol. Cells* 2007, 91, 954-985.
[12] Kroon, R.; Lenes, M.; Hummelen, J. C.; Blom, P. W. M.; De Boer, B. Small Bandgap Polymers for Organic Solar Cells (Polymer Material Development in the Last 5 Years), *Polymer Reviews* 2008, 48, 531–582.
[13] Bharathan, J.; Yang, Y. Polymer electroluminescent devices processed by inkjet printing: I. Polymer light-emitting logo, *Appl. Phys. Lett*. 1998, 72, 2660 -2662.
[14] Shaheen, S. E.; Radspinner, R.; Peyghambarian N., Jabbour G. E., Fabrication of bulk heterojunction plastic solar cells by screen printing, *Appl. Phys. Lett.* 2001, 79, 2996-2998.
[15] Yu, G.; Heeger, A. J. Charge separation and photovoltaic conversion in polymer composites with internal donor/acceptor heterojunctions, *J. Appl. Phys*. 1995, 78, 4510-4515.
[16] Yu, G.; Gao, J.; Hummelen, J.C.; Wudl, F.; Heeger, A.J. Polymer photovoltaic cells: enhanced efficiencies via a network of internal donor–acceptor heterojunctions, *Science* 1995, 270, 1789-1791.
[17] Tada, K.; Hosada, K.; Hirohata, M.; Hidayat, R.; Kawai, T.; Onoda, M.; Teraguchi, M.; Masuda, T.; Zakhidov, A.A.; Yoshino, K. Donor polymer (PAT6)–acceptor polymer (CNPPV) fractal network photocells, *Synth. Met*. 1997, 85, 1305-1306.
[18] Li, G.; Shrotriya, V.; Huang, J.; Yao, Y.; Moriarty, T.; Emery, K.; Yang, Y. High-efficiency solution processable polymer photovoltaic cells by self-organization of polymer blends, *Nat. Mater*. 2005, 4, 864–868.
[19] Ma, W.; Yang, C.; Gong, X.; Lee, K.; Heeger, A. J. Thermally stable, efficient polymer solar cells with nanoscale control of the interpenetrating network morphology, *Adv. Funct. Mater*. 2005, 15, 1617–1622.
[20] Jin, Y. K.; Lee, K.; Coates, N. E.; Moses, D.; Nguyen, T.-Q.; Dante, M.; Heeger, A. J. Efficient tandem polymer solar cells fabricated by all-solution processing, *Science* 2007, 317, 222–225.
[21] Krebs, F.C.; Alstrup, J.; Spanggaard, H.; Larsen, K.; Kold, E. Production of large-area polymer solar cells by industrial silkscreen printing, lifetime considerations and lamination with polyethyleneterephthalate, *Sol. Energy Mater. Sol. Cells* 2004, 83, 293-300.
[22] Krebs, F.C.; Spanggaard, H.; Kjær, T.; Biancardo, M.; Alstrup, J. Large area plastic solar cell modules, *Mater. Sci. Eng*. B 2007, 138,106-111.

[23] Dennler, G.; Lungenschmied, C.; Neugebauer, H.; Sariciftci, N.S.; Labouret, A. Flexible, conjugated polymer-fullerene-based bulk-heterojunction solar cells: basics, encapsulation, and integration, *J. Mater. Res.* 2005, 20, 3224-3233.

[24] Krebs, F.C.; Spanggaard, H. Significant improvement of polymer solar cell stability, *Chem. Mater.* 2005, 17, 5235-5237.

[25] Yang, X.; Loos, J.; Veenstra, S.C.; Verhees, W.J.H.; Wienk, M.; Kroon, J.M.; Michels, M.A.J.; Janssen, R.A.J. Nanoscale morphology of high-performance polymer solar cells, *Nano Lett.* 2005, 5, 579-583.

[26] Katz, E.A.; Gevorgyan, S.; Orynbayev, M.S.; Krebs, F.C. Out-door testing and long-term stability of plastic solar cells, *Eur. Phys. J. Appl. Phys.* 2007, 36, 307-311.

[27] Krebs, F.C.; Norrman, K. Analysis of the failure mechanism for a stable organic photovoltaic during 10,000 h of testing, *Prog. Photovolt.: Res. Appl.* 2007, 15, 697-712.

[28] Blom, P. W. M.; Mihailetchi, V. D.; Koster, L. J. A.; Markov D. E. Device Physics of Polymer:Fullerene Bulk Heterojunction Solar Cells, Adv. Mater. 2007, 19, 1551–1566.

[29] Sievers, D.W.; Shrotriya, V.; Yang, Y. Modeling optical effects and thickness dependent current in polymer bulk-heterojunction solar cells, *J. Appl. Phys.* 2006, 100, 114509 (1-7).

[30] Hoppe, H.; Arnold, N.; Sariciftci, N.S.; Meissner, D. Modeling the optical absorption within conjugated polymer/fullerene-based bulk-heterojunction organic solar cells, *Sol. Energy Mater. Sol. Cells* 2003, 80, 105-113.

[31] Moulé, A. J.; Bonekamp, J.B.; Meerholz, K. The effect of active layer thickness and composition on the performance of bulk-heterojunction solar cells, *J. Appl. Phys.* 2006, 100, 094503 (1-7).

[32] Sylvester-Hvid, K. O.; Ziegler, T.; Riede, M. K.; Keegan, N.; Niggemann, M.; Gombert, A. Analyzing poly(3-hexyl-thiophene):1-(3-methoxy-carbonyl)propyl-1-phenyl-(6,6)C61 bulk-heterojunction solar cells by UV–visible spectroscopy and optical simulations, *J. Appl. Phys.* 2007, 102, 054502 (1-9).

[33] Slooff, L. H.; Veenstra, S. C.; Kroon, J. M.; Moet, D. J. D.; Sweelssen, J.; Koetse, M. M. Determining the internal quantum efficiency of highly efficient polymer solar cells through optical modelling, *Appl. Phys. Lett.* 2007, 90, 143506 (1-3).

[34] Brabec, C. J.; Cravino, A.; Meissner, D.; Sariciftci, N. S.; Fromherz, T.; Rispens, M. T.; Sanchez, L.; Hummelen, J. C. Origin of the open circuit voltage of plastic solar cells, *Adv. Funct. Mater.* 2001, 11, 374-380.

[35] Scharber, M. C.; Mühlbacher, D.; Koppe, M.; Denk, P.; Waldauf, C.; Heeger, A. J.; Brabec C. J. Design rules for donors in bulk-heterojunction solar cells - Towards 10 % energy-conversion efficiency, *Adv. Mater.* 2006, 18, 789-794.

[36] Kooistra, F. B.; Knol, J.; Kastenberg, F.; Popescu, L. M.; Verhees, W. J. H.; Kroon, J. M.; Hummelen, J. C. Increasing the Open Circuit Voltage of Bulk-Heterojunction Solar Cells by Raising the LUMO Level of the Acceptor, *Org. Lett.* 2007, 9, 551-554.

[37] Morvillo, P.; Bobeico, E.; Tuning the LUMO level of the acceptor to increase the open-circuit voltage of polymer-fullerene solar cells: A quantum chemical study, *Sol. Energy Mater. Sol. Cells* 2008, 92, 1192-1198.

[38] Mihailetchi, V.D.; Blom, P.W.M.; Hummelen, J.C.; Rispens, M.T. Cathode dependence of the open-circuit voltage of polymer: fullerene bulk hetero-junction solar cells, *J. Appl. Phys.* 2003, 94, 6849-6854.

[39] Halls, J. M.; Cornill, J.; dos Santos, D. A.; Silbey, R.; Hwang, D.-H.; Holmes, A. B.; Brèdas, J. L.; Friend, R. H. Charge- and energy-transfer processes at polymer/polymer interfaces: A joint experimental and theoretical study, *Phys. Rev.* B 1999, 60, 5721-5727.

[40] Brabec, C. J.; Winder, C.; Sariciftci, N. S.; Hummelen, J. C.; Dhanabalan, A.; van Hal, P. A.; Janssen, R. A. J. A low-bandgap semiconducting polymer for photovoltaic devices and infrared emitting diodes, *Adv. Funct. Mater*. 2002, 12, 709-712.

[41] Kroto, H.W.; Heath, J.R.; O'Brien, S.C.; Curl, R.F.; Smalley, R.E. C_{60}: Buckminsterfullerene, *Nature* 1985, 318, 162-163.

[42] Krätschmer, W.; Lamb, L. D.; Fostiropoulos, K.; Huffman, D. R. Solid C_{60}: A new form of carbon, *Nature* 1990, 347, 354-358.

[43] Nierengarten, J.-F.; Martín, N. Fullerene chemistry, C. R. *Chimie* 2006, 9, 859–860.

[44] Arias, F.; Xie, Q.; Wu, Y.; Lu, Q.; Wilson, S. R.; Echegoyen, L. Kinetic effects in the electrochemistry of fullerene derivatives at very negative potentials, *J. Am. Chem. Soc.* 1994, 116, 6388-6394.

[45] Sariciftci, N. S.; Smilowitz, L.; Heeger, A. J.; Wudl, F. Photoinduced electron transfer from a conducting polymer to buckminsterfullerene, *Science* 1992, 258, 1474-1476.

[46] Fowler, P. W.; Manolopoulos, D. E. *An Atlas of Fullerenes*, Publisher: Clarendon Press, Oxford, 1995.

[47] Hirsch A. (eds.) *Fullerenes and related structures, Topics in Current Chemistry*, Publisher: Springer, Berlin, Heidelberg, 1999, vol. 199.

[48] Hummelen, J. C.; Knight, B. W.; LePeq, F.; Wudl, F.; Yao, J.; Wilkins, C. L. Preparation and Characterization of Fulleroid and Methanofullerene Derivatives, *J. Org. Chem*. 1995, 60, 532-538.

[49] Williams, R. M.; Zwier, J. M.; Verhoeven, J. W. Photoinduced Intramolecular Electron Transfer in a Bridged C_{60} (Acceptor)-Aniline (Donor) System; Photophysical Properties of the First "Active" Fullerene Diad, *J. Am. Chem. Soc.* 1995, 117, 4093-4099.

[50] Wienk, M. M.; Kroon, J. M.; Verhees, W. J. H.; Knol, J.; Hummelen, J. C.; van Hal, P. A.; Janssen, R. A. J. Efficient Methano[70]fullerene/MDMO-PPV Bulk Heterojunction Photovoltaic Cells, *Angew. Chem. Int. Ed.* 2003, 42, 3371 – 3375.

[51] Kooistra, F. B.; Mihailetchi, V. D.; Popescu, L. M.; Kronholm, D.; Blom, P. W. M.; Hummelen, J. C. New C84 Derivative and Its Application in a Bulk Heterojunction Solar Cell, *Chem. Mater*. 2006, 18, 3068-3073.

[52] Popescu, L.M.; van 't Hof, P.; Sieval, A. B.; Jonkman, H.T.; Hummelen, J. C. Thienyl analog of 1-(3-methoxycarbonyl)propyl-1-phenyl-[6,6]-methanofullerene for bulk heterojunction photovoltaic devices in combination with polythiophenes, *Appl. Phys. Lett* 2006, 89, 213507 (1-3).

[53] Riedel, I.; von Hauff, E; Parisi, J.; Martín, N.; Giacolone, F.; Dyakonov, V. Diphenylmethanofullerenes: New and Efficient Acceptors in Bulk-Heterojunction Solar Cells, *Adv. Funct. Mater*. 2005, 15, 1979-1987.

[54] Yang, C.; Kim, J. Y.; Cho, S.; Lee, J. K.; Heeger, A. J.; Wudl, F. Functionalized Methanofullerenes Used as n-Type Materials in Bulk-Heterojunction Polymer Solar Cells and in Field-Effect Transistors, *J. Am. Chem. Soc*. 2008, 130, 6444–6450.

[55] Backer, S. A.; Sivula, K.; Kavulak, D. F.; Fréchet, J. M. J. High Efficiency Organic Photovoltaics Incorporating a New Family of Soluble Fullerene Derivatives, *Chem. Mater.* 2007, 19, 2927-2929.

[56] Wang, X.; Perzon, E.; Delgado, J. L.; de la Cruz, P.; Zhang, F.; Langa, F.; Andersson, M.; Inganäs, O. Infrared photocurrent spectral response from plastic solar cell with low-band-gap polyfluorene and fullerene derivative, *Appl. Phys. Lett.* 2004, 85, 5081-5083.

[57] Wang, X.; Perzon, E.; Mammo, W.; Oswald, F.; Admassie, S.; Persson, N.-K.; Langa, F.; Andersson, M. R.; Inganäs, O. Polymer solar cells with low-bandgap polymers blended with C_{70}-derivative give photocurrent at 1 μm, *Thin Solid Films* 2006, 511 – 512, 576–580.

[58] Wang, X.; Perzon, E.; Oswald, F.; Langa, F.; Admassie, S.; Andersson, M. R.; Inganäs, O. Enhanced Photocurrent Spectral Response in Low-Bandgap Polyfluorene and C_{70}-Derivative-Based Solar Cells, *Adv. Funct. Mater.* 2005, 15, 1665-1670.

[59] Granovsky, A. A. PC GAMESS/ Firefly version 7.1.E, www http://classic.chem.msu.su/gran/gamess/index.html

[60] Schmidt, M. W.; Baldridge, K. K.; Boatz, J. A.; Elbert, S. T.; Gordon, M. S.; Jensen, J.H.; Koseki, S.; Matsunaga, N.; Nguyen, K. A.; Su, S.; Windus, T. L.; Dupuis, M.; Montgomery, General atomic and molecular electronic structure system, *J. A. J. Comput. Chem.* 1993, 14, 1347-1363.

[61] Stewart, J.J.P. Optimization of parameters for semiempirical methods I. Method, *J. Comput. Chem.* 1989, 10, 209-220.

[62] Dobbs, K. D.; Hehre, W. J. Molecular orbital theory of the properties of inorganic and organometallic compounds 4. Extended basis sets for third-and fourth-row, main-group elements, *J. Comput. Chem.* 1986, 7, 359-378.

[63] Pietro, W. J.; Francl, M. M.; Hehre, W. J.; De Frees, D. J.; Pople, J. A.; Binkley, J. S. Self-consistent molecular orbital methods. 24. Supplemented small split-valence basis sets for second-row elements, *J. Am. Chem. Soc.* 1982, 104, 5039-5048.

[64] Becke, A.D. Density-functional thermochemistry. III. The role of exact exchange, *J. Chem. Phys.* 1993, 98, 5648-5652.

[65] Lee, C.; Yang, W.; Parr, R. G. Development of the colle-salvetti correlation-energy formula into a functional of the electron density, *Phys. Rev.* B 1988, 37, 785-789.

[66] Zandler, M.E.; D'Souza, F. The remarkable ability of B3LYP/3-21G(*) calculations to describe geometry, spectral and electrochemical properties of molecular and supramolecular porphyrin–fullerene conjugates, C.R. *Chimie* 2006, 9, 960-981.

[67] Morvillo, P.; Bobeico, E. *Theoretical design of fullerene derivatives as electron acceptors for bulk heterojunction solar cells*, Mold. J. Phys. Sci. 2009, 8, in press.

[68] Suzuki, T.; Maruyama, Y.; Akasaka, T.; Ando, W.; Kobayashi, K.; Nagasel, S. Redox Properties of Organofullerenes, *J. Am. Chem. Soc.* 1994,116, 1359-1363.

[69] Morvillo, P.; Bobeico, E. A fullerene silirane derivative to improve the open circuit voltage in a polymer–fullerene solar cell: a theoretical study, phys. *stat. sol.* (RRL) 2008, 2, 260-262.

[70] Akasaka, T.; Ando, W.; Kobayashi, K.; Nagase, S. Reaction of C_{60} with Silylene, the First Fullerene Silirane Derivative, *J. Am. Chem Soc.* 1993, 115, 1605-1606.

[71] Wakahara, T.; Maeda, Y.; Kako, M.; Akasaka, T.; Kobayashi, K.; Nagase S. Silylation of fullerenes with active species in photolysis of polysilane, *J. Organomet. Chem.* 2003, 685, 177-188.

In: New Trends in Chemical Physics Research
Editor: James N. Lapherty
ISBN: 978-1-61668-535-5

Chapter 12

WOOD PYROLYSIS: INFLUENCE OF PYROLYSIS TEMPERATURE AND HEATING RATE ON CHARCOAL PROPERTIES AND PYROLYSIS PROCESS

Nestor Tancredi*[*1], Andrés Cuña[1] and Maria Irene Yoshida[2]
[1]Cátedra de Fisicoquímica y Laboratorio de Fisicoquímica de Superficies, DETEMA, Facultad de Química, Universidad de la República. Gral. Flores 2124, CC 1157, 11800, Montevideo, Uruguay
[2]Departamento de Química, Universidade Federal de Minas Gerais, Av. Antonio Carlos 6627, Pampulha 31270-901, Belo Horizonte, Minas Gerais, Brazil

ABSTRACT

As in recent decades world petroleum costs are increasing the searching for new alternative energy sources was developed in many countries. Biomass pyrolysis is an important route for charcoal production, a fuel with several advantages in its use compared to biomass or coal combustion. Charcoal has a higher heating value and lesser water content than biomass, is more stable in the presence of microbial agents or plagues and it hardly ever contains sulfur.

Maximum pyrolysis temperature, raw material heating rate up to reaching pyrolysis temperature and pyrolysis time are major parameters in the determination of the resulting charcoal properties. The control of these parameters is hardly performed in industrial charcoal furnaces, but a well controlled pyrolysis process can be carried out at a laboratory level.

In the present work the influence of the maximum pyrolysis temperature and heating rate on the pyrolysis process and charcoal properties was studied. *Eucalyptus grandis* wood was used as the biomass raw material; pyrolysis was performed in a temperature programmable horizontal tubular laboratory furnace in high purity nitrogen atmosphere,

[*] Phone (--598) 2 924 8352, Fax (--598) 2 924 1906, Email: nestor@fq.edu.uy

with maximum pyrolysis temperatures selected in the range 300 – 650 °C and heating rates of 4 and 16 °C min^{-1}. Wood and charcoals characterization was performed through elemental and proximate analysis and determination of high and low heating values by using the calorimetric bomb method. Wood pyrolysis was also studied by TG, at the same conditions used in the laboratory furnace. For those charcoals obtained at 4 °C min^{-1} their combustion in air atmosphere were studied by DSC and TG/DTA; combustion apparent activation energies were determined for charcoals obtained at 4 and 16 °C min^{-1}. Results show that for both heating rates the heating values increase as maximum pyrolysis temperature increases, but there is a different behaviour between 300 and 350°C, where heating value decreases. This seems to be due to a different charcoal composition. In addition, combustion activation energy for the 300°C charcoal is higher than the 350°C one.

Charcoal yield is higher for the lower heating rate being the effect more remarkable for higher maximum pyrolysis temperatures. TG results show that charcoal yield for a 2 hs, 450°C wood pyrolysis is the same that for wood pyrolysis up to 600°C. A similar behaviour is found for a 2 hs, 300°C and wood pyrolysis up to 350°C. DSC and DTA for wood and charcoal combustion show two exothermic peaks due to cellulose, lignin and their pyrolysis products.

INTRODUCTION

As in recent decades world petroleum costs are increasing the searching for new alternative energy sources was developed in many countries. Biomass pyrolysis is an important route for charcoal production, a fuel with several advantages in its use compared to biomass or coal combustion [1]. Charcoal has a higher heating value and lesser water content than biomass, is more stable in the presence of microbial agents or plagues and it hardly ever contains sulfur. In addition, a higher flame combustion temperature is achieved by burning charcoal than biomass materials.

Maximum pyrolysis temperature, raw material heating rate up to reaching pyrolysis temperature and pyrolysis time are major parameters in the determination of the resulting charcoal properties. The control of these parameters is hardly performed in industrial charcoal furnaces, but a well controlled pyrolysis process can be carried out at a laboratory level.

In the present work the influence of the maximum pyrolysis temperature and heating rate on the pyrolysis process and charcoal properties was studied. *Eucalyptus grandis* wood was used as the biomass raw material; pyrolysis was performed in a temperature programmable horizontal tubular laboratory furnace in high purity nitrogen atmosphere, with maximum pyrolysis temperatures selected in the range 300 – 650 °C and heating rates of 4 and 16 °C min^{-1}. Wood and charcoals characterization was performed through elemental and proximate analysis and determination of high and low heating values by using the calorimetric bomb method. Wood pyrolysis was also studied by TG, at the same conditions used in the laboratory furnace. For those charcoals obtained at 4 °C min^{-1} their combustion in air atmosphere were studied by DSC and TG/DTA; combustion apparent activation energies were determined for charcoals obtained at 4 and 16 °C min^{-1}.

Materials and Methods

Eucalyptus grandis wood was used as the biomass raw material. Pyrolysis was performed in a temperature programmable horizontal tubular laboratory furnace (Carbolite CTF) in high purity nitrogen atmosphere, with maximum pyrolysis temperatures selected in the range 300 – 650 °C and heating rates of 4 and 16 °C min^{-1}. Both the wood and the charcoal obtained at the different temperatures were characterized by means of proximate analysis (moisture, ash and volatiles), and elemental analysis (C, H, N and S, Carlo Erba EA 1108 CHNS). Moisture analysis was performed by drying a sample in stove at 105 °C up to constant mass. Ash and volatiles content were determined in agreement to ASTM and ISO norms respectively [2, 3]. To determine the high heating value (HHV) the calorimetric bomb method was used (Parr 1341 Oxigen Bomb Calorimeter), in agreement with Parr Company, instructions sheet 204 M. The low heating value (LHV) was determined from the HHV and the elemental analysis of the samples.

Wood pyrolysis was also studied by TG, at the same conditions used in the laboratory furnace. For those charcoals obtained at 4 °C min^{-1} their combustion in air atmosphere were studied by DSC (Shimadzu DSC-50) and TG/DTA (Netzsch STA 409EP). Combustion apparent activation energies were determined for charcoals obtained at 4 and 16 °C min^{-1}, by applying the Ozawa-Flynn-Wall method.

DSC measurements were carried out in dynamic air atmosphere at 10°C min^{-1} using about 1.0 – 1.5 mg of samples in aluminum pans. The samples were heated from room temperature at 100°C and maintained in isothermal conditions for 10 minutes and, then, heated to 500°C. TG/DTA measurements were carried out in dynamic air atmosphere, in the temperature range from 30 to 750°C using 10°C min^{-1} heating rate, with sample masses between 6.0 – 7.0 mg in alumina crucible. Some additional TG measurements on charcoals prepared at a heating rate of 16°C min^{-1} were carried out in a Shimadzu TG-50 equipment, in dynamic air atmosphere (50 cm^3 STP min^{-1}) using 10°C min^{-1} heating rate, with sample masses between 6.0 – 7.0 mg in platinum pan.

Results

Wood Characterization

Wood characterization results are shown in Table 1. Elemental analysis show that N and S percentages are nil, and C and O percentages similar between them.

Wood Pyrolysis

The effect of the maximum pyrolysis temperature on the properties of charcoals prepared at heating rates of 4 °C min^{-1} and 16 °C min^{-1} are shown in Tables 2 and 3.

Table 1. Wood characterization

Elemental analysis (Weight percentage, dry basis, ash free)					
H (%)	N (%)	C (%)	S (%)	O (%)*	
6.5	0	49.5	0	44.0	
Moisture (%,w.b)	Volatiles (%, d.b)	Ash (%, d.b)	Fixed carbon (%, d.b)	HHV ($kJkg^{-1}$, d.b) ± 200	LHV ($kJkg^{-1}$, d.b) ± 200
20	83.6	0.4	16.0	19000	17600

(*) Determined by difference.

Table 2. Effect of maximum pyrolysis temperature on charcoal properties. Heating rate 4°C min^{-1}

Proximate Analysis (Weight percentage)					
Maximum Temperature (°C)± 1	Moisture (%, w.b)	Volatiles (%, d.b)	Ash (%, d.b)		
300	3.2	46.8	0.7		
350	4.5	35.8	0.9		
450	4.4	21.8	1.2		
600	5.4	9.3	1.3		
650	7.2	7.8	1.3		
Elemental Analysis (Weight percentage, dry basis, ash free)					
Maximum Temperature (°C)± 1	H (%)	N (%)	C (%)	S (%)	O (%) (*)
300	4.1	0.0	67.3	0.0	28.6
350	3.2	0.0	70.2	0.0	26.6
450	2.2	0.0	80.9	0.0	16.9
600	1.1	0.0	87.1	0.0	11.8
650	0.7	0.0	90.0	0.0	9.3
Other properties					
Maximum Temperature (°C)± 1	Charcoal yield (%)	Fixed carbon (%, d.b)	HHV ($kJkg^{-1}$, d.b) ± 200	LHV ($kJkg^{-1}$, d.b) ± 200	
300	47.9	52.5	29600	28600	
350	40.2	63.3	27900	27000	
450	31.5	77.0	30400	29700	
600	29.3	89.4	33300	32900	
650	28.1	90.9	33000	32700	

(*) Determined by difference.

For both heating rates the heating values increase as maximum pyrolysis temperature increases, but there is a different behaviour between 300 and 350°C, where heating value decreases. This seems to be due to a different charcoal composition, as a result of the volatilization of different compounds in each case.

Charcoal yield is higher for the lower heating rate being the effect more remarkable for higher maximum pyrolysis temperatures. This can be explained as at higher heating rates the volatilization is higher as volatile products has lower possibilities to react among them [4]. Also, fixed carbon percentage and carbon percentage slowly increase with maximum pyrolysis temperature increase, being the values a bit higher for 16 °C min^{-1} heating rate.

Table 3. Effect of maximum pyrolysis temperature on charcoal properties. Heating rate 16 °C min^{-1}

Proximate Analysis (Weight percentage)					
Maximum Temperature (°C)± 1	Moisture (%, w.b)	Volatiles (%, d.b)	Ash (%, d.b)		
300	2.7	43.0	0.8		
350	2.9	33.4	1.0		
450	4.6	21.5	1.1		
600	5.2	8.2	1.3		
650	6.1	-	-		
Elemental Analysis (Weight percentage, dry basis, ash free)					
Maximum Temperature (°C)± 1	H (%)	N (%)	C (%)	S (%)	O (%) (*)
300	4.8	0.0	70.8	0.0	24.2
350	4.2	0.0	73.7	0.0	21.9
450	3.5	0.0	81.0	0.0	15.3
600	2.4	0.0	88.6	0.0	8.9
650	-	-	-	-	-
Other properties					
Maximum Temperature (°C)± 1	Charcoal yield (%)	Fixed carbon (%, d.b)	HHV (kJkg^{-1}, d.b) ± 200	LHV (kJkg^{-1}, d.b) ± 200	
300	46.9	56.2	29400	28300	
350	37.5	65.6	28200	27300	
450	29.2	77.4	31100	30300	
600	25.8	90.5	32300	31800	
650	24.7	-	33300	-	

(*) Determined by difference.

TG results for wood pyrolysis are shown in Figures 1 and 2. The profiles are the typical ones for lignocellulosic materials [5, 6]. TG curves show that for both heating rates, charcoal yield for a 2 hs, 450°C wood pyrolysis is the same that for a wood pyrolysis up to 600°C. A similar behaviour is found for a heating rate of 4 °C min^{-1}, for a 2 hs, 300°C wood pyrolysis and a wood pyrolysis up to 350°C. The same charcoal yield could be achieved by different routes, although the properties of the charcoals obtained would not be the same.

Charcoal Combustion

DTG curves comparison for wood and charcoal combustion (Figure 3) shows that the 300°C charcoal prepared at 4°C min^{-1} had a lower autoignition temperature (250°C) than the 350°C one (265). The higher maximum pyrolysis temperature achieved for the 350°C charcoal determine a more important devolatilization, as it is seen in Table 2, and as a consequence a lower quantity of flammable gases is formed when the charcoal is heated up. According with these facts, the autoignition temperature increases with the higher maximum pyrolysis temperature.

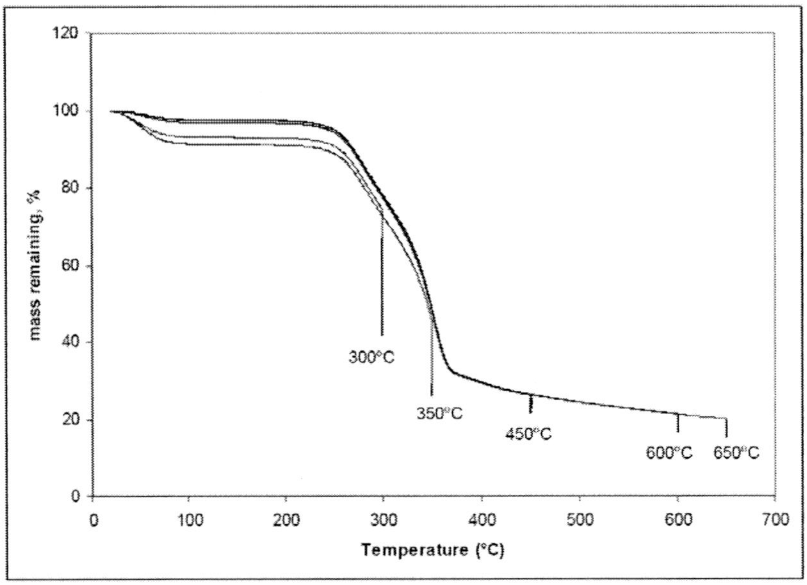

Figure 1. TG curves for E. grandis wood pyrolysis up to different temperatures. Heating rate 4° min^{-1}.

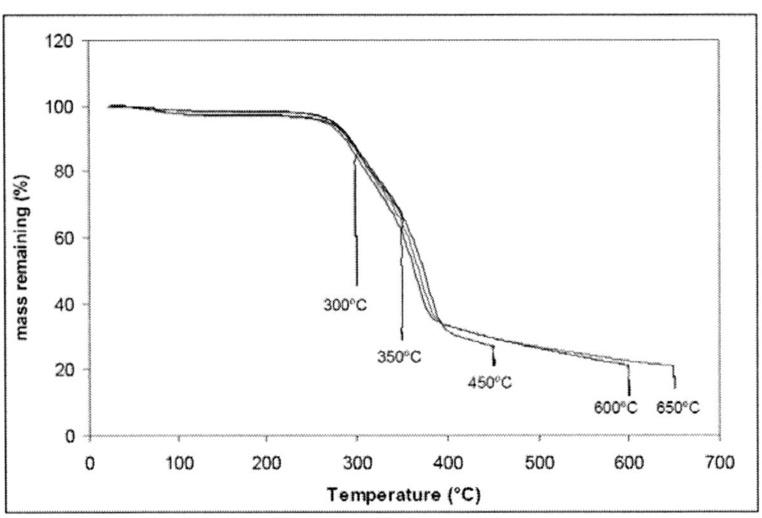

Figure 2. TG curves for E. grandis wood pyrolysis up to different temperatures. Heating rate 16° min^{-1}.

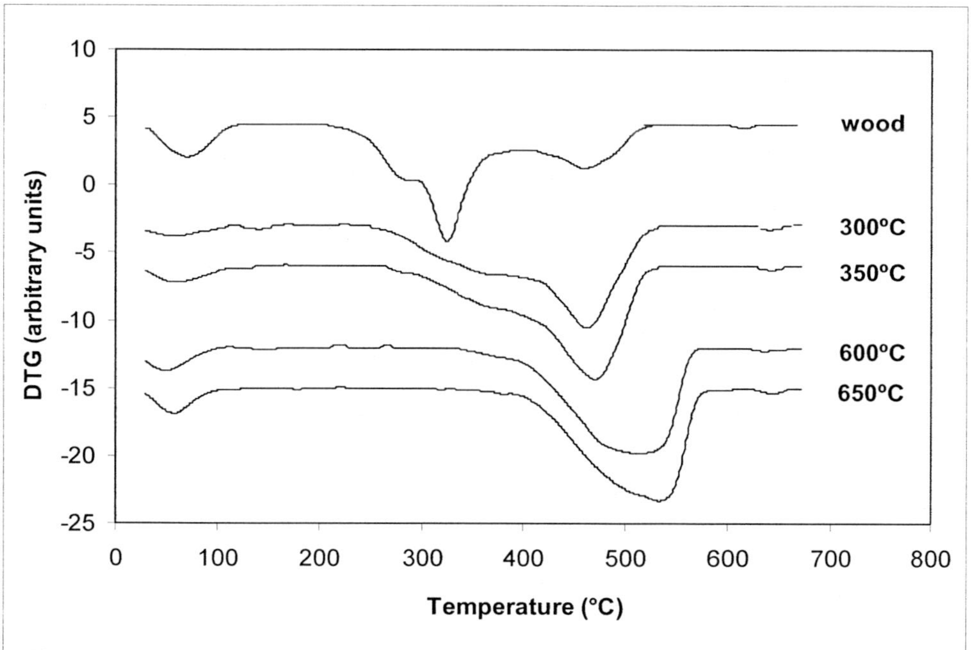

Figure 3. DTG in air for charcoals prepared at a heating rate of 4 °C min^{-1} (TG heating rate 10°C min^{-1}).

DTG curves also shows changes in the number, shape and position of the peaks. All curves present the first peak corresponding to water release at a temperature lower than 100°C. For wood, a second peak with a maximum (absolute value) at 330°C and a shoulder at 290°C is atributtable to combustion of volatile matter coming from hemicellulose (lower temperatures), cellulose (medium-high temperatures) and lignin. A third peak, wider than the second one, placed between 400°C and 530°C, with a maximum at 470°C, would correspond to charcoal combustion. This event has been reported by other researchers [7, 8, 9, 10]. Jiménez et al. (2006) reported a similar profile with a peak at 418°C for charcoal combustion. Also, co-combustion of volatile matter and charcoal would be possible in the temperature range between the second and the third peak.

For charcoals, the second peak disappears as the maximum pyrolysis temperature increases and the content of hemicellulose, cellulose and lignin decreases as a consequence of pyrolysis and devolatilization. The third peak remains and its maximum (absolute value) moves up to higher temperatures.

All these results are in accordance with DTA results (Figure 4)

As the DTA results are expressed in μV mg^{-1} (although the unities in Figure 4 are arbitrary in order to enhance the Figure clearity) the relative combustion enthalpy values for the different processes can be compared. The charcoal combustion enthalpy is clearly higher than that for wood, and shows a tendency to increase with the maximum pyrolysis temperature, as it was seen in Table 2.

When the heating rate was 16°C min^{-1} (TG results not shown), both charcoals had the same autoignition temperature (270°C), a little higher than for 4°C min^{-1}. This result agrees with the need of reaching a higher temperature to obtain a flammable gas mixture for the higher heating rate [7].

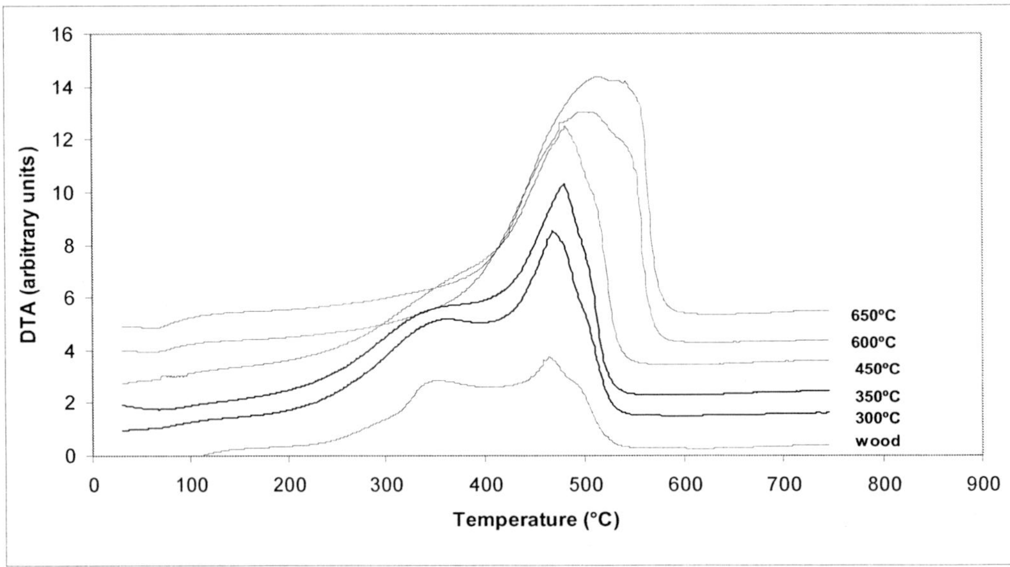

Figure 4. DTA curves in air for wood and charcoals prepared at 4° min^{-1}.

Combustion Apparent Activation Energy Calculations

Taking into account TG and analyses done on charcoals and wood, important differences are found between the charcoals prepared at maximum pyrolysis temperature of 300°C and 350°C. These differences include a decrease in the heating value, in opposition to the major tendency indicating that heating value increases with pyrolysis temperature increase. For these reasons charcoals prepared at 300°C and 350°C were selected to study their combustion properties.

According to the Ozawa-Flynn-Wall method [11], apparent activation energy for charcoal combustion can be determined from thermogravimetric analysis of charcoal in air, carried out at different heating rates. This method is specially convenient for reactions that involve heating mass loss, as combustion ones, and the major supposition is that the activation energy is constant throughout the process. The equation used is

$$\log\left(\frac{\phi_2}{\phi_1}\right) = \frac{0,467}{R} E_{ap}\left(\frac{1}{T_1} - \frac{1}{T_2}\right)$$

where ϕ_1 and ϕ_2 are the heating rates for two thermogravimetric determinations, T_1 and T_2 are the temperatures for a 30 % of mass loss with regarding the initial mass, E_{ap} is the apparent activation energy in J mol^{-1} and R is the gases constant (8,314 J mol^{-1}K^{-1}). The calculated E_{ap} are shown on Table 4. For both heating rates of preparation E_{ap} is lower for the charcoal prepared at 300 °C; therefore combustion rate for this temperature is higher than for the charcoal prepared at 350 °C, according to the alrealdy mentioned different autoignition temperature for these charcoals. When the heating rate of preparation is increased, E_{ap} strongly increase, showing once again that the devolatilization in the 16°C min^{-1} charcoal had a very important consequence on the reactivity of the charcoal, in agreement with other results shown previously in this paper. At the higher heating rate of preparation volatiles produced has a lower residence time, being less probable secondary pyrolytic reactions on the solid surface that lead to a solid product [1]. The facts that the E_{ap} and the autoignition temperature for the charcoal prepared at 4°C min^{-1} are lower than those for the charcoal prepared at 16°C min^{-1} could be explained in terms of a higher combustion reactivity of the volatile matter of that "secondary" charcoal.

Table 4. Combustion apparent activation energy E_{ap} calculated for charcoals prepared from *E.grandis* at two pyrolysis temperatures and two heating rates

Heating rate of preparation (°C min^{-1})	Maximum pyrolysis temperature (°C)	ϕ (K min^{-1})	T (K)	E_{ap} (kJ mol^{-1})
4	300	5	619,7	78,7
		10	647	
	350	5	642,2	83,0
		10	670,0	
16	300	5	634.8	178.6
		10	647.1	
	350	5	657.8	297.7
		10	665.4	

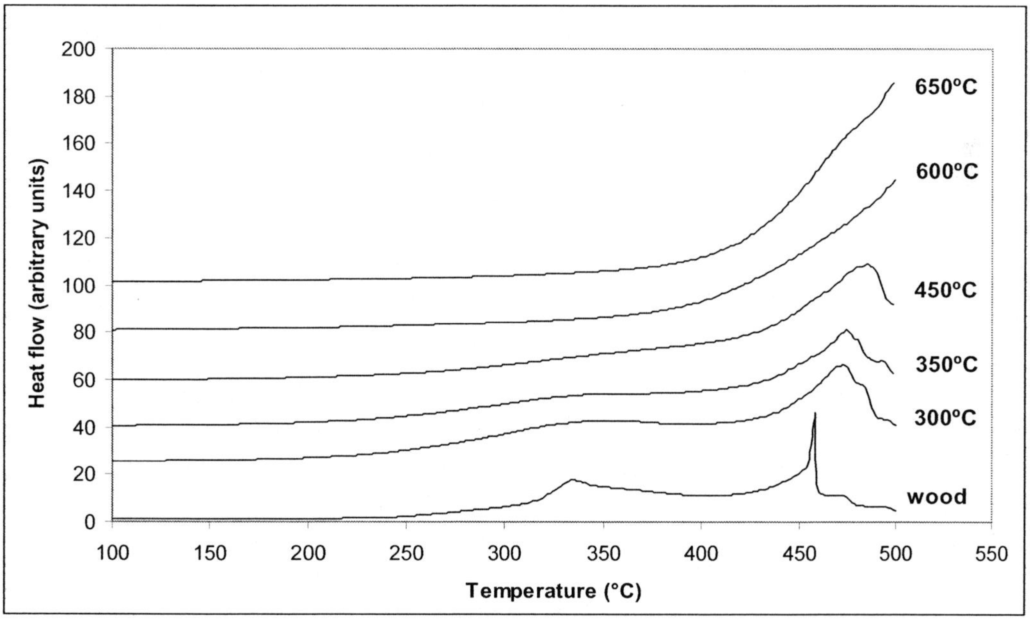

Figure 5. DSC in air for wood and charcoals prepared at 4°C min^{-1}.

DSC results for wood and charcoals prepared from wood at a heating rate of 4°C min^{-1} are shown in Figure 5 The same events described for the case of TG and DTA studies are present here as two peaks appear beyond 100°C. The first peak is placed at 250-380°C, and can be attributed to volatile matter combustion; the second one, narrower and more defined than the first one, is placed at 460°C or higher temperature values, and it is attributable to charcoal combustion. As in the case of DTA curves, the first peak is notorious for wood, it is less clear for 300°C-charcoal and it disappears as the maximum pyrolysis temperature increases. The second peak is always present but it is placed at a higher temperature as the maximum pyrolysis temperature increases.

Conclusion

The influence of the maximum pyrolysis temperature and the heating rate on *E. grandis* wood pyrolysis was studied, for a temperature range of 300-650°C. A clear influence of the maximum pyrolysis temperature on the charcoal mass yield, hygroscopicity, volatile matter content and fixed carbon percentage was found. For the temperature range 300°C-350°C the influence is more important than for the 450°C-650°C one.

Wood pyrolysis TG curves confirm the strong influence of the maximum pyrolysis temperature in the range 300°C-350°C, where a substantial mass loss is observed.

Charcoal and wood combustion studied through TG/DTA and DSC in air atmosphere indicate that the autoignition temperature for the charcoal prepared at 350°C is higher than that of the charcoal prepared at 300°C. The combustion follow different stages revealed as different peaks placed at different temperatures; with increasing maximum pyrolysis temperature the position of the peaks corresponding to solid charcoal combustion moves up while the peaks corresponding to flammable gases combustion tend to disappear. At a constant heating rate of preparation, combustion apparent activation energy is lower for the lower maximum pyrolysis temperature, probably due to a higher content of high heating value volatile matter on the 300°C charcoal. At a constant maximum pyrolysis temperature, combustion apparent activation energy and autoignition temperature are higher for the higher heating rate of preparation of the charcoal, as a consequence of the lower content of "secondary" charcoal.

References

[1] Antal, M. J., Grønli, M., *Ind. Eng. Chem. Res*. 2003, 42,1619-1640.

[2] ASTM D 2866-70 Total Ash Content of Activated Carbon. ASTM International, West Conshohocken, PA, www.astm.org.

[3] ISO 562-1981 Determination of the volatile matter content of a hard coal and coke.

[4] Jiménez, L. F.; Baquero, M. C.; Díaz, J., *Rev. Colomb. Quím*., 2006, 35, 177-190.

[5] Wang, S.; Liu, Q.; Luo, Z.; Wen, L.; Cen, K., *Front. Energy Power Eng. China* 2007, 1, 1–7.

[6] Worasuwannarak, N.; Potisri, P.; Tanthapanichakoon, W., The Joint International Conference on "*Sustainable Energy and Environment* (SEE)"1-3 December 2004, Hua Hin, Thailand.

[7] Liang, X. H; .Kozinski, J. A. *Fuel* 2000; 79, 1477-86.

[8] Lapuerta, M.; Hernández, J. J.; Rodríguez, *J. Biomass and Bioenergy*, 2004, 27, 385 – 391.

[9] Bilbao, R.; Mastral, J. F.; Aldea, M. E.; Ceamanos, J., *J. Anal. Appl. Pyrol.* 1997, 39, 53–64.

[10] Grammelis, P.; Basinas, P.; Malliopoulou, A.; Sakellaropoulos, G., *Fuel,* 2009, 88, 195–205.

[11] Hatakeyama, T.; Liu, Z. *Handbook of Thermal Analysis*, Second Edition, John Wiley and Sons, Chichester, EN, 2000, pp. 47-48.

In: New Trends in Chemical Physics Research
Editor: James N. Lapherty

ISBN: 978-1-61668-535-5

Chapter 13

NITRATION OF CELLULOSE AND PERFORMANCE OF CELLULOSE NITRATE BASED ADHESIVES FOR SELECTED SUBSTRATES

F. N. Emengo, J. K. C. Nduka* and E. O. Onwu

Pure and Industrial Chemistry Department
Nnamdi Azikiwe University, P.M.B. 5025 Awka,
Anambra State, Nigeria.

ABSTRACT

Background, the effects of composition of mixed nitric and sulphuric acids on the nitration of cellulose and the effects of average molar mass, degree of nitration and unperturbed polymer dimension of cellulose nitrate on bonding strength and tack of joints of selected substrates have been investigated. Nitration of cellulose was non-stoichiometric, the degree of nitration decreasing from 14 to 7 percent as the mole fraction of water in mixed acid increased. Degree of nitration increased non-linearly with average molar mass. Bond strength increased to a maximum and then decreased for the three substrates as average molar mass, degree of nitration and unperturbed polymer dimension increased.

Results indicate that maximum bond strength was obtained at a degree of nitration of about 12 percent and unperturbed polymer dimension of 1.20 x 10^{-12} cm^2 for the three substrates; at average molar mass of 6.87 x 10^3 for wood and leather and of 6.87 x 10^3 to 8.25 x 10^3 for paper. The tack of cellulose nitrate joints showed visco elastic response to applied stress. Onset of visco elasticity occurred at degree of nitration of about 13 percent, unperturbed polymer dimension of about 1.40 x 10^{-12} cm^2 and average molar mass of about 7 x 10^3 for the three substrates.

It is concluded that bond strengths of these joints for paper, wood and leather are influenced by degree of nitration, average molar mass and unperturbed polymer dimension of cellulose nitrate.

* E-mail: johnnduka2000@yahoo.co.uk

Keywords: A. solvent-based, B. Fibres. C. Wood, D. Tack.

INTRODUCTION

Cellulose is a naturally occurring long chain polymer containing a series of glucosidic rings joined by oxygen bridges. All adhesives based on cellulose are derivatives formed by reacting one or more of three hydroxyl groups on each ring structure with appropriate chemicals. It has been suggested that some free hydroxyl groups are necessary for optimum adhesion. Thus the best cellulose nitrate adhesives are those nitrated to give 11.2 to 11.6 percent nitrogen [1] However, the nitration of cellulose to a required degree is uncertain principally because of lack of a comprehensive mechanism for the reaction and numerous sources of degradation of cellulose nitrate during reaction and processing [2]. It has been suggested by Clark (1981) [3] that the nitration of cellulose nitrate is not stoichiometric. The irregular nitration was attributed to impurities in chemicals used in the nitration of cellulose, combination, quality and quantity of processing additive [4,5], the reversibility of the nitration reaction whereby the nitrous oxides by-products degrade the cellulose residues by oxidation [6], inhomogenieties in the structure of cellulose or inhomogenieties in the nitrating mix, presence of cellulose sulphate esters, and trapped free acids [2]. Because of the complexity of sources of degradation, a direct relationship would not be expected between the adhesive properties and degree of nitration. In this paper, the relationships between the degree of nitration on the one hand and composition of mixed acids and average molar mass of cellulose nitrate on the other are reported. The effects of average molar mass, degree of nitration and unperturbed polymer dimension of cellulose nitrate on bonding strength and tack for paper wood and leather substrates have also been reported.

EXPERIMENTALS

Material

Cotton Linters

Cotton linters were obtained from Nigerian Cotton Mill Ltd. Onitsha, Nigeria. The cellulose content was found to be 85% using Klason's method [7]. Sulphuric acid, 98.99% purity was a BDH product. Nitric acid, 69.72% purity was a BDH product. Absolute alcohol, 95.0% purity, amyl acetate, acetone and butyl acetate were analar grade.

Methods

Nitration of Cellulose

Cotton linters was degreased with carbon tetrachloride and scoured. It was then washed and dried with alcohol until the water content was 2% [4] 100g of cotton linters was immersed for 30 minutes in a large excess of mixed acid of various ratio of sulphuric and nitric acids, contained in a 500ml beaker placed in a thermostat bath at 30°C in a fume cupboard. The mixed acids were drained out and the nitrated cellulose was then quenched in a large excess of water at 30°C. The water was centrifuged out. The nitrated cellulose was carefully washed. Esters of sulphate and nitrosulphate type, which cause instability of the cellulose nitrate, are hydrolyzed readily by weak mineral acids and were removed by boiling with dilute acid solutions, followed by thorough washing with water at room temperature until no free acid was trapped in the nitrated cellulose. The cellulose nitrate was dried with alcohol until the water content was 2% and was then stored in stoppered plastic containers [4]

Determination of Degree of Nitration of Cellulose Nitrate

The degree of nitration of cellulose nitrate was measured by the gas volumetric method [8]. The method is based on the volumetric measurement of nitric oxide liberated from the reaction:

$$2\ (\text{Cellulose} - NO + 4H_2SO_4 + 3Hg \longrightarrow (\text{Cellulose})_2SO_4 + 3HgSO_4 + 2NO + 4H_2O$$

5g cellulose nitrate was added to 20ml of concentrated sulphuric acid and 1g of mercury in a round-bottomed flask. At the end of the reaction, the volume of nitric oxide collected over water at atmospheric pressure and room temperature (743mmHg and 25°C) was measured [9]. The volume of NO at Standard Temperature and Pressure (STP) was calculated and the percentage of nitrogen based on 5g of cellulose nitrate used was obtained from it. The final result recorded was the average of two measurements on each sample.

Measurement of Average Molar Mass of Cellulose Nitrate and Estimation of Unperturbed Polymer Dimension

Average molar mass of cellulose nitrate was determined by a solution method for cellulose nitrate having degree of nitration from 10 to 14. The relative viscosities of polymer solutions were determined using the Ubbelohde Suspended - level Viscometer, which is a U – tube viscometer. The measured quantity is the time required for the meniscus of the solution to pass between the two etches marks on either side of the measuring bulb. The instrument was immersed in a vertical position in a thermostatted bath at 30°C, and then filled with sufficient solution or solvent. The viscometer was operated by stopping the end of the right hand tube, sucking the solution into the measuring tube, unstopping the right hand tube, releasing the pressure and then taking the time for the meniscus to pass between the two etch marks on either side of the measuring bulb by means of a stop watch. For each solution (or solvent), the mean of three such readings was taken. Measurements were made on three solutions of known concentrations and also on the solvent.

Neglecting the kinetic energy correction, dynamic viscosity, η = At where t is the flow time and A is a constant equal to 0.009203 mm^2/s. The intrinsic viscosity, $[\eta]$ was obtained by plotting $(t - t_O)/ct_O$ against c and extrapolating to infinite dilution.

As the viscosity method for average molar mass of a polymer is not an absolute one, Mark – Houwink equation was employed.

$$[\eta] = KM^{\alpha} \qquad (1)$$

The parameters K and α are constants for a given polymer – solvent system at a given temperature. For cellulose nitrate - acetone system at $25^{O}C$, K = $1.42 \times 10^{-4} dl/g$ and α = 0.9338 [8,10] The unperturbed polymer dimensions were also estimated from the intrinsic viscosities obtained. The method of Flory and Fox (1951)[11] was used.

$$\bar{r}_o^2 = \frac{[[\eta]M]^{2/3}}{\theta} \qquad (2)$$

$\bar{r}_o^2$ is the unperturbed mean -square end to end separation of a polymer chain, θ is a universal constant equal to 2.87×10^{21} if $\bar{r}_o^2$ is expressed in cm and [η] is in dl/g.

Formulation of the Cellulose Nitrate Adhesives

Cellulose nitrate adhesives were formulated by adding the constituents as shown in Table 1 and mixing with a motorized stirrer at 30°C. Cellulose nitrate was added to the mixture of the solvents and camphor. The cellulose formed a homogenous paste with the other constituents.

Measurement of Tack

The tack of cellulose nitrate on paper was measured[12] using typing paper cut into strips measuring 20 x 2cm. The strips were marked into divisions measuring 2cm. The divisions were numbered consecutively I, 2, 3 ..... minutes. A uniform smear of Ig of the adhesive to be tested was applied on the back of the substrate by means of a glass rod. The treated paper was immediately placed on another paper of the same dimension. Pressure was applied over the surface to ensure a good lap. A stopwatch was started. The two strips of paper were peeled from each other by the application of a steady force at the rate of one minute per division. The time, t observed was recorded. This corresponds to the setting speed. The reciprocal of this time was taken as the tack of paper.

The tack of wood and leather were determined using samples measuring 10 x 5cm. 2g of the adhesive to be tested was uniformly smeared on the back of the substrate using a glass rod. The treated substrate was immediately placed on the corresponding substrate forming wood/wood and leather/leather pairs. A sliding force was applied at intervals of time to check the resistance to movement. The time, t, a resistance to movement was observed and recorded. This was regarded as the setting speed and the tack was taken as the reciprocal of this time.

Measurement of Bonding Strength

Samples measuring 10 x 5cm were used for the measurement of bonding strength. A uniform smear of 2g of the adhesive was applied on the substrate using a glass rod. The treated substrate was placed on the corresponding substrate to give paper/paper, wood/wood and leather/leather pairs. The leather and also the wood pairs were each clamped together and allowed to stand for 24 hours at room temperature. In case of paper, a load of 1kg was placed to cover the paper totally and allowed to stand for 24 hours at room temperature. At the end of 24 hours, the substrates were separated and the type of failure was recorded. No bonding between the substrates was recorded as very low; slight fibre failure was recorded as low; pronounced fibre failure was recorded as high and intermediate failure was recorded as medium.

RESULTS AND DISCUSSION

Results have been obtained for the following investigations.

(1) The dependence of the degree of nitration of cellulose on the composition of mixed sulphuric and nitric acids, Table 2 and Figure I.
(2) The dependence of degree of nitration of cellulose nitrate on average molar mass, figure 2.
(3) The effects of average molar mass, degree of nitration and unperturbed molar mass on bonding strength of cellulose nitrate for selected substrates, Table 3.
(4) The effects of average molar mass, degree of nitration and unperturbed polymer dimension of cellulose nitrate on tack for selected substrates, Figures 3, 4 and 5.

Table 1. Formulation of Cellulose nitrate adhesives

Substance	Weight percent
Cellulose nitrate	15.00
Camphor	6.00
Acetone	11.00
Ethanol	44.00
Amyl acetate	24.00

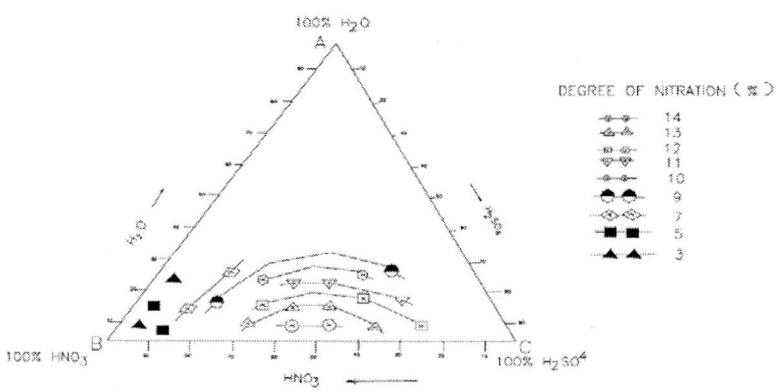

Figure 1. Effects of composition of nitrate mixture on degree of nitration of cellulose.

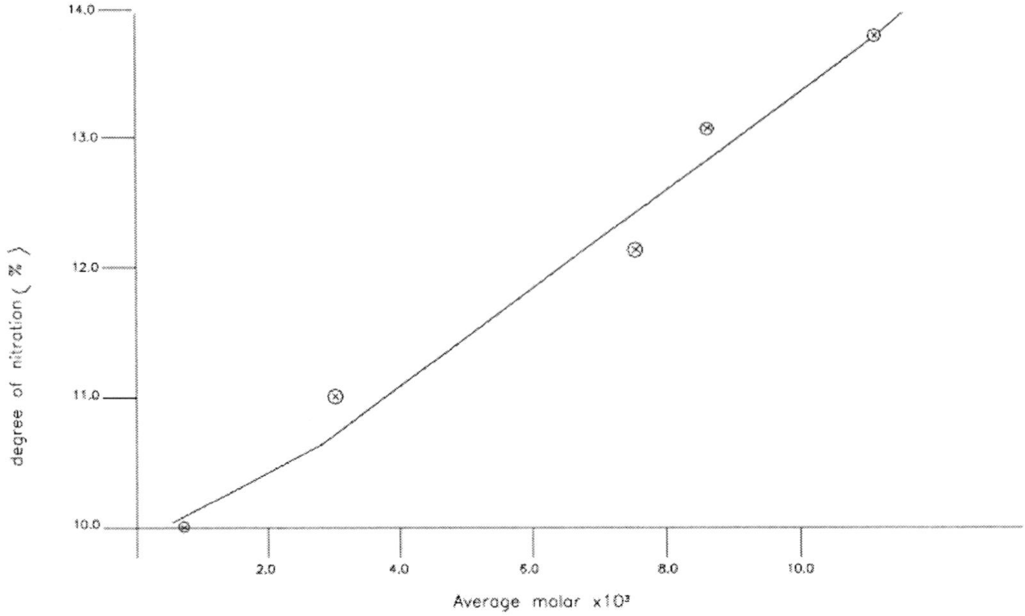

Figure 2. Effects of average molar mass of cellulose nitrate on degree of nitration.

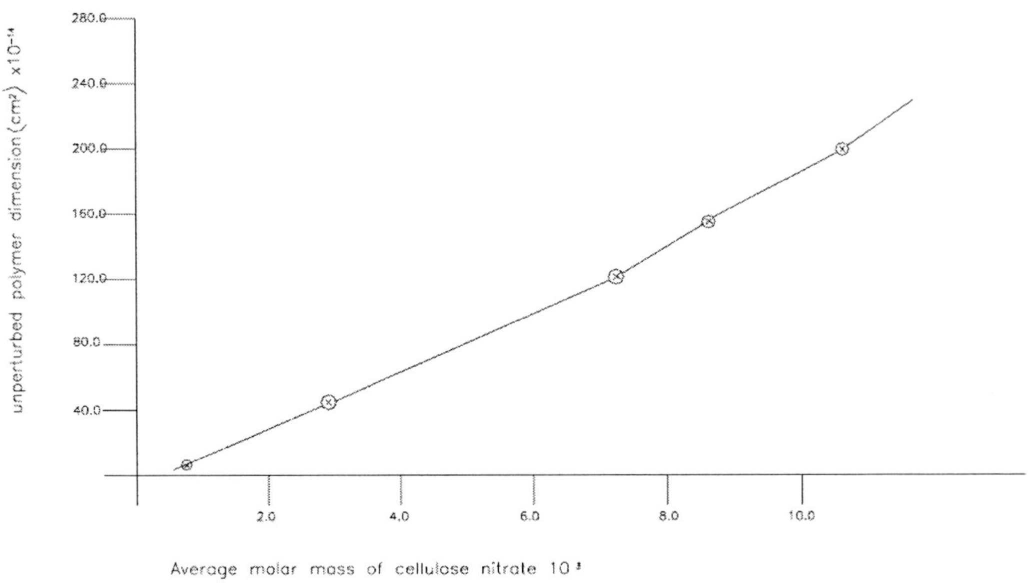

Figure 3. Effects of average molar mass of cellulose nitrate on unpertubed polymer.

Effects of Composition of Mixed Acids on Degree of Nitration of Cellulose

Under the same conditions of temperature and time, cellulose has been nitrated using widely varying compositions of sulphuric and nitric acids shown in the ternary diagram in fig. 1 and also in Table 2.

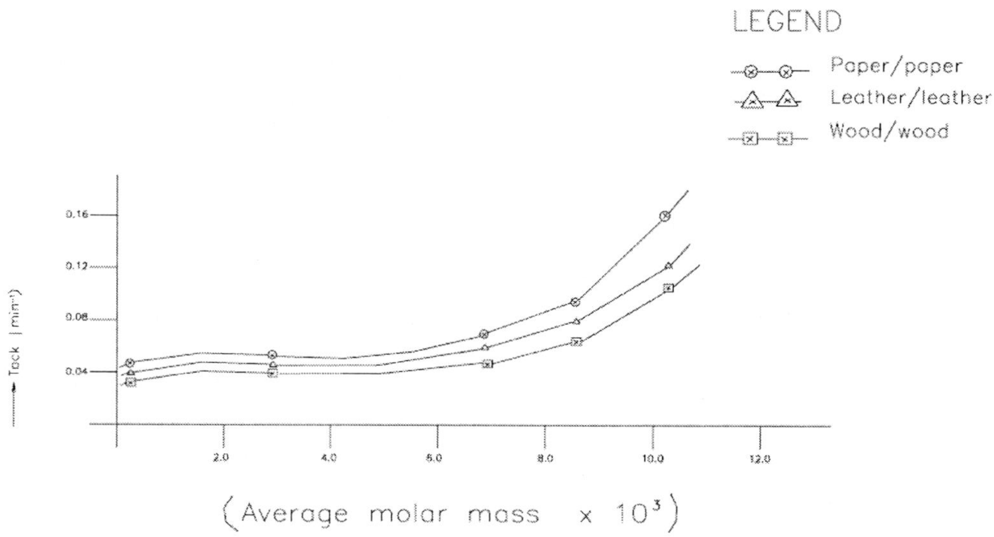

Figure 4. Effects of degree of nutrition of cellulose nitrate on tack for paper, wood and leather substrate.

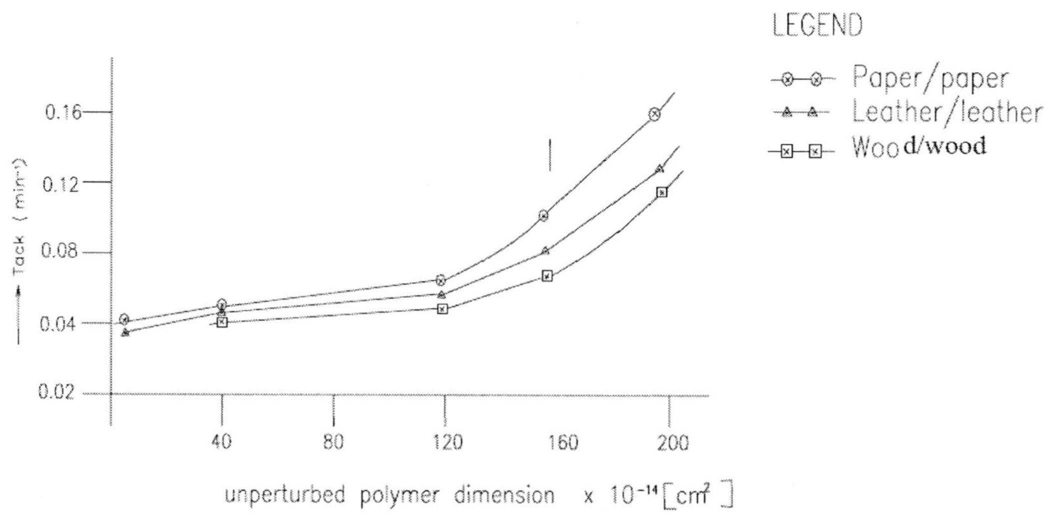

Figure 5. Effects of unpertubed polymer dimension on tack of cellulose nitrate for paper, wood and leather substrate.

The results show that the degree of nitration of cellulose nitrate varies from 3 to 14 percent. Theoretically, complete replacement of the hydroxyl groups in cellulose should give a product containing 14.15 percent nitrogen. Thus the results show that the nitration of cellulose proceed non-stoichiometrically. This observation is in agreement with earlier observations [13,3]. The irregular nitration was explained to arise from the inhomogeneities in the original cellulose fibres. It was believed that the amorphous and crystalline regions of the cellulose fibres reacted differently in the nitration mix. However Clark (1981)[3] using electron spectroscopy chemical analysis demonstrated that the physical structure of cellulose cannot be responsible for the irregular nitration of cellulose.

Table 2. Effect of composition of nitrating mixture on degree of nitration

Composition of nitrating mixture			Degree of nitration (% nitrogen content)
$\%H_2SO_4$	$\%HNO_3$	$\%H_2O$	
43.0	52.0	5.0	14%
52.0	43.0	5.0	14%
33.0	62.0	5.0	13%
62.0	33.0	5.0	13%
38.0	47.0	15.0	13%
47.0	38.0	15.0	13%
71.0	24.0	5.0	12%
30.0	55.0	15.0	12%
55.0	30.0	15.0	12%
64.0	21.0	15.0	11%
33.0	42.0	25.0	11%
42.0	33.0	25.0	11%
26.0	49.0	25.0	10%
49.0	26.0	25.0	10%
21.0	64.0	15.0	9%
11.0	64.0	15.0	9%
56.0	19.0	25.0	9%
13.0	72.0	15.0	7%
19.0	56.0	25.0	7%
14.0	81.0	5.0	5%
4.0	81.0	15.0	5%
5.0	90.0	5.0	3%
4.0	71.0	25.0	3%
11.0	64.0	25.0	3%

The net degree of nitration is controlled by a number of factors. These include the concentration of mixed acids, the reversibility of the nitration reaction and the corrosive power of the deterioration products. The major factor of the effect of concentration on degree of nitration is the mole fraction of water. Figure 1 shows that the degree of nitration decreases from 14 to 7 as the mole fraction of water is increased. Furthermore, high acid concentration i.e, low mole fraction of water increases degree of nitration by minimizing degradation by hydrolysis of nitro and sulphate esters and oxidation of cellulose residues. High acid concentration can minimize degradation, evidently because of lowered hydrogen ion concentration, with corresponding reduction of hydrolytic rupture of the cellulose chains. The nitration of cellulose proceeds according to the reversible reaction [6]

$$\text{Cellulose} - OH + 2HNO_3 \rightleftharpoons \text{Cellulose} - NO_2 + HNO_3 + H_2O$$

This reaction is reversible in the presence of nitric acid and water. Nitric acid oxidizes the cellulose residues and produce nitrous oxides that catalyse further degradative reactions. Perhaps, nitration of cellulose proceeds stoichiometrically but degradative reactions whose intensity depends on several factors alter the structure of cellulose nitrate.

The Dependence of the Observed Degree of Nitration of Cellulose on Average Molar Mass

The results in Figure 2 show that degree of nitration increases with average molar mass in a non-linear manner. It is unexpected that the degree of polymerization of a polymer chain would have such a profound effect on its degree of nitration. The structure of incompletely nitrated cellulose remains inexplicable. X ray diffraction patterns have been claimed [13] (Lewis et al, 1949) to indicate mixtures of completely nitrated cellulose with cellulose, which is un-nitrated or relatively so. A possible explanation of this is that in a given glucopyranose radical, the introduction of a single nitrate group renders the two remaining hydroxyls sufficiently sensitive to reaction so that complete replacement occurs in that radical. This would visualize a partly nitrated cellulose molecule as consisting of a chain of certain glucopyranose radicals completely nitrated and others unattacked. However solvent fractionation of a given nitrated product shows little or no variation in nitrogen content but wide differences in viscosity. The variation of degree of nitration with average molar mass can be explained in terms of side reactions or changes in structure, which are determined by the strength of the mixed acids and by the time and temperature of reaction. Thus traces of sulphuric and nitrosulphuric esters undoubtedly form and nitric acid exerts an oxidizing action on the cellulose, while both acids catalyse the hydrolysis of both the cellulose itself and its esters. In the absence of these degradative reactions, degree of nitration would not vary with average molar mass for the same structure of nitrated cellulose. For lower concentrations of the acids implying higher molar fractions of water and higher concentration of hydrogen ions, these degradative reaction would be more important and could alter the relationships between degree of nitration and average molar mass. However because the structure of cellulose nitrated in the presence of nitric and sulphuric acids is uncertain, the relationship between degree of nitration and average molar mass cannot be predicted.

Bonding Strengths of Cellulose Nitrate-Based Adhesives for Paper, Wood and Leather Substrates

Table 3 shows bonding strengths of joints made by cellulose nitrate adhesives of different average molar masses, degrees of nitration and unperturbed polymer dimensions for paper, wood and leather substrates. Adhesion is a complex phenomenon and the total bond strength is usually made up of contributions from various factors. Chemical bonds, both covalent and electrostatic can contribute to bond strength. Ionic bonds, usually of the donor-acceptor, types are closely associated with the electrostatic theory of adhesion, [14, 15,16,17].Other factors that could contribute to adhesive bond strength include molecular diffusion, adsorption and mechanical contacts. Molecular diffusion is important in the autohesion of polymers [18], formation of films from latices, heat sealing of thermoplastics, polymer to metal adhesion and in the combination of films of adhesives above their glass transition temperature.

Table 3. Effects of molar mass, degree of nitration and unperturbed polymer dimension of cellulose nitrate-based adhesives on bonding strength of paper, wood and leather joints

Average molar mass x 10^3	Degree of nitration, (%)	Unperturbed polymer dimension (cm^2 x 10^{-14})	Bonding strength paper/paper	Bonding strength Wood/wood	Bonding strength Leather/leather
5.36	10.00	4.40	Low	Very low	Very low
27.53	11.00	37.00	Medium	Low	Low
68.73	12.00	120.00	High	High	High
82.46	13.00	150.00	High	Medium	Medium
104.74	14.00	200.00	Medium	Low	Low

Diffusion is also important in the wetting of surfaces by liquids, which is an essential precondition for adhesive action. The identification of diffusion as the mechanism of adhesion for rubbery polymers and resins above the glass transition temperature was due to Voyutskii(1960) [19] ,but its quantitative development and its general acceptance was due to Vasenin (1960, 1961, 1963, 1965) [20-23],. Additionally adsorption of the polymer could conceivably contribute to adhesion in various ways, namely reduction of surface energy when a liquid is allowed to wet the surface of a solid and orientation of the polymer molecule on the surface including those cases where particularly strong adhesive bonds could be associated with chemisorption. Adsorptions in the sense of ``oriented polymer molecules have been successfully applied to explain adhesion on high-energy solids [24-25]. The identification of adsorption as the mechanism of adhesion is due to Anand (1969, 1970, 1973) [26-28] who held the opinion that diffusion does take place during autohesion but that its contribution to bond strength seems insignificant compared with the physical adsorption brought about by van der waals forces. He proposed that firstly contact must be made, when made adsorption is instantaneous. The growth in bond strength is due, according to Anand, to the process of making contact, and not to diffusion; mechanical factors also play important role in explaining bond strength of many systems. These factors generally enhance systems based on adsorption or electrostatic action [29-32]. Because adhesion is a complex phenomenon, it is not comprehensible with a single model. Explaining the total bond strength of two adhesively bonded substrates requires the application of several adhesion theories at one and the same time. This fact is borne out in the various substrates bonded with cellulose nitrate based adhesives investigated.

Effects of Various Factors on Bond Strength Produced by Cellulose Nitrate

Results in Table 3 show that as the average molar mass of cellulose nitrate is increased bond strength increased to a maximum and then decreased for paper, wood and leather substrates. Whereas increases to maximum occurred gradually for wood and leather substrates, it was appreciably more rapid for the paper substrate. Surface chemical groups on paper and wood are likely to be similar but different from the surface chemical groups on leather. The close similarities between bond strength data for leather and wood and the differences between wood and paper with surface groups derived from the glucopyranose units indicate that wetting of the substrates which is an adsorption process does not differ significantly among the three surfaces.

Mechanical factors are known to be important in adhesion of fibrous materials such as textile [31] and leather [32]. Mechanical factors are expected to be important in the adhesion of the substrates bonded with cellulose nitrate. Fibre end protrude from each substrate and embed themselves in the adhesive. The bond strength would be expected to be proportional to the number of fibre ends originally present and a function of these broken when the bonded substrates are separated. Of course, for fibre ends to remain embedded implies some specific adhesion between the cellulose nitrate and the fibre material. Fibres of paper, wood and leather would differ in their specific adhesions to cellulose nitrate. Such differences would be expected to result in differences in the bond strengths of paper/paper, wood/wood and leather substrates joined together with cellulose nitrate.

Results in Table 3 and Figure 3 also show that unperturbed polymer dimension increases as the average molar mass is increased. Figure 3 shows that increase of unperturbed polymer dimension become very sharp at a molar mass of around 6.0×10^3. Unperturbed polymer dimension is a direct measure of the extent of spread of the cellulose nitrate on the substrate. It would therefore be expected that embedding of fibre ends in the cellulose nitrate will increase in bonding strength. This is shown by the results in Table 3 and Figure 3. However, the large and progressive increase in unperturbed polymer dimension at high molar mass does not lead to progressive increase in bond strength. A maximum in bond strength was reached at a molar mass of 6.9×10^3 for wood and leather substrates and at 6.9×10^3 to 8.3×10^3 for paper substrates.

Vasenin (1960, 1961, 1963, 1965) [20-23], applying the diffusion theory of adhesion obtained an equation for the force required to separate two substrates bonded by an adhesive. The force should be directly proportional to the rate of separation, inversely proportional to two thirds root of the average molar mass and one quarter root of time of contact. This equation predicts that the bond strength, which is related to the force of separation, would increase with average molar mass.

The results in Table 3 also show that bond strength increases as the degree of nitration of cellulose nitrate is increased, reaches a maximum at a degree of nitration of 12.0 for wood and leather substrates and at a degree of nitration of 12.0 to 13.0 for paper. A degree of nitration of 12.0 has been stated as optimal for cellulose nitrate adhesives[1]. A possible explanation of this observation: is that a balance of NO_3 and OH groups is essential for optimal adhesion. The OH groups in the glucopyranose units are capable of forming hydrogen bonds. The hydrogen bonds would enhance attractive forces across the interface. These donor-acceptor interfacial interactions all involve redistribution of electrons, and as they are confined to the interface involve a change in the electrical state and is a form of electrostatic adhesion.

Tack of Cellulose Nitrate Based Adhesives for Paper, Wood and Leather Substrates

Results in Table 4 show the tack of cellulose nitrate adhesives of different average molar masses, degrees of nitration and unperturbed polymer dimensions for paper/paper, wood/wood and leather/leather substrates.

Table 4. Effects of molar mass, unperturbed polymer dimension, degree of nitration on tack of cellulose nitrate-based adhesives of paper, wood and leather substrates

Average Molar Mass x 10^3	Degree of nitration, (%)	Unperturbed Polymer Dimension ($cm^2 \times 10^{-14}$)	*Track (mm^{-1})*		
			Paper/paper	Wood/wood	Leather/leather
5.306	10.000	4.400	0.040	0.004	0.039
27.530	11.000	37.000	0.050	0.046	0.048
68.730	12.00	120.000	0.070	0.056	0.063
82.460	13.000	150.000	0.100	0.067	0.077
104.740	14.000	200.000	0.170	0.110	0.125

Tack is that property of the adhesive whereby light contact with the surface of another body brings about a condition requiring force to restore the original separated state. Voyutsjkii (1960)[19] specifically connects tack with spinnability. This ability to pull out into filaments implies that long chain molecules lubricated by any solvents present, uncoil and partially orient along the axis of the filament. At some degree of orientation the stress required to produce further orientation exceeds that necessary for viscous flow and the molecules slide over each other. This process can continue until thinning of the filament eventually causes rupture, or alternatively, failure takes place at the substrate surface and filaments retract, partly by elastic recoil, and partly by surface tension forces, to their origin. However considering adhesion, Wake (1961) [33] showed that tack implies a continuous appreciation of energy to a maximum before separation. This does not apply in the case of purely viscous phenomena. Dahlquist (1959, 1969) [34-35] explained tack of a pressure sensitive adhesive as the response of a viscoelastic material to dynamic deformation, a response, which therefore, involves the complex rather than the elastic modulus. The viscoelasticity of the bulk polymer used as the adhesive plays a major part in resisting separation by peeling even when separation is apparently at the Interface and surface factors might be expected to dominate. Dahlqulst (1969) [35]determined the time – temperature mechanical properties using the Williams, Landel and Ferry, WLF (1955) [36] relationship and showed that these were correlated with tack behaviour using a shift factor analogous to the WLF transform: Bates had subsummed by a single expression data generated by experiments covering all the variables known to determine the tack energy of a given material[37]. The expression Bates gave is

$$Y = A \exp [m \log_{10} (\text{probe separation rate})] \quad (3)$$

In this equation A and m are constants. The constant A occurs as the result of an integration and is an integration constant which includes the compliance of the adhesive and the geometric form of the probe; m is a material constant which appears dependent on the loss modulus of the adhesive.

The measurement of tack is an energy measurement. It involves wetting and separation. In addition when made with a probe there is contact deformation. All three components are time dependent. This mutual dependence results in the possibility of subsuming results from experiments at different loads, dwell times, temperatures, film thickness and rates of separation in a single curve. The tack of a viscoelastic polymer exhibits to itself or a closely similar polymer is a diffusion process.

Effects of Various Factors on Tack of Cellulose Nitrate Adhesives

Table 4 and Figure 4 show that as the average molar mass of cellulose nitrate is increased, there was a progressive increase in tack for the three substrates. At a molar mass of 7.0 10^3 for the paper/paper and about 9.0 10^3 for leather/leather and wood/wood, the tack increased sharply. Although the surface groups of paper and wood would be expected to be similar being derived from cellulose and would be different from those of leather which is derived from proteins, the behaviour of leather was closer to paper than that of wood. This indicates that wetting characteristics of cellulose nitrates on the substrates do not contribute significantly to tack.

Where the curves in Figure 4 show gradual build of tack would correspond to regions in which the joint was failing cohesively [38]. The gradual build up of force required to separate the substrates occurs as purely viscous behaviour; gives way to visco elasticity and the detachment of cellulose nitrate polymer from substrate occurs when the force required to extend the polymer from the substrate reached a value greater than that which would balance attachment to the polymer. The results shown in Table 4 and Figure 4. shows that the onset of viscoelasticity occurs earlier in the paper/paper system than in the wood/wood and leather/leather systems. In other words, the average molar mass of cellulose nitrate required for the deformation to change from viscous to visco elastic behaviour is lower for the paper/paper system than for the wood/wood and leather/leather systems. Adhesion of these systems is therefore determined by both bulk and interfacial factors. The significance and nature of the interfacial contributions to adhesion are shown by the effects of degree of nitration and unperturbed polymer dimension on tack. Figure 5. Shows that tack builds up as the unperturbed polymer dimension is increased and increased sharply at a particular unperturbed polymer dimension. The behaviour of the three systems are similar. The unperturbed polymer dimension at which visco elastic behaviour sets is identical for the three systems. This might be taken to indicate that the contributions of bulk effects to tack are similar for the three systems. However the tack at each unperturbed polymer dimension is highest for the paper/paper system followed by the leather/leather and wood/wood system in that order. These differences reflect the significance of interfacial contribution in the adhesion of these systems. The effects of unperturbed polymer dimension bring out the contributions of interfacial contribution more distinctively than the effects of average molar mass.

The contributions of interfacial factors decrease in the order paper/paper, leather/leather and wood/wood.

Unperturbed polymer dimension is a measure of the extent of spread of the cellulose nitrate on the substrate. Diffusion of the polymer occurs by movements of individual segments (say glucopyranose units) resulting in movements of the ends of the molecule. Under equilibrium conditions, the centre of gravity will not move appreciably but if there is a concentration, stress or temperature gradient then as the molecule alters its configuration by thermal agitation, its center of gravity moves and the effect of this is that the molecule diffuses through its neighbours. Vasenin(1960, 1961, 1963, 1965) [20-23], connected diffusion of a long-chain molecule with the force required to remove it from a substrate by considering the interchange of momentum between that part of the molecule which has diffused into the matrix itself (i.e. polymer and substrate) and the matrix itself. This interchange of momentum would appear to increase in the order; wood, leather and paper, that is, in the order of increasing unperturbed polymer dimension on these substrates.

Figure 6 shows that tack builds up as the degree of nitration is increased and increased sharply at a degree of nitration of about 13 percent. The behaviour of the three systems are again similar. The degree of nitration at which viscoelastic deformation sets in are similar for the three systems. However, the tack at each degree of nitration decreases in the order paper, leather and wood.

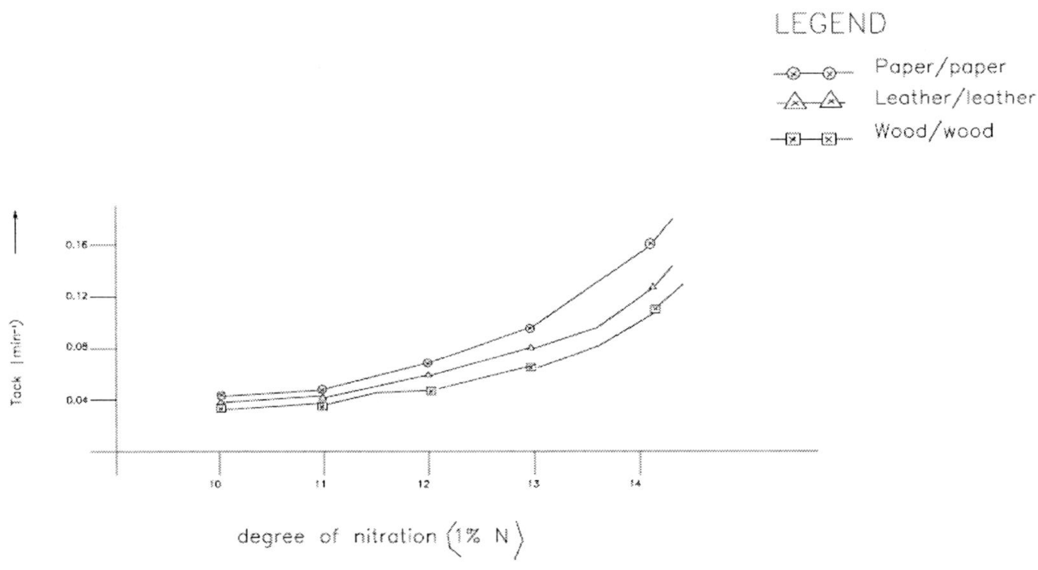

Figure 6. Effects of degree of nutrition of cellulose nitrate on tack for paper, wood and leather substrate.

The effects of degree of nitration on tack indicate the significance of specific chemical forces in the adhesion of these systems, and point out that electrostatic theory is relevant [14]. The electrostatic interactions are associated with the degree of nitration and hydrogen bonding. The effects of such interactions are to enhance attractive forces across the interface beyond that which would be expected from a consideration of the surface free energies of the two phases considered separately. These donor-acceptor interfacial interactions involve redistribution of electron and as they are confined to the interface, involve a change in electrical state and are forms of electrostatic adhesion.

CONCLUSION

The nitration of cellulose by mixed nitric and sulphuric acids proceeds non-stochiometrically. The degree of nitration decreases as the mole fraction of the water in the mixed acid is increased and increases non-linearly with molar mass. Bond strength of cellulose nitrate joint is made up of contributions from mechanical, diffusional and electrostatic factors. Bond strengths of these joints for paper, wood and leather are influenced by degree of nitration, average molar mass and unperturbed polymer dimension of cellulose nitrate. Tack of cellulose nitrate joint shows viscoelastic response to applied stress. Tack is influenced by average molar mass, degree of nitration and unperturbed polymer dimension of cellulose nitrate and is determined by bulk and interfacial factors from electrostatic interactions associated with degree of nitration and hydrogen bonding.

REFERENCES

[1] Chalmers I and Bathe p, Chemical Specialties , *Domestic and Industrial*, 2nd edn, vol. 2 George Godwin, London (1979).
[2] Reilly JA, (JAIC), 30 (2), 145 – 162 (1991).
[3] Clark DT, *American Chemical Society Series*, 162, 247-249 (1981).
[4] Worden, *Nitrocellulose Industry*. Vols 1 and 2 New York: Van Nostrand (1911)
[5] Pettitpas P and Mathieu M, *Transaction of the Faraday Society*, 42B (17), 525 (1946).
[6] Miles FD, *Cellulose Nitrate Oliver and Boyd*, London (1955).
[7] Seven A and Rydholm P, Pulping Processes, Interscience Publishers, England (1967).
[8] Urbaniski J, Czerwinski W, Janicka K, Majewska F and Zowall H, *Handbook of Analysis of synthetic polymers*, John Wiley and Sons Inc. New York, p.424 (1977).
[9] Brady JE and Humiston GE 3rd edn , General Chemistry Principles and Structure, p 609 John Wiley an Sons. Inc. New York (1982).
[10] Blackley D C, *Addition Polymers : Formation and Characterisation*. Smith, D A, (ed) pp131- 219,Butterworths, London (1968).
[11] Flory PJ and Fo TG, *J. Am. Chem*, Soc., 73, (1951).
[12] Emengo FN. Chukwu SER, Mozie J, *Int. J. Adhesion Adhesives*, 22(2), 93 -100 (2002).
[13] Lewis K. , Squares L and Boughton G, *Chemistry of Colloidal and Amorphous Materia*, 367 – 373. The Macmillan Company, New York (1949).
[14] Denyagin BV, N A Krotova, V V Karasev, YM Krillova and A N AleiNokova, *Proc 2nd Int. Congr. Surface Activity*, Div. VI (C)BP 595 (1957).
[15] Roberts A, J. *Phys D. Appl. Phys*. 10, 1802 (1977).
[16] Fowkes FM, *Recent Advances in Adhesion in Adhesion* (ed. L. H. Lee), Gordon and Beach, New York, p 39 (1973).
[17] Fowkes FM, *Amer. Chem. Soc. Div. Org. Coat Plastic Chem*., Honolulu, Hawii, p. 13 (1979).
[18] Wake WC, *Adhesion and Adhesives, Fundamentals and Practice*, pp 67 –114 Society of Chemical Industries, London (1954).
[19] Voyutskii SS, *Autohesion and Adhesion of High Polymer*, Moscow, and (transl by S. Kaganoff and ed. By V. Vakula), Interscience, New York (1960).

[20] Vasenin RM, Vys. Soed, 2, 851 (RAPRA TransI I005, R. J. Moseley) Vasenin, Vys. Soed, 2, 875 (RAPRA Transl. 1006, R. J. Moseley) (1960).

[21] Vasenin RM, *Vys Soed.*, 3, 679 (RAPRA Trnsl. 1006, R. J. Moseley) (1961).

[22] Vasenin RM, *Nauchn Tr. Mosk Tekhol. Inst. Legoi Prom* 26, (RAPRA Transl, 1010, J. Moseley) (1963).

[23] Vasenin RM, *Adhesives Age*, 8 (5) 21 and (6), 30 (1965).

[24] Glazer J, *J. Polym Sci.* 13, 355 (1954).

[25] Schonhorn H, *J. Polym Sci.*, Al, 2243 and 3523 (1963).

[26] Anand J N, J. *Adhesion 1*(1) 31 (1969).

[27] Anand J N, J. *Adhesion 2* (1) 23 (1970).

[28] Anand J N, J. *Adhesion* (1973)

[29] Narcus H, *Trans. J. Plastics lnst.*, 35 (117) 529 (1967).

[30] Perrins LE and Pettet K, *Plastics and Polymers* 39 (144), (1971)

[31] Boreoff E M and Wake W C, (a) Trans. Inst. Rubber Ind. 25, 199.(b) *Trans.Inst.Rubber Ind,* 25, 210 (1949)

[32] Haines BM, *Aspects of Adhesion* - 3 (ed. D. J. Alner) University of London Press, London, p 40 (1967).

[33] Wake WC, Adhesion (ed D. D. Eley) Clarendon Press, Oxford, p 191 (1961)

[34] Dahlquist C, *Adhesives Age*, 2, 25 (1959).

[35] Dahlquist C, Adhesion: Fundamentals and Practices (ed. U.K. Ministry of Technology). Maclaren, London. p 143 (1969).

[36] Williams ML, Landel RF and Ferry JD, J. Amer. Chem. Soc., 77, 3701 (1955).

[37] Bates R, *Studies in the Nature of Adhesives Tack* .Thesis. The City Universiy,London vol. 2. George Godwin, London (1974).

[38] Gent AN and Petieh, *Proc. Roy. Soc.*, A 310, 433 (1969).

In: New Trends in Chemical Physics Research
Editor: James N. Lapherty

ISBN: 978-1-61668-535-5

Chapter 14

PHONON DYNAMICS OF FE-BASED GLASSY ALLOYS

*Aditya M. Vora**
Parmeshwari 165, Vijaynagar Area, Hospital Road,
Bhuj – Kutch, 370 001, Gujarat, India

ABSTRACT

The well recognized model potential is used to investigate the vibrational properties of four Fe-based binary glassy alloys viz. $Fe_{90}Zr_{10}$, $Fe_{80}B_{20}$, $Fe_{83}B_{17}$ and $Fe_{80}P_{20}$. The thermodynamic and elastic properties are also computed from the elastic limits of the PDC. Three theoretical approaches given by Hubbard-Beeby (HB), Takeno-Goda (TG) and Bhatia-Singh (BS) are used in the present study to compute the phonon dispersion curves (PDC). Five local field correction functions proposed by Hartree (H), Taylor (T), Ichimaru-Utsumi (IU), Farid *et al.* (F) and Sarkar *et al.* (S) are employed to see the effect of exchange and correlation in the aforesaid properties.

Keywords: pseudopotential; pair potential; phonon dispersion curves (PDC); Fe-based binary glassy alloys; static properties; vibrational properties.

PACS: 63.50.+x, 65.60.+a.

1. INTRODUCTION

The mankind has been manufacturing glassy materials for several thousands years. Compared to that, the scientific study of amorphous materials has a much shorter history. And only recently, there has been an explosion of interest to these studies as more promising materials are produced in the amorphous form. The range of applications of metallic glasses

* Tel. : +91-2832-256424, E-mail address : voraam@yahoo.com.

is vast and extends from the common window glass to high capacity storage media for digital devices [1-9].

$Fe_{90}Zr_{10}$ glass is a good example of ferromagnetic glassy material and it is the member of transition metal- transition metal (TM-TM) group. The vibrational dynamics of this glass was studied by Bhandari *et al.* [10] employing Hubbard-Beeby (HB) [11] and Bhatia-Singh (BS) [12] approaches. They have also calculated the temperature dependence electrical resistivity of $Fe_{90}Zr_{10}$. Also, the structural study of $Fe_{90}Zr_{10}$ glass was made by Hausleitner and Hafner [13] using the MD technique. Recently Wildes *et al.* [14] have studied the magnetic behaviour of $Fe_{90}Zr_{10}$ glass, experimentally. The $Fe_{80}B_{20}$, $Fe_{83}B_{17}$ and $Fe_{80}P_{20}$ metallic glasses are the most important candidate of transition metal-metalloid (TM-M) group. Self-consistent spin-polarized electronic structure of $Fe_{80}B_{20}$ glass has been extensively studied by Hafner *et al.* [15, 16]. The PDC of $Fe_{80}B_{20}$ glass has been studied by Gupta *et al.* [17] employing Hubbard-Beeby (HB) [11] approach. They have compared their findings with available experimental results [17] and found to be in good agreement with them. The vibrational dynamics of $Fe_{83}B_{17}$ glass was studied by Saxena [18] using Hubbard-Beeby (HB) [11] approach using model potential formalism. The atomic dynamics of $Fe_{80}P_{20}$ glass has been studied by Hausleitner and Hafner [16]. Recently, we have been reported the vibrational properties of some metallic glasses using model potential formalism [1-9].

Looking to the advantages of metallic glasses, the present paper emphasizes the vibrational dynamics of four Fe-based binary glassy alloys, viz. $Fe_{90}Zr_{10}$, $Fe_{80}B_{20}$, $Fe_{83}B_{17}$ and $Fe_{80}P_{20}$ using well recognized model potential [1-9]. The thermodynamics and elastic properties of such as longitudinal sound velocity υ_L, transverse sound velocity υ_T, isothermal bulk modulus B_T, modulus of rigidity G, Poisson's ratio σ, Young's modulus Y and Debye temperature θ_D are computed from the elastic limit of the dispersion relation. Five different types of local field correction functions proposed by Hartree (H) [19], Taylor (T) [20], Ichimaru-Utsumi (IU) [21], Farid *et al.* (F) [22] and Sarkar *et al.* (S) [23] are used to study the exchange and correlation effects in the aforesaid studies. The phenomenological theories of Hubbard-Beeby (HB) [11], Takeno-Goda (TG) [24, 25] and Bhatia-Singh (BS) [12, 26] are employed to generate the phonon dispersion curves (PDC). The most important ingredient of the PDC is pair potential computed theoretically in Wills-Harrison (WH) [27] form from well recognized model potential [1-9].

2. Theoretical Methodology

The fundamental ingredient, which goes into the calculation of the phonon dynamics of metallic glasses, is the pair potential. In the present study, for TM-TM and TM-M glasses, the pair potential is computed using [1-9, 27],

$$V(r) = V_S(r) + V_b(r) + V_r(r). \tag{1}$$

The *s*-electron contribution to the pair potential $V_S(r)$ is calculated from,

$$V_s(r) = \left(\frac{Z_S^{\,2} e^2}{r}\right) + \frac{\Omega_O}{\pi^2} \int F(q) \left[\frac{Sin(qr)}{qr}\right] q^2 \, dq \,. \qquad (2)$$

Here $Z_S \sim 1.5$ is found by integrating the partial *s*-density of states resulting from self-consistent band structure calculation for the entire 3d and 4d series [27], while Ω_O the effective atomic volume of the one component fluid.

The energy wave number characteristics appearing in the Eq. (2) is written as [1-9, 27]

$$F(q) = \frac{-\Omega_O \, q^2}{16\pi} |W_B(q)|^2 \frac{[\varepsilon_H(q) - 1]}{\{1 + [\varepsilon_H(q) - 1][1 - f(q)]\}} \,. \qquad (3)$$

Here, $W_B(q)$ is the effective bare ion potential, $\varepsilon_H(q)$ the Hartree dielectric response function and $f(q)$ the local field correction function to introduce the exchange and correlation effects.

The well recognized model potential $W_B(q)$ [1-9] used in the present computation of phonon dynamics of binary metallic glasses is of the form

$$W_B(q) = \frac{-4\pi e^2 Z}{\Omega_O q^2} \left[\begin{array}{l} \left\{ \begin{array}{l} -1 + \frac{12}{U^2} + \frac{U^2}{1+U^2} + \frac{6U^2}{(1+U^2)^2} + \frac{18U^2}{(1+U^2)^3} - \frac{6U^4}{(1+U^2)^3} \\ + \frac{24U^2}{(1+U^2)^4} - \frac{24U^4}{(1+U^2)^4} \end{array} \right\} \cos(U) \\ + \\ \left\{ \begin{array}{l} \frac{6}{U} - \frac{12}{U^3} + \frac{U}{1+U^2} + \frac{3U}{(1+U^2)^2} - \frac{3U^3}{(1+U^2)^2} + \frac{6U}{(1+U^2)^3} \\ - \frac{18U^3}{(1+U^2)^3} + \frac{6U}{(1+U^2)^4} - \frac{36U^3}{(1+U^2)^4} + \frac{6U^5}{(1+U^2)^4} \end{array} \right\} \sin(U) \\ + \\ 24U^2 \exp(1) \left\{ \frac{U^2 - 1}{(1+U^2)^4} \right\} \end{array} \right] . \qquad (4)$$

here $U = qr_C$. r_C is the model potential parameter. This form has feature of a Coulombic term outside the core and varying cancellation due to repulsive and attractive contributions to the potential within the core in real space. The detailed information of this potential is given in the literature [1-9]. The model potential parameter r_C is calculated from the well known formula [28] as follows:

$$r_C = \left[\frac{0.51\, r_S}{(Z)^{1/3}}\right]. \tag{5}$$

Here r_S is the Wigner-Seitz radius of the amorphous alloys.

The d-electron contributions to the pair potential are expressed in terms of the number of d-electron Z_d, the d-state radii r_d and the nearest-neighbour coordination number N_C as follows:

$$V_b(r) = -Z_d\left(1-\frac{Z_d}{10}\right)\left(\frac{12}{N_C}\right)^{1/2}\left(\frac{28.06}{\pi}\right)\frac{2r_d^{\,3}}{r^5}, \tag{6}$$

and

$$V_r(r) = Z_d\left(\frac{450}{\pi^2}\right)\frac{r_d^{\,6}}{r^8}. \tag{7}$$

The theories of Hubbard-Beeby (HB) [11], Takeno-Goda (TG) [24, 25] and Bhatia-Singh (BS) [12, 26] have been employed in the present computation. The expressions for longitudinal phonon frequency ω_L and transverse phonon frequency ω_T as per HB, TG and BS approaches are given below [11, 12, 24-26].

According to the Hubbard-Beeby (HB) [11], the expressions for longitudinal phonon frequency ω_L and transverse phonon frequency ω_T are,

$$\omega_L^2(q) = \omega_E^2\left[1-\frac{\sin(q\sigma)}{q\sigma}-\frac{6\cos(q\sigma)}{(q\sigma)^2}+\frac{6\sin(q\sigma)}{(q\sigma)^3}\right], \tag{8}$$

and

$$\omega_T^2(q) = \omega_E^2\left[1-\frac{3\cos(q\sigma)}{(q\sigma)^2}+\frac{3\sin(q\sigma)}{(q\sigma)^3}\right]. \tag{9}$$

with $\omega_E^2 = \left(\frac{4\pi\rho}{3M}\right)\int_0^\infty g(r)V''(r)r^2dr$ is the maximum frequency.

The theory for computing the phonon dynamics in amorphous solids, approach proposed by Takeno-Goda (TG) [24, 25] have been employed in the present computation. The expressions for longitudinal phonon frequency ω_L and transverse phonon frequency ω_T are

$$\omega_L^2(q)=\left(\frac{4\pi\rho}{M}\right)\int_0^\infty dr\, g(r)\left[\left\{rV'(r)\left(1-\frac{\sin(qr)}{qr}\right)\right\}+\left\{r^2V''(r)-rV'(r)\right\}\right.$$
$$\left.\left(\frac{1}{3}-\frac{\sin(qr)}{qr}-\frac{2\cos(qr)}{(qr)^2}+\frac{2\sin(qr)}{(qr)^3}\right)\right], \tag{10}$$

and

$$\omega_T^2(q)=\left(\frac{4\pi\rho}{M}\right)\int_0^\infty dr\, g(r)\left[\left\{rV'(r)\left(1-\frac{\sin(qr)}{qr}\right)\right\}+\left\{r^2V''(r)-\right.\right.$$
$$\left.\left.rV'(r)\right\}\left(\frac{1}{3}+\frac{2\cos(qr)}{(qr)^2}+\frac{2\sin(qr)}{(qr)^3}\right)\right]. \tag{11}$$

Here M, ρ are the atomic mass and the number density of the glassy alloy while $V''(r)$ is the second derivative of the pair potential, respectively.

Recently Bhatia-Singh (BS) [12] was modified by Shukla and Campnaha [26], they introduced screening effects. Then, with the above assumptions and modification, the dispersion equations for an amorphous material can be written as [12, 26]

$$\rho\omega_L^2(q)=\frac{2N_C}{q^2}(\beta I_0+\delta I_2)+\frac{k_e k_{TF}^2 q^2\varepsilon(q)|G(qr_S)|^2}{q^2+k_{TF}^2\varepsilon(q)}, \tag{12}$$

and

$$\rho\omega_T^2(q)=\frac{2N_C}{q^2}\left(\beta I_0+\frac{1}{2}\delta(I_0-I_2)\right). \tag{13}$$

Other used constants in Eqs. (12) and (13) can be found in Refs. [12, 26]. Here M is the effective atomic mass, ρ is the number density and N_C is the coordination number of the glassy system, respectively. In the long wavelength limit of the frequency spectrum, both phonon frequencies viz. the longitudinal ω_L and transverse ω_T phonon frequencies are proportional to the wave vectors and obey the relationships [1-9, 11, 12, 24-26],

$$\omega_L \propto q \text{ and } \omega_T \propto q\,,\ \therefore\ \omega_L=\upsilon_L q \text{ and } \omega_T=\upsilon_T q\,. \tag{14}$$

where, υ_L and υ_T are the longitudinal and transverse sound velocities of the glassy alloys, respectively. Detail expressions of the long wavelength limit of the frequency spectrum are narrated in our earlier papers [1-9].The present study also includes isothermal bulk modulus B_T, modulus of rigidity G, Poisson's ratio σ, Young's modulus Y and Debye temperature θ_D from the elastic limit of the PDC. All the quantities are computed from the longitudinal and transverse sound velocities (υ_L and υ_T). The bulk modulus B_T, modulus of rigidity G, Poisson's ratio σ, Young's modulus Y and Debye temperature θ_D are obtained using the expressions [1-9],

$$B_T = \rho_M\left(\upsilon_L^2 - \frac{4}{3}\upsilon_T^2\right), \tag{15}$$

$$G = \rho_M\,\upsilon_T^2 . \tag{16}$$

With ρ_M is the isotropic number density of the solid.

$$\sigma = \frac{1-2\left(\dfrac{\upsilon_T^2}{\upsilon_L^2}\right)}{2-2\left(\dfrac{\upsilon_T^2}{\upsilon_L^2}\right)}, \tag{17}$$

$$Y = 2G(\sigma+1), \tag{18}$$

and

$$\theta_D = \frac{\hbar\,\omega_D}{k_B} = \frac{\hbar}{k_B}2\pi\left[\frac{9\rho}{4\pi}\right]^{1/3}\left[\frac{1}{\upsilon_L^3}+\frac{2}{\upsilon_T^3}\right]^{(-1/3)} . \tag{19}$$

The low temperature specific heat C_V can be calculated from the following expressions [29],

$$C_V = \frac{\Omega_O\hbar^2}{k_B T^2}\sum_{\lambda=L,T}\int\frac{d^3q}{(2\pi)^3}\frac{\omega_\lambda^2(q)}{\left[\exp\left(\dfrac{\hbar\omega_\lambda(q)}{k_B T}\right)-1\right]\left[1-\exp\left(-\dfrac{\hbar\omega_\lambda(q)}{k_B T}\right)\right]} . \tag{20}$$

Here, $\hbar$, k_B ω_D are the Plank's constant, Boltzmann's constant and Debye frequency, respectively.

3. Results and Discussion

The input parameters and other related constants used in the present computations are tabulated in table 1, which are taken from the literature [27].

The present pair potentials $V(r)$ of $Fe_{90}Zr_{10}$ glass alongwith those of Bhandari *et al.* [10] are shown in Figure 1(a). It is apparent from the Figure that the position of the first minimum of the pair potential is greatly affected by the different types of screening functions. The first zero for pair potential $V(r = r_0)$ due to all screening functions occurs at $r_0 \approx 2.9$ au. The results obtained from S-screening give maximum well depth. Furthermore, well depth of presently computed pair potentials moves towards the left and give deep potential as compared to Bhandari *et al.* [10].

Table 1. Input parameters and other constants

Glasses	Z	Ω_O (au)	r_C (au)	Z_d	N_C	r_d (au)
$Fe_{90}Zr_{10}$	3.10	85.39	0.6565	6.10	8.40	1.63
$Fe_{80}B_{20}$	3.00	72.46	0.6363	5.40	8.20	1.41
$Fe_{83}B_{17}$	3.00	73.45	0.6389	5.57	8.17	1.42
$Fe_{80}P_{20}$	3.40	91.88	0.6315	5.40	8.20	1.41

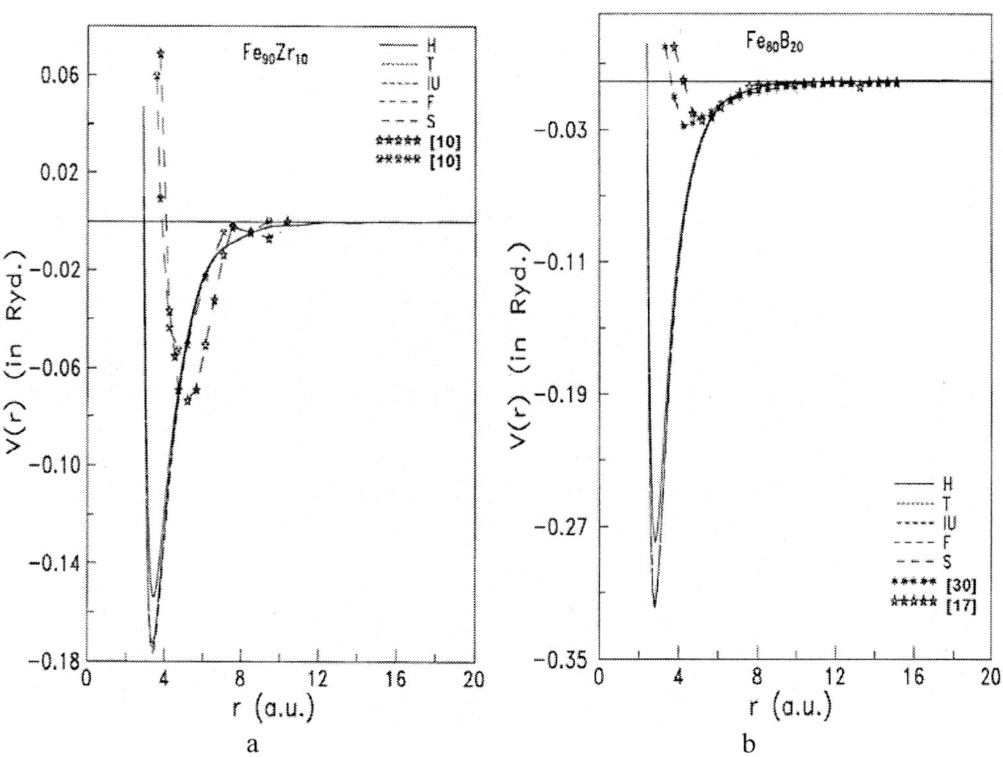

Figure 1 Continued

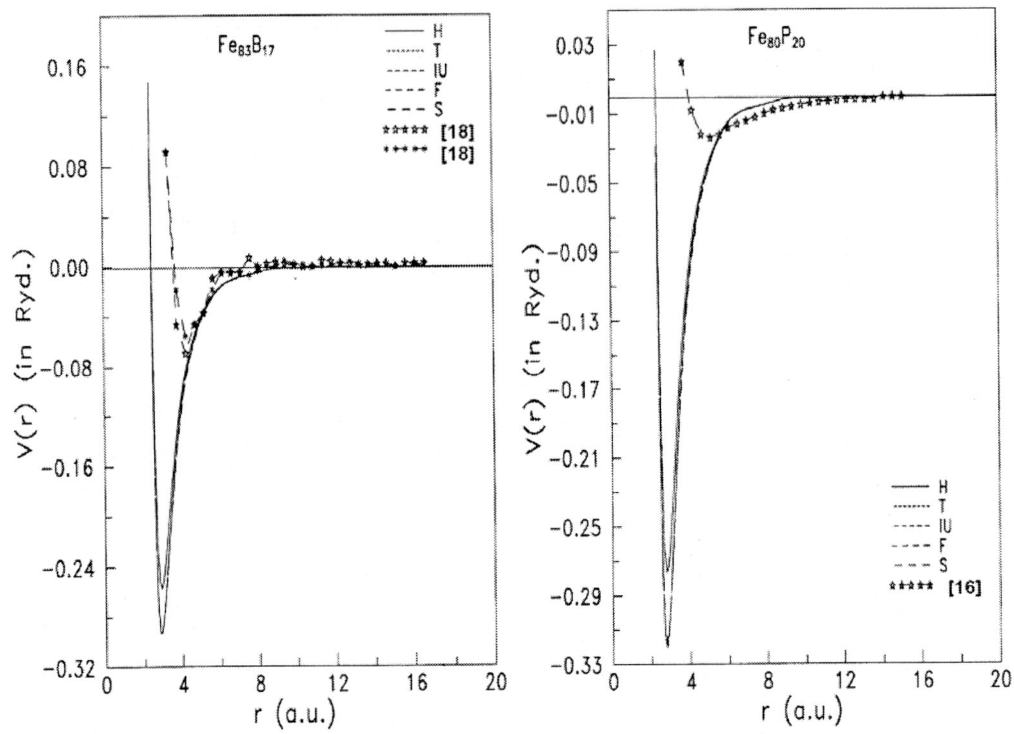

c d

Figure 1. Pair potentials for (a) $Fe_{90}Zr_{10}$, (b) $Fe_{80}B_{20}$, (c) $Fe_{83}B_{17}$ and (d) $Fe_{80}P_{20}$ metallic glasses.

The presently computed pair potentials $V(r)$ of $Fe_{80}B_{20}$ glass are shown in Figure 1(b) with other theoretical data [17, 30]. The first zero for pair potential $V(r = r_0)$ due to H-sceening occurs at $r_0 = 2.4$ au, while the exchange and correlation effect suppresses this zero to $r_0 \leq 2.3$ au. It is noticed that the position of the well depth of computed pair potentials $V(r)$ moves towards the left as compared to that of Gupta *et al.* [17] and Hausleitner-Turek [30] which show higher results than the computed one.

The present yielding is not showing any oscillatory behaviour. All pair potentials $V(r)$ converge towards the zero at large r-region without any oscillatory behaviour. The pair potentials $V(r)$ of binary $Fe_{83}B_{17}$ metallic glass are shown in Figure 1(c).

It is seen that the effect of exchange and correlation functions does not change nature of the pair potential. The first zero position of pair potential at $r = r_0$ due to all screening functions occurs at $r_0 \approx 3.7$ au. Thus, the effect of inclusion of exchange and correlation on the pair potential $V(r = r_0)$ is not seen substantially. It is observed that well depth of computed pair potentials moves towards the left and also high as compare to that of Saxena [18]. The computed pair potentials $V(r)$ of $Fe_{80}P_{20}$ glass are shown in Figure 1(d). The first zero for pair potential $V(r = r_0)$ due to all the screening functions occurs at $r_0 \approx 3.8$ au. The maximum depth in the pair potential is obtained for S-function. It is observed that well depth of presently computed potentials moves towards the left as compared to that of Hausleitner and Hafner [16], whose results are very shallow. The present yielding as well as those of Hausleitner and Hafner [16] do not show any significant oscillations and almost convergent in the large r-region.

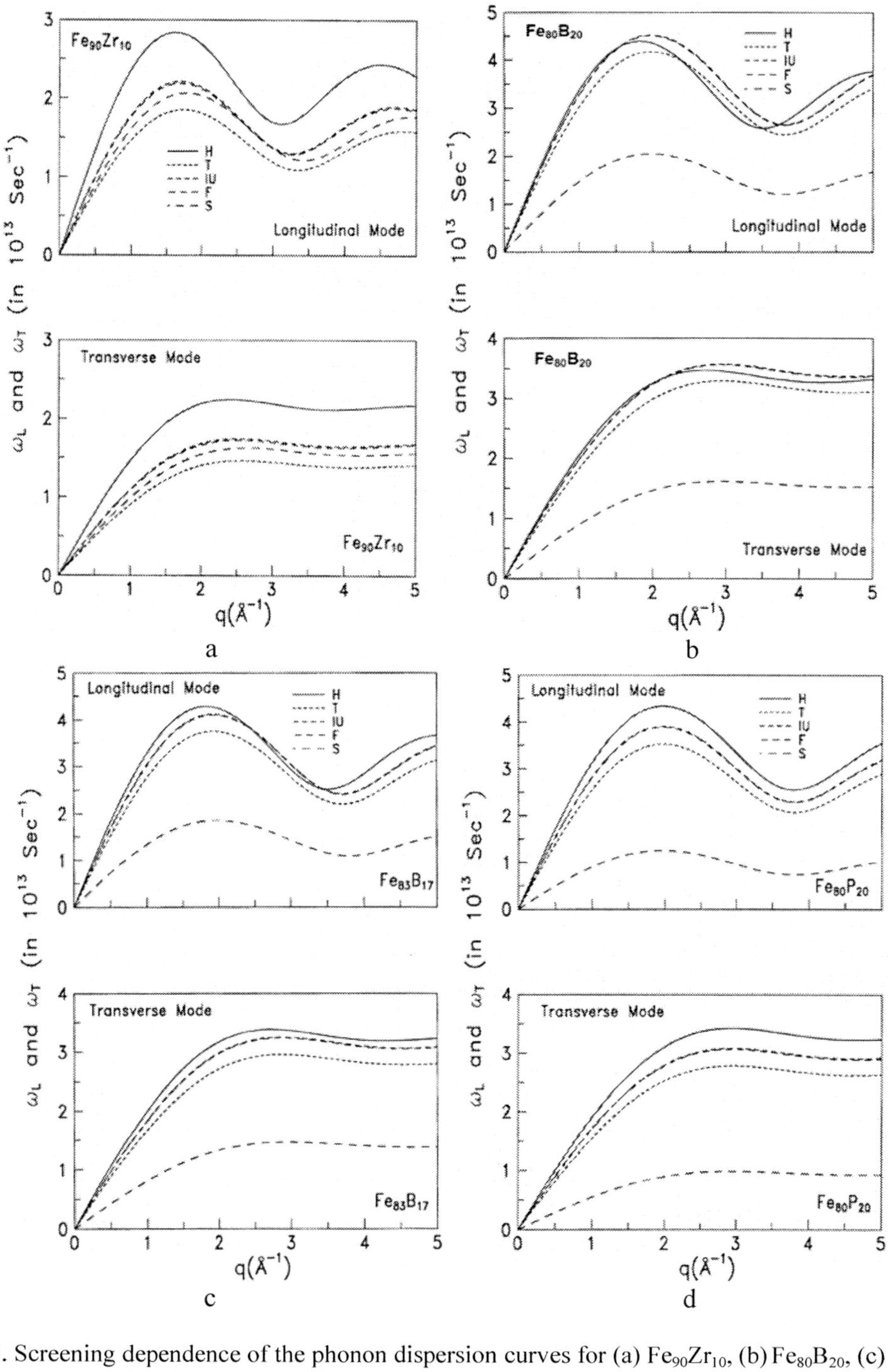

Figure 2. Screening dependence of the phonon dispersion curves for (a) $Fe_{90}Zr_{10}$, (b) $Fe_{80}B_{20}$, (c) $Fe_{83}B_{17}$ and (d) $Fe_{80}P_{20}$ metallic glasses using HB approach.

From the Figure 1(a)-(d), it is observed that the shifting of the pair pot pair potentials $V(r)$ with respect to atomic volume Ω_O of the amorphous alloys. The repulsive coulomb interaction and the attractive interactions represented by oscillatory nature are observed. It is also noticed that when volume Ω_O of the glassy alloys decreases (Ω_O is less for $Fe_{80}B_{20}$ glass), the potential depth deepens. It means that the pair potentials $V(r)$ for $Fe_{80}B_{20}$ glass show higher depth in comparison with other metallic glasses. All the pair potentials $V(r)$ show the combined effect of the s- and d-electrons. Bretonnet and Derouiche [24] observed that the repulsive part of $V(r)$ is drawn lower and its attractive part is deeper due to the d-electron effect. When we go from $Fe_{83}B_{17} \rightarrow Fe_{90}Zr_{10}$, the net number of d-electron r_d decreases, hence the pair potentials $V(r)$ are shifted towards the lower r-values. Therefore, the present results support the d-electron effect as noted by Bretonnet and Derouiche [31]. Also, from the Figure 1(a)-(d), it can be noted that the Coulomb repulsive potential part dominates the oscillations due to ion-electron-ion interactions, which shows the waving shape oscillation of the potential after $r \approx 10$ au. Hence, the pair potentials are converged towards a finite value instead of zero in repulsive region. The phonon modes for longitudinal and transverse branches of $Fe_{90}Zr_{10}$, $Fe_{80}B_{20}$, $Fe_{83}B_{17}$ and $Fe_{80}P_{20}$ metallic glasses calculated using HB approach with the five screening functions are shown in Figure 2(a)-(d). The PDC of $Fe_{90}Zr_{10}$ glass obtained using HB approach with all five screening functions is displayed in Figure 2(a). The behaviour of both the phonon modes have been suppressed after the inclusion of the screening effects. The first minimum in the longitudinal mode for H, T, IU, F and S-local field correction functions is around at $q \approx 3.1$Å^{-1}, 3.3Å^{-1}, 3.3Å^{-1}, 3.3Å^{-1} and 3.4Å^{-1}, respectively.

The influence of various screening on longitudinal frequency with respect to H-screening is 34.87% for T, 22.17% for IU, 23.02% for F and 27.41% for S-screening. While such influence on ω_T at $q \approx 1.0$Å^{-1} position is 38.40% for T, 15.94% for IU, 25.82% for F and 32.74% for S-screening function. In the present study the phonon eigen frequencies for longitudinal and transverse phonon modes of $Fe_{80}B_{20}$ glass calculated using HB approach with all five screening functions are shown in Figure 2(b).

It is evident that the influence of exchange and correlation effects suppresses both phonon frequencies for low wave vector region. The first minimum occurs in the longitudinal branch at $q \approx 3.5$Å^{-1}, 3.8Å^{-1}, 3.8Å^{-1}, 3.8Å^{-1} and 3.9Å^{-1} for H, T, IU, F and S-local field correction functions, respectively. The influence of various screening functions at first maximum of ω_L with respect to H-screening is 4.94% for T, 3.00% for IU, 2.76% for F and 53.55% for S-screening. Such screening influence observed at $q \approx 1.0$Å^{-1} in transverse branch for T, IU, F and S-function is 11.20%, 3.63%, 4.00% and 56.41% in comparison to static H-dielectric function, respectively. The phonon eigen frequencies of $Fe_{83}B_{17}$ glass calculated using HB approach with the five screening functions are displayed in Figure 2(c) which help us to study screening influence on the aforesaid properties. The first depth in the longitudinal branch is around at 1.7Å^{-1} for H, 2.6Å^{-1} for T as well as IU, 2.5Å^{-1} for F and 1.7Å^{-1} for S-local field correction function. The screening influence at first maximum of ω_L for T is 9.07%, for IU is 0.00%, for F is 0.44% and for S-screening is 55.43% with respect to H-screening which does not include any exchange and correlation effects. Such influence on ω_T at $q \approx 1.0Å^{-1}$ position for T is 16.27%, for IU is 7.92%, for F is 8.32% and for S-screening is 59.53%, respectively. The PDC of $Fe_{80}P_{20}$ glass is not reported by others till our literature survey. To show the screening influence on the eigen frequencies Figure 2(d) is drawn where the results of HB approach are used. It is apparent from the Figure that the influence of exchange and correlation effects suppresses the phonon frequencies in both longitudinal as well as transverse branches. The first minimum in the longitudinal branch for H, T, IU, F and S-local field correction functions is around at $q \approx 3.8Å^{-1}$, $3.8Å^{-1}$, $3.8Å^{-1}$, $3.8Å^{-1}$ and $3.8Å^{-1}$, respectively. As compared to previously reported metallic glasses, the influence of T, IU and F on ω_L is small. Such influence is ranging from 9.94% to 18.05%. The S-function enhances the ω_L of $Fe_{80}P_{20}$ glass by $\approx$ 71.74% in comparison of H-dielectric function. While the influence of screening at $q \approx 1.0Å^{-1}$ point on transverse mode is 18.65% for T, 9.94% for IU, 10.37% for F and 71.14% for S-screening with respect to H-screening.

The PDC obtained from three approaches viz. HB, TG and BS with S-function of $Fe_{90}Zr_{10}$ glass are plotted in Figure 3(a) and show the characteristic behaviour of dispersion relation in a disordered system. The present outcome due to HB and TG approaches are low in comparison with the reported data [10]. The first deepness in the longitudinal branch is observed at $q \approx 1.9Å^{-1}$ for HB, $q \approx 3.4Å^{-1}$ for TG and $q \approx 1.9Å^{-1}$ for BS approach. The first crossing position of ω_L and ω_T in the HB, TG and BS approaches is found, respectively, at $2.6Å^{-1}$, $1.2Å^{-1}$ and $1.1Å^{-1}$.

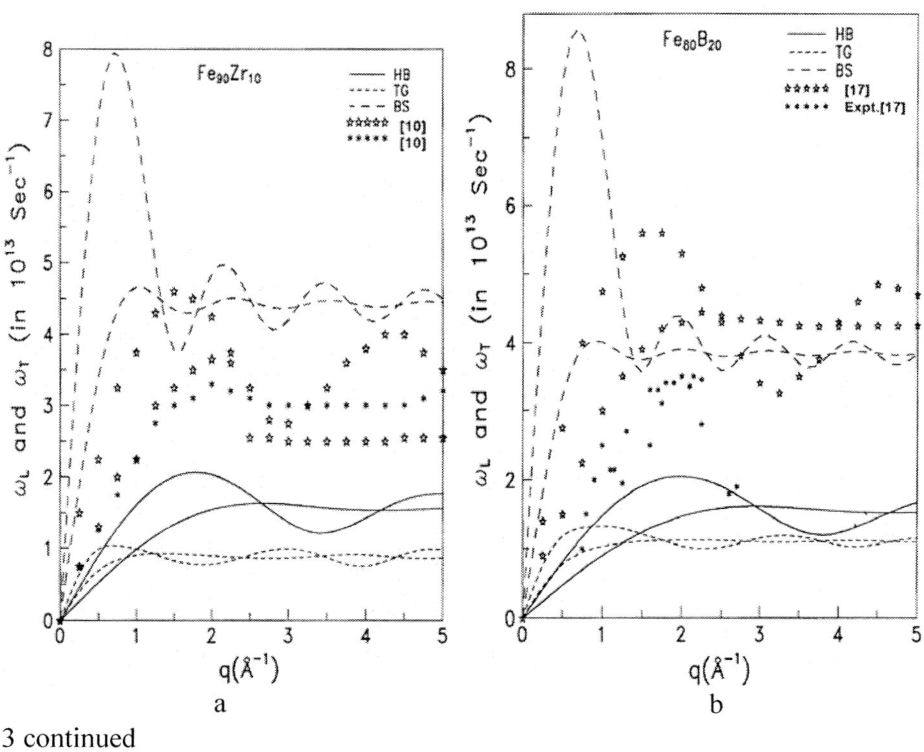

Figure 3 continued

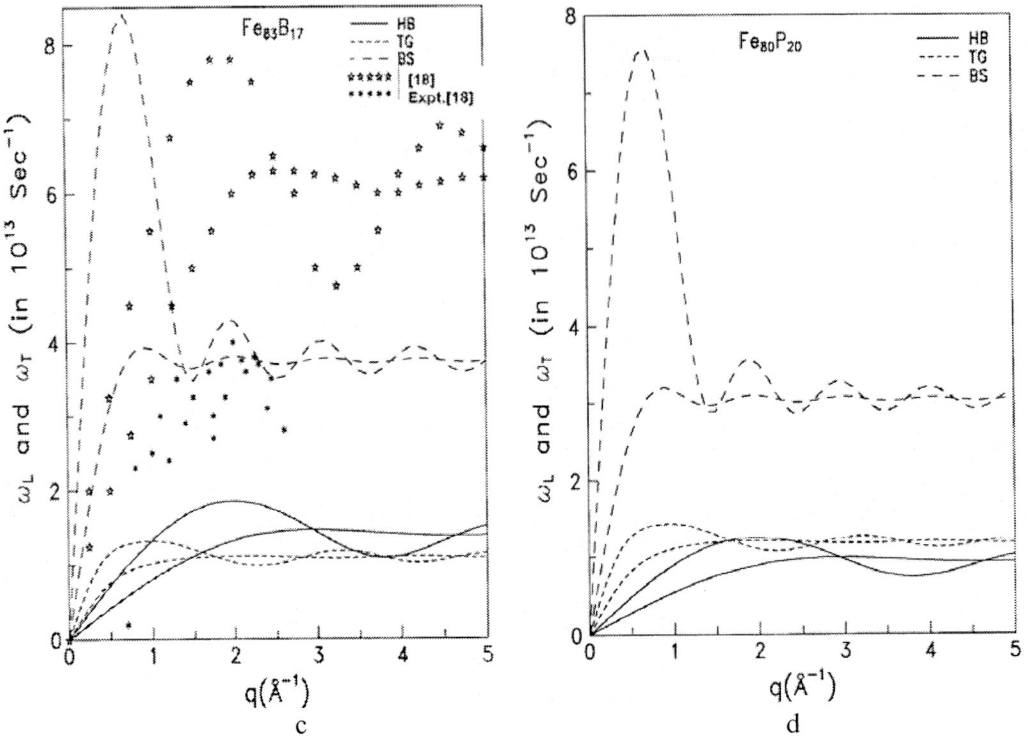

Figure 3. Phonon dispersion curves for a) $Fe_{90}Zr_{10}$, (b) $Fe_{80}B_{20}$, (c) $Fe_{83}B_{17}$ and (d) $Fe_{80}P_{20}$ metallic glasses using HB, TG and BS approaches with S-local field correction function.

The PDC obtained due to three approaches of $Fe_{80}B_{20}$ glass are shown in Figure 3(b). The first depth in the longitudinal branch falls at $q \approx 3.9Å^{-1}$ for HB, $q \approx 2.3Å^{-1}$ for TG and $q \approx 1.5Å^{-1}$ for BS approach. The first crossing position of ω_L and ω_T in the HB, TG and BS approaches is observed, respectively, at $2.9Å^{-1}$, $1.7Å^{-1}$ and $1.4Å^{-1}$. The results due to BS approach are very high in contrast to TG and HB approaches. The present yielding is found in line with other theoretical and experimental data [17]. The PDC obtained from three approaches viz. HB, TG and BS with S-function of $Fe_{83}B_{17}$ glass are apparent in Figure 3(c) and show the characteristic behaviour of dispersion relation in a disordered system. The present outcome due to HB and TG approaches are very low in comparison with the reported theoretical and experimental data [18]. The first minimum in the longitudinal branch lies at $q \approx 3.8Å^{-1}$ for HB, $q \approx 2.3Å^{-1}$ for TG and $q \approx 1.5Å^{-1}$ for BS approach. The first intersect point of both phonon frequencies in the HB, TG and BS approaches is seen at $2.9Å^{-1}$, $1.8Å^{-1}$ and $1.4Å^{-1}$, respectively.

The PDC due to three approaches (HB, TG, BS) of $Fe_{80}P_{20}$ glass with S-local field correction function are shown in Figure 3(d). It is seen that the oscillations are prominent in the longitudinal phonon modes as compared to the transverse modes. The present outcomes of PDC due to BS approach are higher than those due to HB and TG approaches. The first minimum in the longitudinal branch occurs at $q \approx 3.8Å^{-1}, q \approx 2.2Å^{-1}$ and $q \approx 1.5Å^{-1}$, for HB, TG and BS approaches, respectively. The first crossover location of both phonon branches is seen at $2.9Å^{-1}$, $1.7Å^{-1}$ and $1.4Å^{-1}$, respectively, in the HB, TG and BS approaches.

It is observed from the Figure 3(a)-(d) that, the first peak position of longitudinal branch of $Fe_{80}B_{20}$ glass is higher while that of $Fe_{80}P_{20}$ glass is lower in comparison with other metallic glasses. The lower dip in the longitudinal branch justifies the correctness and stability of the pair potential. The same results are observed in transverse branch also. Moreover, it is observed from the Figure 4 that, the oscillations are more prominent in the longitudinal phonon modes as compared to the transverse modes. This shows the existence of collective excitations at larger momentum transfer due to longitudinal phonons only and the instability of the transverse phonons due to the anharmonicity of the atomic vibrations in the metallic systems. Also in the high wave vector region, damping of phonons dominates the transverse mode which is indicating the fluid characteristic of the glass i.e. transverse phonon behaviour is monotonic. Here in transverse branch, the frequencies increase with the wave vector q and then saturates at $\approx q = 2.0Å^{-1}$, which supports the well known Thorpe model [32] in which, it describes a glass like a solid containing finite liquid cluster. The transverse phonons are absorbed for frequencies larger than the smallest eigen frequencies of the largest cluster.

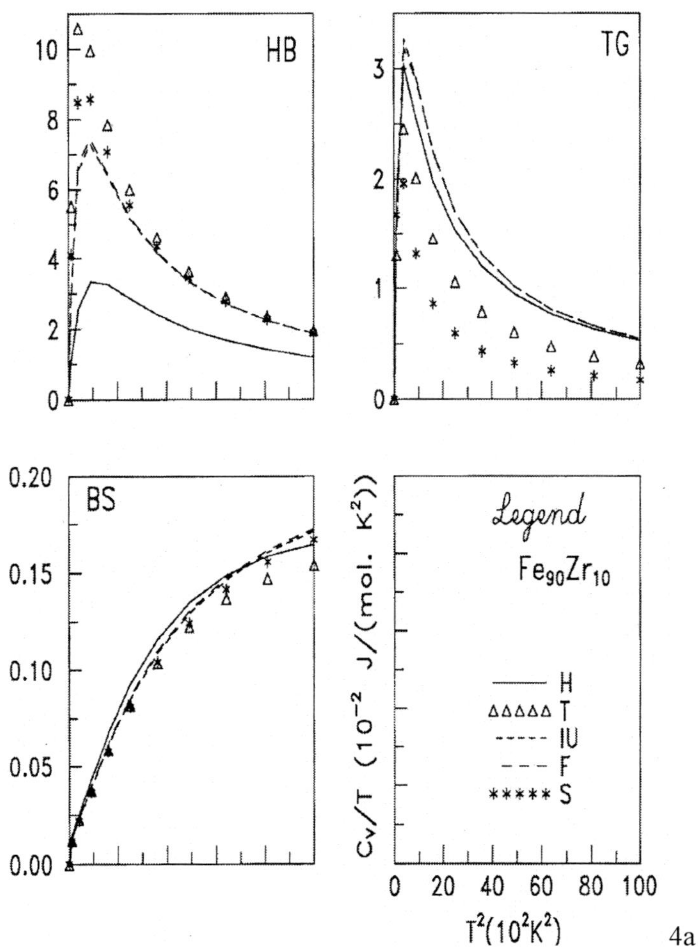

Figure 4 continued

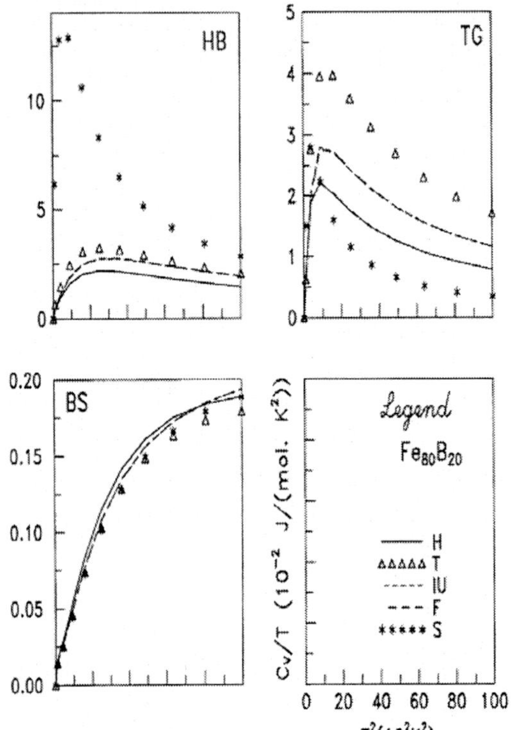

4b

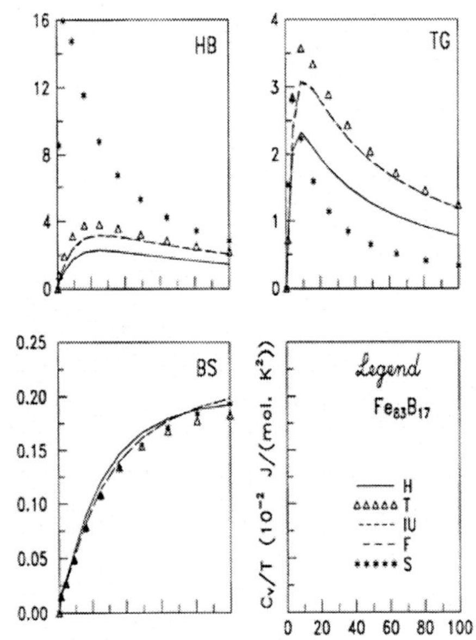

4c

Figure 4. Screening dependence of the low temperature specific heat for a) $Fe_{90}Zr_{10}$, (b) $Fe_{80}B_{20}$, (c) $Fe_{83}B_{17}$ and (d) $Fe_{80}P_{20}$ metallic glasses computed from HB, TG and BS approaches.

As shown in Figure 4(a)-(d) for $Fe_{90}Zr_{10}$, $Fe_{80}B_{20}$, $Fe_{83}B_{17}$ and $Fe_{80}P_{20}$ metallic glasses, the exchange and correlation functions also affected in the anomalous behaviour (i.e. deviation from the T^3 law), which is observed in the vibrational part of the low temperature specific heat C_V. The reason behind the anomalous behaviour may be due to the low frequency modes modify the generalized vibrational density of states of the glass with that of the polycrystal.

These modes are mainly responsible for the difference in the temperature dependence of the vibrational part of the specific heat which departs from the normal behaviour. The existence of a portion of the spectrum with 'softer phonons' (resembling rotons in liquid helium) may be cause of anomalous behaviour of low temperature specific heat C_V. In the low temperature region, a contribution to the low temperature specific heat C_V is made by phonons of the initial part of the frequency spectrum.

When the temperature reaches a value at which the energy of the thermal motions becomes comparable to the energy of 'softer phonons' minimum, an additional contribution appears to heat capacity from the roton portion of the phonon frequency $\omega(q)$. Here also initial rise in C_V/T is obtained for low temperature range in the HB and TG approaches and then fall to low values of C_V/T for further increase in T. Such a high fluctuation in $C_V/T \to T^2$ curve at low temperature is absent in the BS approach.

Table 2. Thermodynamic and Elastic Properties of $Fe_{90}Zr_{10}$Glass

App.	SCR	v_L x 10^5 cm/sec	v_T x 10^5 cm/sec	B_T x 10^{11} dyne/cm^2	G x 10^{11} dyne/cm^2	σ	Y x 10^{11} dyne/cm^2	θ_D (K)
HB	H	2.8073	1.6208	3.4113	2.0468	0.2499	5.1170	230.01
	T	1.7072	0.9856	1.2616	0.7569	0.2500	1.8924	139.87
	IU	2.0871	1.2050	1.8857	1.1314	0.2499	2.8285	171.00
	F	2.0645	1.1920	1.8450	1.1070	0.2499	2.7675	169.15
	S	1.8562	1.0717	1.4915	0.8949	0.2499	2.2373	152.09
TG	H	3.7509	1.6695	8.0669	2.1717	0.3765	5.9785	240.80
	T	3.3855	1.9186	5.1063	2.8682	0.2634	7.2476	272.71
	IU	3.5793	1.8849	6.2908	2.7684	0.3081	7.2428	269.41
	F	3.6178	1.9227	6.3573	2.8805	0.3032	7.5076	274.63
	S	2.8751	1.6987	3.4428	2.2484	0.2318	5.5394	240.56
BS	H	8.7030	3.7767	44.1979	11.1135	0.3840	30.7620	545.29
	T	8.8698	3.9556	45.0447	12.1914	0.3759	33.5475	570.48
	IU	8.8507	3.9295	44.9946	12.0312	0.3772	33.1399	566.83
	F	8.8301	3.9118	44.8551	11.9231	0.3779	32.8580	564.33
	S	8.9135	4.0007	45.2779	12.4709	0.3739	34.2667	576.83
Expt. [10]		–	–	–	–	–	–	16.66
Others [10]			4.8 6.2	2.81 2.35	9.60 23.8	6.05 4.23	0.23 0.41	15.01 12.00

Furthermore, the thermodynamic and elastic properties of Fe-based glassy alloys estimated from the elastic limit of the PDC are tabulated in Tables 2-5. From the Table2, it is noted that the present outcomes of these properties of $Fe_{90}Zr_{10}$ glass are found on the line with other such reported data [10].

It is seen from the Table 3 that the screening theory plays an important role in the prediction of the thermodynamic and elastic properties of $Fe_{80}B_{20}$ glass. As phonon modes obtained for BS approach is very high compared to HB and TG approaches, the thermodynamic and elastic properties obtained for BS approach is also higher in magnitude compared to HB and TG approaches. The present results are found in qualitative agreement with the available experimental [17] or theoretical [17] data.

The thermodynamic and elastic properties are calculated from the elastic part of the PDC and narrated in Table 4 for $Fe_{83}B_{17}$ glass. It is noticed that the properties are very sensitive to screening functions used. The υ_L and υ_T due to S-screening show low values in HB and TG approaches while high values in BS approach as compared to other screening.

Table 3. Thermodynamic and Elastic Properties of $Fe_{80}B_{20}$ Glass

App.	SCR	v_L x 10^5 cm/sec	v_T x 10^5 cm/sec	B_T x 10^{11} dyne/cm^2	G x 10^{11} dyne/cm^2	σ	Y x 10^{11} dyne/cm^2	θ_D (K)
HB	H	3.8570	2.2268	5.9849	3.5909	0.2500	8.9773	333.79
	T	3.3861	1.9550	4.6128	2.7677	0.2499	6.9192	293.04
	IU	3.6747	2.1216	5.4324	3.2594	0.2499	8.1486	318.01
	F	3.6605	2.1134	5.3905	3.2343	0.2499	8.0857	316.78
	S	1.6619	0.9595	1.1111	0.6666	0.2499	1.6666	143.82
TG	H	4.8530	2.0830	12.8652	3.1420	0.3871	8.7164	317.80
	T	4.6937	2.2569	11.0351	3.6885	0.3496	9.9563	342.58
	IU	4.9114	2.2747	12.4718	3.7470	0.3635	10.2178	345.93
	F	4.9432	2.3088	12.5474	3.8602	0.3605	10.5035	350.98
	S	3.2967	1.8910	4.4175	2.5895	0.2548	6.4987	283.61
BS	H	9.4947	3.5570	53.0644	9.1621	0.4184	25.9906	545.01
	T	9.6119	3.7067	53.6360	9.9496	0.4127	28.1106	567.51
	IU	9.5753	3.6666	53.4138	9.7351	0.4141	27.5327	561.47
	F	9.5725	3.6662	53.3784	9.7331	0.4141	27.5263	561.41
	S	9.6453	3.7405	53.8594	10.1317	0.4115	28.6016	572.59
Expt. [17]			4.773	–	1.41	0.65	0.3	1.69
Others [17]			5.15	3.05	1.04 1.38	0.688 0.50	0.229	1.693

Moreover, both the velocities give the fair results due to BS approach compare to others [18] while B_T gives good agreement in the TG approach compare to experimental and others results [18]. Also BS approach provides good results for Y as compared to experimental and others data [18]. The thermodynamic and elastic properties of $Fe_{80}P_{20}$ glass are also computed and reported in Table 5. Here for BS approach, the results obtained for υ_L, υ_T, B_T, G, σ, Y and θ_D are very high compared to TG and HB approaches. The screening influence on PDC is reflected on these properties, also.

Table 4. Thermodynamic and Elastic Properties of $Fe_{83}B_{17}$ Glass

App.	SCR	v_L x 10^5 cm/sec	v_T x 10^5 cm/sec	B_T x 10^{11}dyne/cm^2	G x 10^{11}dyne/cm^2	σ	Y x 10^{11} dyne/cm^2	θ_D (K)
HB	H	3.7563	2.1687	5.7613	3.4568	0.2500	8.6419	323.60
	T	3.1207	1.8018	3.9765	2.3859	0.2499	5.9648	268.85
	IU	3.4322	1.9816	4.8099	2.8860	0.2499	7.2149	295.68
	F	3.4170	1.9728	4.7674	2.8604	0.2500	7.1511	294.37
	S	1.5026	0.8675	0.9219	0.5532	0.2500	1.3829	129.45
TG	H	4.7506	2.0302	12.5478	3.0293	0.3883	8.4110	308.39
	T	4.4761	2.2065	9.9540	3.5783	0.3395	9.5861	332.96
	IU	4.6931	2.2194	11.3607	3.6201	0.3560	9.8176	335.65
	F	4.7259	2.2538	11.4368	3.7332	0.3528	10.1006	340.70
	S	3.2662	1.8710	4.4105	2.5727	0.2558	6.4617	279.37
BS	H	9.3156	3.4901	51.8441	8.9522	0.4184	25.3949	532.34
	T	9.4301	3.6364	52.3994	9.7186	0.4127	27.4582	554.23
	IU	9.3938	3.5965	52.1792	9.5065	0.4141	26.8867	548.26
	F	9.3911	3.5962	52.1445	9.5048	0.4141	26.8811	548.20
	S	9.4631	3.6696	52.6193	9.8971	0.4115	27.9395	559.21
Expt. [18]		–	–	17.0	–	–	20.6	555 ± 90
Others [18]		6.0	4.0	10.2	–	–	24.447	599

The dielectric function plays an important role in the evaluation of potential due to the screening of the electron gas. For this purpose in the present investigation, the local filed correction functions due to H, T, IU, F and S are used. Reason for selecting these functions is that H-function does not include exchange and correlation effect and represents only static dielectric function, while T-function cover the overall features of the various local field correction functions proposed before 1972. IU, F and S-functions are recent one among the existing functions and not exploited rigorously in such study. This helps us to study the relative effects of exchange and correlation in the aforesaid properties. Hence, the five different local field correction functions show variations up to an order of magnitude in all the properties.

In all three approaches, it is very difficult to judge which approach is the best for computations of vibrational dynamics of Fe-based binary metallic glass, because each approach has own identity. The HB approach is simplest and older one theory, which generating consistent results of the phonon data of the glassy systems, because this approach needs minimum number of input parameters. While TG approach is developed upon the quasi-crystalline approximation in which effective force constant depends upon the correlation function for the displacement of atoms and correlation function of displacement itself depends on the phonon frequencies. The BS approach is retained the interatomic interactions of the atoms, effective between the first nearest neighbours only hence, the disorder of the atoms in the formation of metallic glasses is more which show deviation in magnitude of the PDC and their related vibrational properties. From the present study we are concluded that, all three approaches are suitable for studying the vibrational dynamics of the amorphous materials. Hence, successful application of the model potential with three approaches is observed from the present study.

Table 5. Thermodynamic and Elastic Properties of $Fe_{80}P_{20}$ Glass

App.	SCR	v_L x 10^5 cm/sec	v_T x 10^5 cm/sec	B_T x 10^{11} dyne/cm^2	G x 10^{11} dyne/cm^2	σ	Y x 10^{11} dyne/cm^2	θ_D (K)
	H	3.5222	2.0335	4.2746	2.5648	0.2500	6.4119	281.62
	T	2.8654	1.6543	2.8290	1.6974	0.2500	4.2434	229.10
HB	IU	3.1721	1.8314	3.4670	2.0802	0.2500	5.2005	253.62
	F	3.1569	1.8227	3.4340	2.0604	0.2499	5.1510	252.41
	S	1.0164	0.5868	0.3560	0.2136	0.2500	0.5340	81.27
	H	4.7945	2.1277	10.5135	2.8078	0.3774	7.7348	299.52
	T	4.5387	2.3845	8.0744	3.5265	0.3094	9.2351	332.64
TG	IU	4.7384	2.3882	9.2092	3.5372	0.3297	9.4073	334.03
	F	4.7800	2.4282	9.2950	3.6568	0.3261	9.6986	339.46
	S	3.5755	2.0416	4.4819	2.5852	0.2581	6.5048	283.00
	H	9.0302	3.0378	42.9431	5.7234	0.4362	16.4399	431.05
	T	9.1310	3.1774	43.3614	6.2616	0.4311	17.9222	450.57
BS	IU	9.0970	3.1397	43.1741	6.1138	0.4324	17.5148	445.29
	F	9.1036	3.1492	43.1989	6.1509	0.4320	17.6165	446.62
	S	9.1448	3.1909	43.4469	6.3148	0.4307	18.0690	452.45

Conclusions

Lastly it is concluded that, in the study of phonon dynamics of metallic glasses, the pair potentials and its derivatives as well as pair correlation function plays an important role. In the present computation, the WH form is adopted to generate the pair potentials, which ignores the angular interaction due to partially filled d-bands in transition metals. Most recent model potential with WH model and HB approach produces consistent results of phonon dynamics for all metallic glasses. Hence, the present model potential is suitable for studying the phonon dynamics of four Fe-based metallic glasses, which confirms applicability of the model potential in the aforementioned study. Such study on phonon dynamics of other binary as well as ternary liquid alloys and metallic glasses is in progress, which will be communicated in near future.

References

[1] A. M. Vora, M. H. Patel, P. N. Gajjar and A. R. Jani, *Solid State Phys.* 46 (2003) 315.
[2] P. N. Gajjar, A. M. Vora and A. R. Jani, in: *Proceedings of the 9th Asia Pacific Physics Conference Hanoi*, The Gioi Publication, Vietnam (2006) 429.
[3] Aditya M. Vora, *Chinese Phys. Lett.* 23 (2006) 1872.
[4] Aditya M. Vora, *J. Mater. Sci.* 43 (2007) 935.
[5] Aditya M. Vora, *Acta Phys. Pol.* A111 (2007) 859.
[6] Aditya M. Vora, *J. Non-Cryst. Sol.* 352 (2006) 3217.
[7] Aditya M. Vora, *Front. Mater. Sci. China* 1 (2007) 366.
[8] Aditya M. Vora, *FIZIKA A*16 (2007) 187.

[9] Aditya M. Vora, Rom. *J. Phys*. 3-4 (2008) – in press.
[10] D. Bhandari, A. Prasad, A. Gupta and N. S. Saxena, in *The Physics of Disordered Materials*, M. P. Saksena, N. S. Saxena, D. Bhandari (Eds.), NISCOM, New Delhi, 1997, pp. 86-90.
[11] J. Hubbard and J. L. Beeby, *J. Phys. C: Solid State Phys*. 2 (1969) 556-571.
[12] A. B. Bhatia and R. N. Singh, *Phys. Rev* B31 (1985) 4751.
[13] Ch. Hausleitner and J. Hafner, *J. Non-Cryst. Sol*. 144 (1992) 175.
[14] A. R. Wildes, J. R. Stewart, N. Cowlam, S. Al-Heniti, L. F. Kiss and T. Kemény, *J. Phys.: Condens. Matter* 15 (2003) 675.
[15] J. Hafner, M. Tegze and Ch. Becker, *Phys. Rev*. B49 (1994) 285.
[16] Ch. Hausleitner and J. Hafner, *Phys. Rev*. B47 (1993) 5689.
[17] A. Gupta, D. Bhandari and N. S. Saxena, *Ind. J. Pure* and *Appl. Phys*. 36 (1998) 366.
[18] N. S. Saxena, *J. Non-Cryst. Sol*. 196 (1996) 37.
[19] W. A. Harrison, Elementary Electronic Structure, World Scientific, Singapore (1999).
[20] R. Taylor, *J. Phys. F: Met. Phys*. 8 (1978) 1699.
[21] S. Ichimaru and K. Utsumi, *Phys. Rev*. B24 (1981) 7385.
[22] B. Farid, V. Heine, G. Engel and I. J. Robertson, *Phys. Rev*. B48 (1993) 11602.
[23] A. Sarkar, D. S. Sen, S. Haldar and D. Roy, *Mod. Phys. Lett*. B12 (1998) 639.
[24] S. Takeno and M. Goda, *Prog. Thero. Phys*. 45 (1971) 331.
[25] S. Takeno and M. Goda, *Prog. Thero. Phys*. 47 (1972) 790.
[26] M. M. Shukla and J. R. Campanha, *Acta Phys. Pol*. A94 (1998) 655.
[27] J. M. Wills and W. A. Harrison, *Phys. Rev*. B28 (1983) 4263.
[28] J. Hafner and V. Heine, J. Phys. F: *Met. Phys*. 13 (1983) 2479.
[29] N. P. Kovalenko and Y. P. Krasny, *Physica* B162 (1990) 115.
[30] Ch. Hausleitner and I. Turek, *J. Non-Cryst. Sol*. 156-158 (1993) 210.
[31] J. L. Bretonnet and A. Derouiche, *Phys. Rev*. B43 (1990) 8924.
[32] M. F. Thorpe, *J. Non-Cryst. Sol*. 57 (1983) 355.

In: New Trends in Chemical Physics Research
Editor: James N. Lapherty
ISBN: 978-1-61668-535-5

Chapter 15

VIBRATIONAL DYNAMICS OF ZR-BASED BULK METALLIC GLASSES

*Aditya M. Vora**
Parmeshwari 165, Vijaynagar Area, Hospital Road,
Bhuj–Kutch, 370 001, Gujarat, India

ABSTRACT

The phonon dynamics of some Zr-Ni bulk metallic glasses has been studied at room temperature in terms of phonon eigen frequencies of longitudinal and transverse modes employing three different approaches proposed by Hubbard-Beeby (HB), Takeno-Goda (TG) and Bhatia-Singh (BS). The well recognized model potential is employed successfully to explain electron-ion interaction in the metallic glass. The present findings of phonon dispersion curve are found in fair agreement with available theoretical as well as experimental data. The thermodynamic properties obtained by HB and TG approaches are found very lower than those obtained by BS approach.

Keywords: Pair potential, Zr-Ni bulk metallic glasses, phonon dispersion curves, thermal properties, elastic properties.

PACS:63.50. +x, 65.60. +a.

1. INTRODUCTION

The mankind has been manufacturing glassy materials for several thousands years. Compared to that, the scientific study of amorphous materials has a much shorter history. And only recently, there has been an explosion of interest to these studies as more promising

* Tel. : +91-2832-256424, E-mail address : voraam@yahoo.com.

materials are produced in the amorphous form. The range of applications of metallic glasses is vast and extends from the common window glass to high capacity storage media for digital devices [1-17].

The theoretical calculations of phonon dispersion curves (PDC) of $Zr_{67}Ni_{33}$ glass were studied by Gupta *et al.* [11] for the first time. They adopted Hubbard-Beeby (HB) [18] and Bhatia-Singh (BS) [19] approaches in their study and compared the results with MD outcomes of Aihara *et al.* [14] and found their results are in good agreement with reported experimental data of Aihara *et al.* [14] with different temperatures. Experimentally, PDC of series of Zr-Ni amorphous alloys has been extensively studied by Aihara *et al.* [15]. Also, Otomo *et al.* [16] have studied the dynamical structure factor as well as collective excitations of this glass using neutron-in-elastic-scattering (NIS) technique. The vibrational dynamics of $Ni_{50}Zr_{50}$ metallic glass have been studied by Gupta *et al.* [12] using Takeno-Goda (TG) [20] and Bhatia-Singh (BS) [19] approaches. Recently, Lad and Arun Pratap [17] have studied the PDC of five Zr-Ni alloys with model potential formalism using Takeno-Goda (TG) [20] and Bhatia-Singh (BS) [19] approaches with Wills-Harrison (WH) form of pair potential [21]. Also, they have noted that, their results are found to be very close to those derived from MD and to the available experimental values [15]. They have fitted the parameter of the potential with the experimental data in most of the cases.

This article introduces pseudopotential based theory to address the problems of phonon dynamics of bulk metallic glasses and their related elastic as well as thermodynamic properties with the method of computing the properties under investigation. Three main theoretical approaches given by Hubbard-Beeby (HB) [18], Takeno-Goda (TG) [20] and Bhatia-Singh (BS) [19, 22] are used in the present study for computing the phonon frequencies of the bulk non-crystalline or glassy alloys. Five local field correction functions viz. Hartree (H) [23], Taylor (T) [24], Ichimaru-Utsumi (IU) [25], Farid *et al.* (F) [26] and Sarkar *et al.* (S) [27] are used for the first time in the present investigation to study the screening influence on the aforesaid properties of bulk metallic glasses. Besides, the thermodynamic properties such as longitudinal sound velocity υ_L, transverse sound velocity υ_T and Debye temperature θ_D, low temperature specific heat capacity C_V and some elastic properties viz. the isothermal bulk modulus B_T, modulus of rigidity G, Poisson's ratio σ and Young's modulus Y are also calculated from the elastic part of the phonon dispersion curves (PDC).

2. Theoretical Methodology

The fundamental ingredient, which goes into the calculation of the vibrational dynamics of bulk metallic glasses, is the pair potential. In the present study, for bulk metallic glasses, the pair potential is computed using Wills-Harrison (WH) form [1-12, 21],

$$V(r) = V_S(r) + V_b(r) + V_r(r). \qquad (1)$$

The s-electron contribution to the pair potential $V_S(r)$ is calculated from [1-10],

$$V_s(r) = \left(\frac{Z_S{}^2 e^2}{r}\right) + \frac{\Omega_O}{\pi^2} \int F(q) \left[\frac{Sin(qr)}{qr}\right] q^2 \, dq \, . \tag{2}$$

Here $Z_S \sim 1.5$ is found by integrating the partial s-density of states resulting from self-consistent band structure calculation for the entire 3d and 4d series [21], while Ω_O is the effective atomic volume of the one component fluid.

The energy wave number characteristics appearing in the equation (2) is written as [1-10]

$$F(q) = \frac{-\Omega_O \, q^2}{16\pi} |W_B(q)|^2 \frac{[\varepsilon_H(q)-1]}{\{1+[\varepsilon_H(q)-1][1-f(q)]\}} \, . \tag{3}$$

Here, $W_B(q)$ is the bare ion potential $\varepsilon_H(q)$ the modified Hartree dielectric function, which is written as [23]

$$\varepsilon(q) = 1 + (\varepsilon_H(q) - 1)(1 - f(q)) \, . \tag{4}$$

While, $\varepsilon_H(X)$ is the static Hartree dielectric function and the expression of it is given by [23],

$$\varepsilon_H(q) = 1 + \frac{m e^2}{2\pi k_F \hbar^2 \eta^2} \left(\frac{1-\eta^2}{2\eta} \ln \left| \frac{1+\eta}{1-\eta} \right| + 1 \right); \; \eta = \frac{q}{2k_F} \tag{5}$$

here $m, e, \hbar$ are the electronic mass, the electronic charge, the Plank's constant, respectively and $k_F = (3\pi^2 Z/\Omega_O)^{1/2}$ is the Fermi wave vector, in which Z the valence. While $f(q)$ is the local field correction function. In the present investigation, the local field correction functions due to Hartree (H) [23], Taylor (T) [24], Ichimaru-Utsumi (IU) [25], Farid *et al.* (F) [26] and Sarkar *et al.* (S) [27] are incorporated to see the impact of exchange and correlation effects on the aforesaid properties. The details of all local field correction functions are given below:

The Hartree (H) screening function [23] is purely static, and it does not include the exchange and correlation effects. The expression of it is,

$$f(q) = 0 \, . \tag{6}$$

Taylor (T) [24] has introduced an analytical expression for the local field correction function, which satisfies the compressibility sum rule exactly. This is the most commonly used local field correction function and covers the overall features of the various local field correction functions proposed before 1972. According to Taylor (T) [24],

$$f(q)=\frac{q^2}{4k_F^2}\left[1+\frac{0.1534}{\pi k_F^2}\right]. \tag{7}$$

The Ichimaru-Utsumi (IU) local field correction function [25] is a fitting formula for the dielectric screening function of the degenerate electron liquids at metallic and lower densities, which accurately reproduces the Monte-Carlo results as well as it also, satisfies the self consistency condition in the compressibility sum rule and short range correlations. The fitting formula is

$$f(q)=A_{IU}\left(\frac{q}{k_F}\right)^4+B_{IU}\left(\frac{q}{k_F}\right)^2+C_{IU}+\left[A_{IU}\left(\frac{q}{k_F}\right)^4+\left(B_{IU}+\frac{8A_{IU}}{3}\right)\left(\frac{q}{k_F}\right)^2-C_{IU}\right]\left\{\frac{4-\left(\frac{q}{k_F}\right)^2}{4\left(\frac{q}{k_F}\right)}\ln\left|\frac{2+\left(\frac{q}{k_F}\right)}{2-\left(\frac{q}{k_F}\right)}\right|\right\}. \tag{8}$$

On the basis of Ichimaru-Utsumi (IU) local field correction function [25] local field correction function, Farid *et al.* (F) [26] have given a local field correction function of the form

$$f(q)=A_F\left(\frac{q}{k_F}\right)^4+B_F\left(\frac{q}{k_F}\right)^2+C_F+\left[A_F\left(\frac{q}{k_F}\right)^4+D_F\left(\frac{q}{k_F}\right)^2-C_F\right]\left\{\frac{4-\left(\frac{q}{k_F}\right)^2}{4\left(\frac{q}{k_F}\right)}\ln\left|\frac{2+\left(\frac{q}{k_F}\right)}{2-\left(\frac{q}{k_F}\right)}\right|\right\}. \tag{9}$$

Based on equations (8-9), Sarkar *et al.* (S) [27] have proposed a simple form of local field correction function, which is of the form

$$f(q)=A_S\left\{1-\left(1+B_S\left(\frac{q}{k_F}\right)^4\right)\exp\left(-C_S\left(\frac{q}{k_F}\right)^2\right)\right\}. \tag{10}$$

The parameters A_{IU}, B_{IU}, C_{IU}, A_F, B_F, C_F, D_F, A_S, B_S and C_S are the atomic volume dependent parameters of IU-, F- and S-local field correction functions. The mathematical expressions of these parameters are narrated in the respective papers of the local field correction functions [25-27].

The well recognized model potential $W_B(r)$ [1-10] (in r-space) used in the present computation is of the form,

$$W(r) = \frac{-Z e^2}{r_C^3}\left[2-\exp\left(1-\frac{r}{r_C}\right)\right]r^2 ; \qquad r \le r_C$$
$$= \frac{-Z e^2}{r} \qquad ; \qquad r \ge r_C \tag{11}$$

This form has feature of a Coulombic term out side the core and varying cancellation due to a repulsive and an attractive contribution to the potential within the core. Hence it is assumed that the potential within the core should not be zero nor constant but it should very as a function of r. Thus, the model potential has the novel feature of representing varying cancellation of potential within the core over and above its continuity at $r=r_C$ and weak nature [1-10]. Here r_C is the parameter of the model potential of bulk metallic glasses. The model potential parameter r_C is calculated from the well known formula [28, 29] as follows :

$$r_C = \left[\frac{0.51\, r_S}{(Z)^{1/3}}\right]. \tag{12}$$

Here r_S is the Wigner-Seitz radius of the bulk metallic glasses.

The d-electron contributions to the pair potential are expressed in terms of the number of d-electron Z_d, the d-state radii r_d and the nearest-neighbour coordination number N as follows [1-12, 21]:

$$V_b(r) = -Z_d\left(1-\frac{Z_d}{10}\right)\left(\frac{12}{N}\right)^{1/2}\left(\frac{28.06}{\pi}\right)\frac{2\, r_d^{\,3}}{r^5}, \tag{13}$$

and

$$V_r(r) = Z_d\left(\frac{450}{\pi^2}\right)\frac{r_d^{\,6}}{r^8}. \tag{14}$$

Here, $V_b(r)$ takes into account the Friedel-model band broadening contribution to the transition metal cohesion and $V_r(r)$ arises from the repulsion of the d-electron muffin-tin orbital on different sites due to their non-orthogonality. Wills and Harrison (WH) [21] have studied the effect of the s- and d-bands.

The phonon theories of Hubbard-Beeby (HB) [18], Takeno-Goda (TG) [20] and Bhatia-Singh (BS) [19, 22] have been employed in the present computation. The expressions for longitudinal phonon frequency ω_L and transverse phonon frequency ω_T as per HB, TG and BS approaches are given below [12, 13, 18, 19].

According to the Hubbard-Beeby (HB) [18] theory, the expressions for longitudinal and transverse phonon frequencies are as follows,

$$\omega_L^2(q) = \omega_E^2 \left[1 - \frac{\sin(q\sigma)}{q\sigma} - \frac{6\cos(q\sigma)}{(q\sigma)^2} + \frac{6\sin(q\sigma)}{(q\sigma)^3}\right] \tag{15}$$

$$\omega_T^2(q) = \omega_E^2 \left[1 - \frac{3\cos(q\sigma)}{(q\sigma)^2} + \frac{3\sin(q\sigma)}{(q\sigma)^3}\right] \tag{16}$$

with $\omega_E^2 = \left(\frac{4\pi\rho}{3M}\right)\int_0^\infty g(r)V''(r)r^2 dr$ is the maximum frequency.

Following to Takeno-Goda (TG) [20] approach, the wave vector (q) dependent longitudinal and transverse phonon frequencies are written as

$$\omega_L^2(q) = \left(\frac{4\pi\rho}{M}\right)\int_0^\infty dr\, g(r)\left[\left\{rV'(r)\left(1 - \frac{\sin(qr)}{qr}\right)\right\} + \left\{r^2V''(r) - rV'(r)\right\}\right.$$
$$\left.\left(\frac{1}{3} - \frac{\sin(qr)}{qr} - \frac{2\cos(qr)}{(qr)^2} + \frac{2\sin(qr)}{(qr)^3}\right)\right], \tag{17}$$

$$\omega_T^2(q) = \left(\frac{4\pi\rho}{M}\right)\int_0^\infty dr\, g(r)\left[\left\{rV'(r)\left(1 - \frac{\sin(qr)}{qr}\right)\right\} + \left\{r^2V''(r) - \right.\right.$$
$$\left.\left. rV'(r)\right\}\left(\frac{1}{3} + \frac{2\cos(qr)}{(qr)^2} + \frac{2\sin(qr)}{(qr)^3}\right)\right]. \tag{18}$$

According to modified Bhatia-Singh (BS) [19] approach, the phonon frequencies of longitudinal and transverse branches are given by Shukla and Campanha [22],

$$\omega_L^2(q) = \frac{2N_C}{\rho q^2}(\beta I_0 + \delta I_2) + \frac{k_e k_{TF}^2 q^2 \varepsilon(q)|G(qr_S)|^2}{q^2 + k_{TF}^2 \varepsilon(q)} \tag{19}$$

$$\omega_T^2(q) = \frac{2N_C}{\rho q^2}\left(\beta I_0 + \frac{1}{2}\delta(I_0 - I_2)\right) \tag{20}$$

Other used constants in Eqs. (19) and (20) can be found in Refs. [19, 22]. Here M is the effective atomic mass, ρ is the number density and N_C is the effective coordination number of the glassy system, respectively. While $V'(r)$ and $V''(r)$ are the first and second derivative of the pair potential, respectively.

In the long-wavelength limit of the frequency spectrum, the both the frequencies i.e. transverse and longitudinal are proportional to the wave vectors and obey the relationships,

$$\omega_L \propto q \text{ and } \omega_T \propto q,$$
$$\therefore\ \omega_L = \upsilon_L q \text{ and } \omega_T = \upsilon_T q. \tag{21}$$

where υ_L and υ_T are the longitudinal and transverse sound velocities of the glassy alloys, respectively. The mathematical expressions of υ_L and υ_T are given in earlier papers [1-13, 18, 19]. In the long-wavelength limit of the frequency spectrum, transverse and longitudinal sound velocities υ_L and υ_T are computed. The isothermal bulk modulus B_T, modulus of rigidity G, Poisson's ratio σ, Young's modulus Y and the Debye temperature θ_D are found using the expressions [1-12, 18-20, 22],

$$B_T = \rho\left(\upsilon_L^2 - \frac{4}{3}\upsilon_T^2\right), \tag{22}$$

$$G = \rho\upsilon_T^2. \tag{23}$$

With ρ is the isotropic number density of the solid.

$$\sigma = \frac{1-2\left(\frac{\upsilon_T^2}{\upsilon_L^2}\right)}{2-2\left(\frac{\upsilon_T^2}{\upsilon_L^2}\right)}, \tag{24}$$

$$Y = 2G(\sigma+1), \tag{25}$$

$$\theta_D = \frac{\hbar\omega_D}{k_B} = \frac{\hbar}{k_B}2\pi\left[\frac{9\rho}{4\pi}\right]^{1/3}\left[\frac{1}{\upsilon_L^3}+\frac{2}{\upsilon_T^3}\right]^{(-1/3)}. \tag{26}$$

here ω_D is the Debye frequency.

Now a day it is firmly established that for all amorphous solids, the universal temperature behaviour of vibrational contributions to the heat capacity (C_V) differs essentially from that of crystal. In the thermodynamic limit ($N_O \to 0$, $\Omega_0 \to 0$ and $N_O/\Omega_0 =$ constant, with N_O is the number of atoms in the unit cell), one can obtain [30]

$$C_V = \frac{\Omega_0 \hbar^2}{k_B T^2} \sum_\lambda \int \frac{d^3 q}{(2\pi)^3} \frac{\omega_\lambda^2(q)}{\left[\exp\left(\frac{\hbar\omega_\lambda(q)}{k_B T}\right)-1\right]\left[1-\exp\left(-\frac{\hbar\omega_\lambda(q)}{k_B T}\right)\right]} . \quad (27)$$

where, T is the temperature of the binary system. The basic features of temperature dependence of C_V are determined by the behavior of $\omega_L(q)$ and $\omega_T(q)$.

3. RESULTS AND DISCUSSION

The input parameters and other related constants used in the present computations are written in Table 1.

Table 1. Input parameters and constants used in the present computation.

Glasses	Z	Ω_O (au)	r_C (au)	Z_d	N_C	r_d (au)
$Zr_{17}Ni_{83}$	3.66	142.73	0.7084	3.52	12	2.44
$Zr_{33}Ni_{67}$	3.34	129.54	0.7104	4.48	12	2.23
$Zr_{50}Ni_{50}$	3.00	115.52	0.7386	5.50	12	2.00
$Zr_{67}Ni_{33}$	2.66	101.50	0.7148	6.52	12	1.78
$Zr_{83}Ni_{17}$	2.34	88.31	0.7168	7.48	12	1.57

The computed data of pair potentials of $Zr_{17}Ni_{83}$, glass are shown in Figure 1(a) alongwith available theoretical results [17]. The present results are found to be in qualitative agreement with the available theoretical data [17]. It is seen that well depth of presently computed potentials move towards the left and found to be in qualitative agreement with available pair potential [17]. The first zero for $V(r=r_0)$ due to all the screenings occurs at $r_0 \approx 3.8$ au. The presently calculated pair potentials $Zr_{33}Ni_{67}$ glassy alloy of are shown in Figure 1(b). It is seen that the exchange and correlation effects affect largely the behaviour of the pair potential. The first zero for $V(r=r_0)$ due to all the screenings occurs at $r_0 \approx 3.8$ au. It is seen that well depth of presently computed potentials move towards the left and found to be in qualitative agreement with available pair potential [17]. The presently computed pair potentials of $Zr_{50}Ni_{50}$ glass are drawn in Figure 1(c). It is seen that the exchange and correlation functions affect the nature of the pair potential as well as the first zero of $V(r)$

appreciably. The first zero for $V(r = r_0)$ due to all the screenings occurs at $r_0 \approx 3.6$ au. It is seen that well depth of presently computed potentials move towards the left and found to be in qualitative agreement with available pair potential [12, 13, 17]. The computed pair potentials of $Zr_{67}Ni_{33}$ glass are displayed in Figure 1(d).

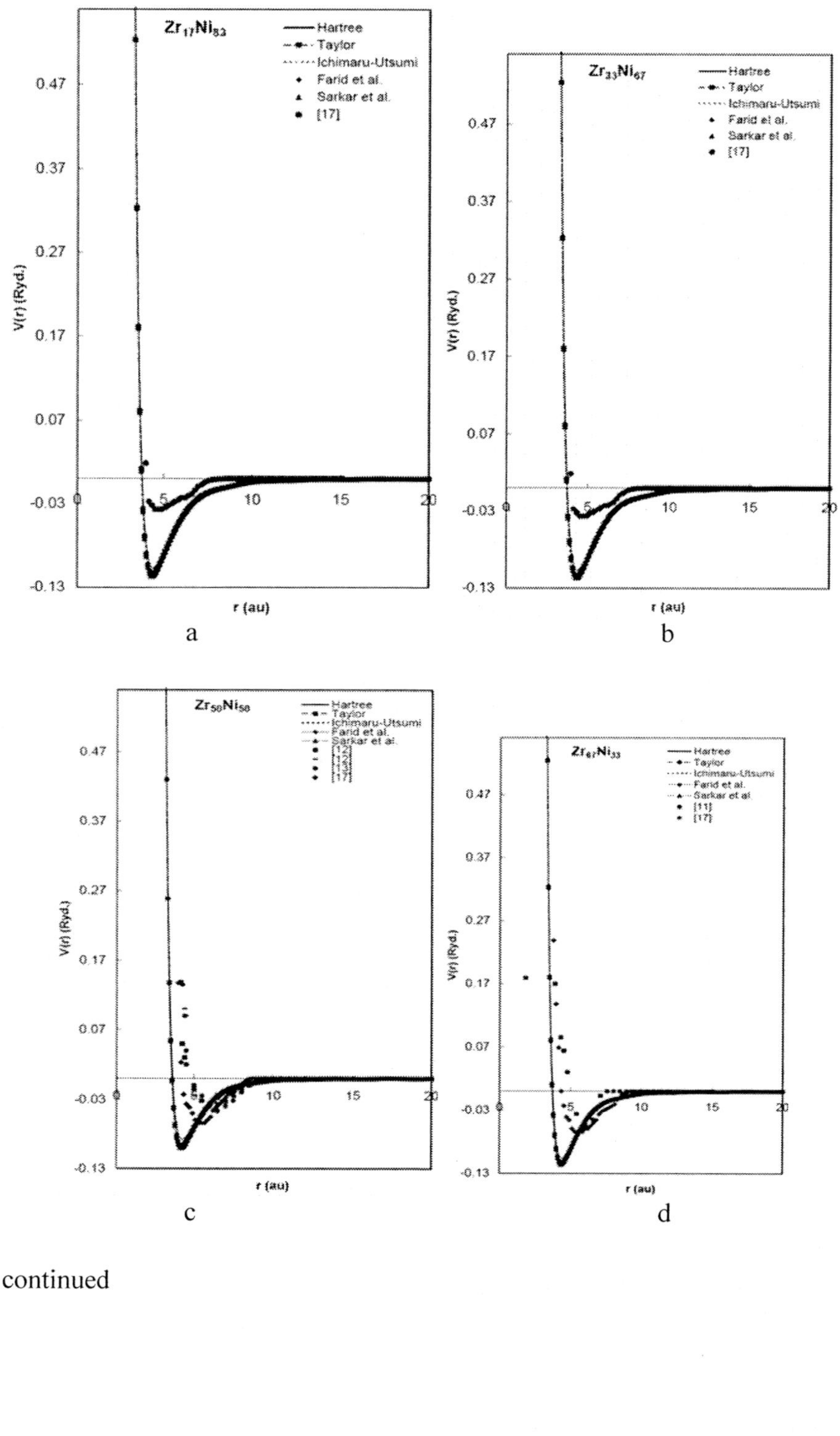

Figure 1 continued

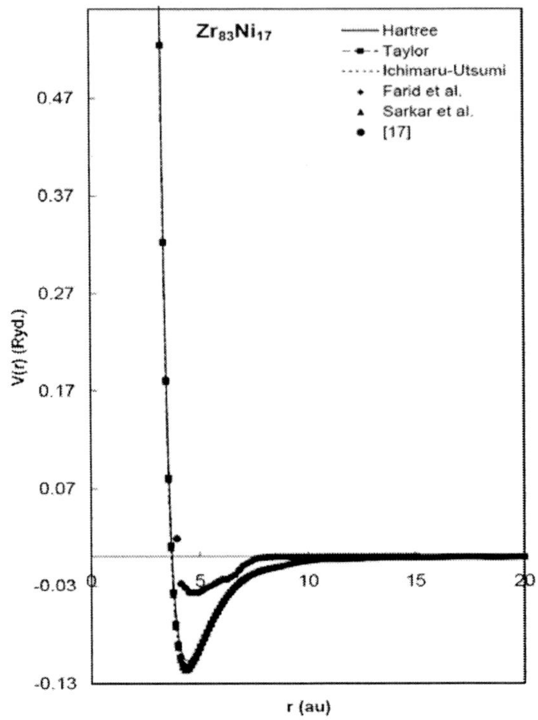

e

Figure 1. Screening dependence pair potentials for (a) $Zr_{17}Ni_{83}$, (b) $Zr_{33}Ni_{67}$, (c) $Zr_{50}Ni_{50}$, (d) $Zr_{67}Ni_{33}$ and (e) $Zr_{83}Ni_{17}$ metallic glasses.

It is observed that the nature of the pair potentials change appreciably with inclusion of exchange and correlation effects. The first zero for $V(r = r_0)$ due to all the screenings occurs at $r_0 \approx 3.8$ au. It is seen that well depth of presently computed potentials move towards the left and found to be in qualitative agreement with available pair potential [11, 17]. The WH form of pair potentials of $Zr_{83}Ni_{17}$ glass have been computed to see the atomic interactions and displayed in Figure 1(e).

It is apparent that the position of the first minimum, well width and well depth of the pair potential is highly affected by various screening functions. The first zero for $V(r = r_0)$ due to all the screenings occurs at $r_0 \approx 3.8$ au. It is seen that well depth of presently computed potentials move towards the left and found to be in qualitative agreement with available pair potential [17]. The presently calculated pair potentials of all glassy alloys elucidate that the inclusion of screening functions hardly changes the nature of the pair potentials, except around the first minimum. The well depth is slightly increases due to influence of various screening functions compared to H-screening. The maximum depth in the pair potential is obtained for S-function. The pair potentials due to remaining T, IU and F-screening functions are lying between those of H and S-screening function. Also, the present results of pair potentials of all Zr-based glassy alloys do not show any oscillatory behaviour and almost constant in the large r -region. The presently computed pair potentials from H, T IU, F and S-local field correction functions for most of the amorphous alloys are overlapped with each other. The present results of the pair potentials for most of the glassy alloys do not show oscillatory behaviour and potential energy remains negative and constant in the large r-region. Thus, the Coulomb repulsive potential part dominates the oscillations due to ion-electron-ion interactions, which show the waving shape of the pair potential after $r = 10.0$ au. From the Figures 1(a)-(e), it is observed that the shifting of the pair potentials with respect to atomic volume Ω_O of the amorphous alloys. The repulsive coulomb interaction and the attractive interactions represented by oscillatory nature are observed. It is also noticed that when volume Ω_O of the glassy alloys increases (Ω_O is more for $Zr_{17}Ni_{83}$ glass), the potential depth deepens. It means that the pair potential for $Zr_{17}Ni_{83}$ glass shows higher depth in comparison with other metallic glasses. Hence, the pair potentials converge towards a finite value instead of zero in attractive region. All the pair potentials show the combined effect of the s- and d-electrons. Bretonnet and Derouiche [31] are observed that the repulsive part of $V(r)$ is drawn lower and its attractive part is deeper due to the d-electron effect and the $V(r)$ is shifted towards the lower r-values. Therefore, the present results are supported the d-electron effect as noted by Bretonnet and Derouiche [31].

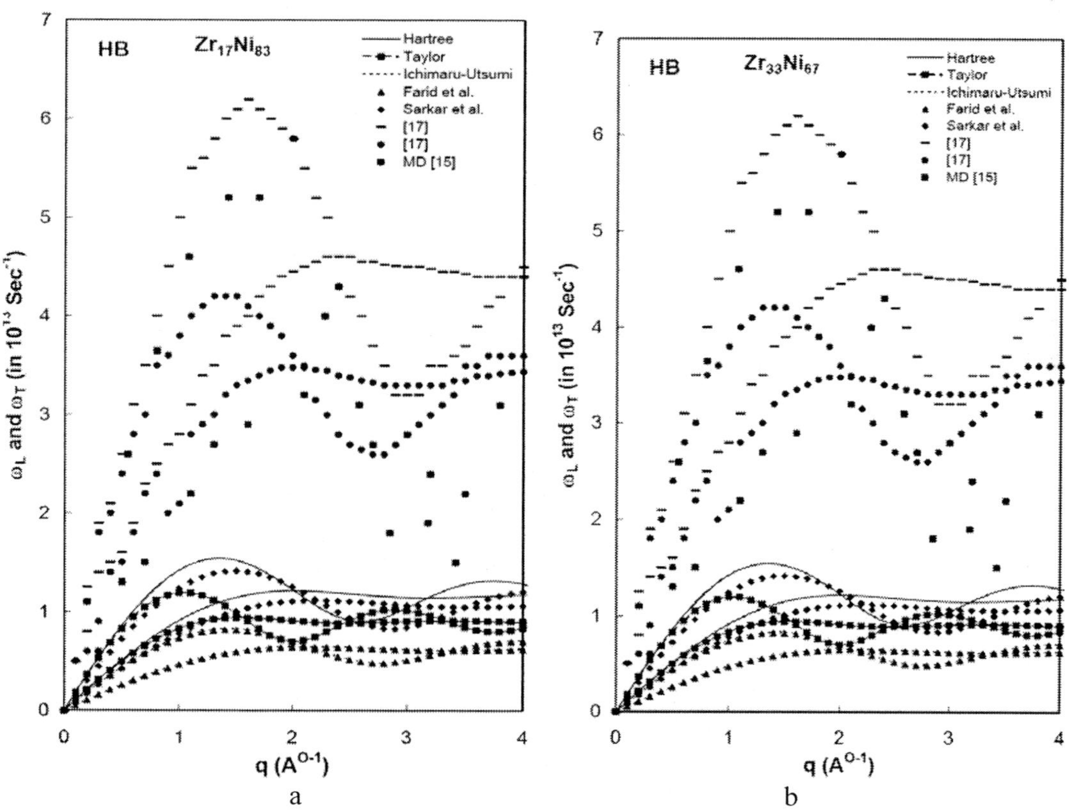

Figure 2. Continued on next page.

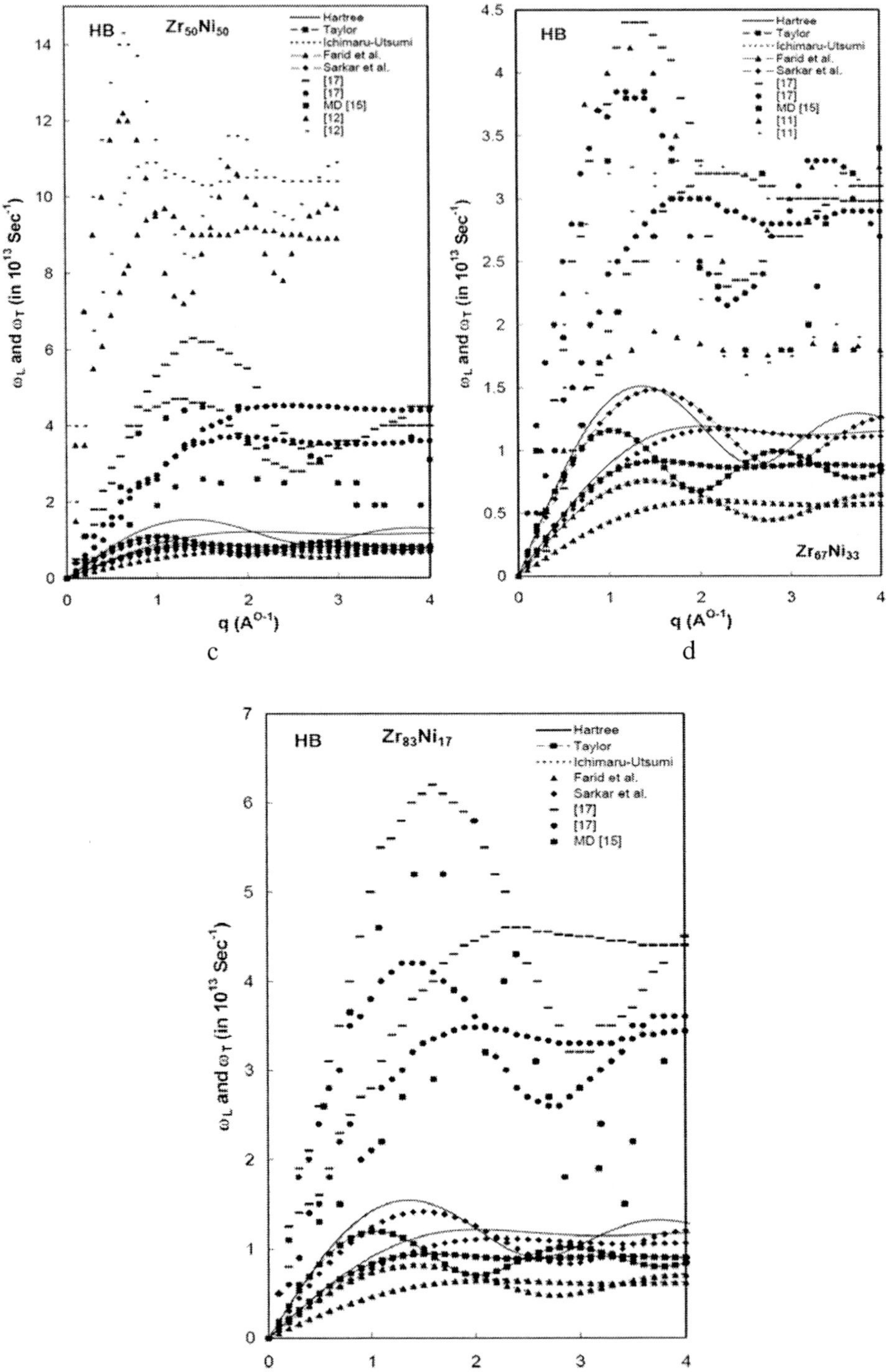

Figure 2. Screening dependence of the phonon dispersion curves for (a) $Zr_{17}Ni_{83}$, (b) $Zr_{33}Ni_{67}$, (c) $Zr_{50}Ni_{50}$, (d) $Zr_{67}Ni_{33}$ and (e) $Zr_{83}Ni_{17}$ metallic glasses using HB approach.

The phonon modes for longitudinal and transverse branches of $Zr_{17}Ni_{83}$, $Zr_{33}Ni_{67}$, $Zr_{50}Ni_{50}$, $Zr_{67}Ni_{33}$ and $Zr_{83}Ni_{17}$ metallic glasses calculated using HB approach with the five screening functions are shown in figures 2(a)-(e). To study the screening influence on the phonon eigen frequencies, the longitudinal and transverse phonon modes obtained using HB approach are shown in Figure 2(a) for $Zr_{17}Ni_{83}$ glass. The enhancement in both the phonon branches is concluded due to the effect of exchange and correlation functions. The present results of PDC due to H, T and F are lying between those due to IU and S-screening. The first minimum in the longitudinal branch for H, T, IU, F and S-local field correction functions occurs at $q \approx 1.3Å^{-1}$, $1.0Å^{-1}$, $1.3Å^{-1}$, $1.3Å^{-1}$ and $1.47Å^{-1}$, respectively.

Figure 2(b) shows the phonon frequencies for $Zr_{33}Ni_{67}$ glass obtained for various local field correction functions. It is seen that the first minimum in the longitudinal branch for H, T, IU, F and S-local field correction functions occurs at $q \approx 1.3Å^{-1}$, $1.1Å^{-1}$, $1.3Å^{-1}$, $1.3Å^{-1}$ and $1.4Å^{-1}$, respectively. The longitudinal and transverse phonon frequencies of $Zr_{50}Ni_{50}$ glass calculated using HB approach to study the screening influence is shown in Figure 2(c). It is seen that the influence of exchange and correlation effect lower down the phonon frequencies except for S-local field correction function. The first minimum in the longitudinal branch for H, T, IU, F and S-screening is occurs at $q \approx 1.4Å^{-1}$, $1.0Å^{-1}$, $1.0Å^{-1}$, $1.0Å^{-1}$ and $1.0Å^{-1}$, respectively. The screening influence on phonon frequencies of $Zr_{67}Ni_{33}$ glass is shown in Figure 2(d) where the results due to HB approach are included. It is apparent that both the longitudinal and transverse phonon frequencies are lower down by the inclusion of exchange and correlation effect. The first minimum in the longitudinal branch for H, T, IU, F and S-local field correction functions lies at $q \approx 1.4Å^{-1}$, $1.0Å^{-1}$, $1.4Å^{-1}$, $1.4Å^{-1}$ and $1.5Å^{-1}$, respectively. In the present study the phonon frequencies of $Zr_{83}Ni_{17}$ glass, calculated using HB approach with the five screening functions are shown in Figure 2(e). It is noticed that both the phonon modes are very sensitive to the screening effect. The first depth in the longitudinal branch for H, T, IU, F and S-local field correction functions is around at $q \approx 1.4Å^{-1}$, $1.0Å^{-1}$, $1.5Å^{-1}$, $1.5Å^{-1}$ and $1.5Å^{-1}$, respectively. The present results are found to be in qualitative agreement with the available theoretical data [11, 12, 15, 17].

The phonon modes for longitudinal and transverse branches of $Zr_{17}Ni_{83}$, $Zr_{33}Ni_{67}$, $Zr_{50}Ni_{50}$, $Zr_{67}Ni_{33}$ and $Zr_{83}Ni_{17}$ metallic glasses calculated using TG approach with the five screening functions are shown in figures 3(a)-(e). To study the screening influence on the phonon eigen frequencies, the longitudinal and transverse phonon modes obtained using TG approach are shown in Figure 3(a) for $Zr_{17}Ni_{83}$ glass. The enhancement in both the phonon branches is concluded due to the effect of exchange and correlation functions. The present results of PDC due to H, T and F are lying between those due to IU and S-screening. The first minimum in the longitudinal branch for H, T, IU, F and S-local field correction functions occurs at $q \approx 1.0Å^{-1}$, $0.7Å^{-1}$, $0.6Å^{-1}$, $0.6Å^{-1}$ and $1.1Å^{-1}$, respectively. Figure 3(b) shows the phonon frequencies for $Zr_{33}Ni_{67}$ glass obtained for various local field correction functions. It is seen that the first minimum in the longitudinal branch for H, T, IU, F and S-local field correction functions occurs at $q \approx 1.0Å^{-1}$, $0.7Å^{-1}$, $0.6Å^{-1}$, $0.6Å^{-1}$ and $1.1Å^{-1}$, respectively. The longitudinal and transverse phonon frequencies of $Zr_{50}Ni_{50}$ glass calculated using TG approach to study the screening influence is shown in Figure 3(c). It is seen that the influence of exchange and correlation effect lower down the phonon frequencies except for S-local field correction function.

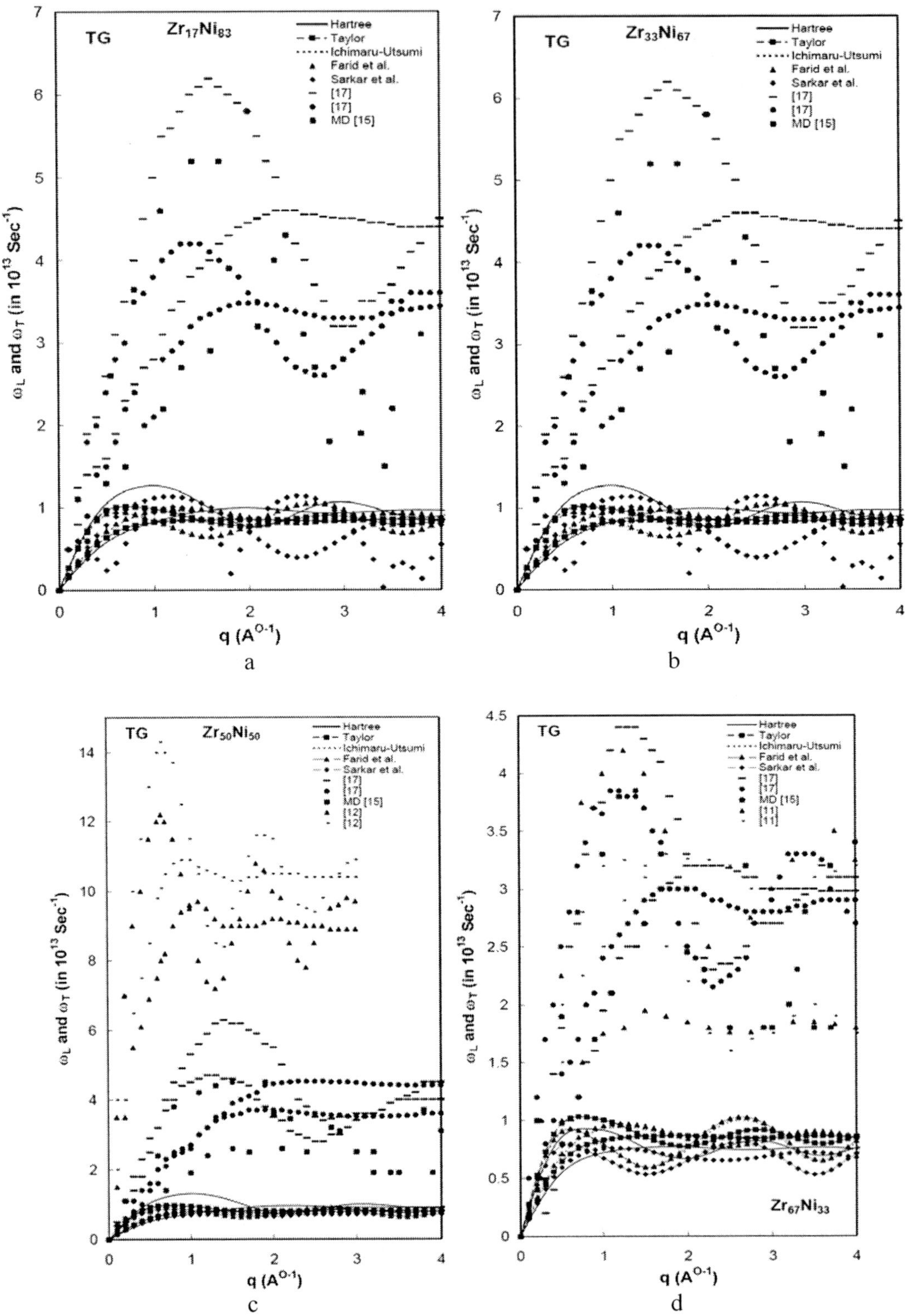

Figure 2 continued

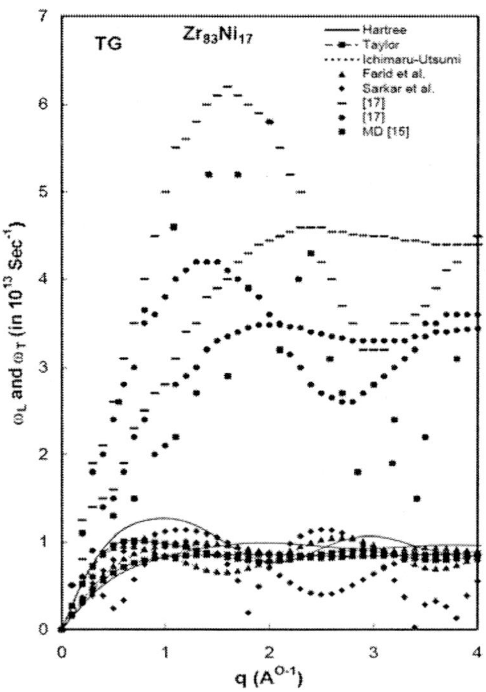

e

Figure 3. Screening dependence of the phonon dispersion curves for (a) $Zr_{17}Ni_{83}$, (b) $Zr_{33}Ni_{67}$, (c) $Zr_{50}Ni_{50}$, (d) $Zr_{67}Ni_{33}$ and (e) $Zr_{83}Ni_{17}$ metallic glasses using TG approach.

The first minimum in the longitudinal branch for H, T, IU, F and S-screening is occurs at $q \approx 0.8$Å^{-1}, 0.7Å^{-1}, 0.6Å^{-1}, 0.6Å^{-1} and 0.8Å^{-1}, respectively. The screening influence on phonon frequencies of $Zr_{67}Ni_{33}$ glass is shown in Figure 3(d) where the results due to TG approach are included. It is apparent that both the longitudinal and transverse phonon frequencies are lower down by the inclusion of exchange and correlation effect. The first minimum in the longitudinal branch for H, T, IU, F and S-local field correction functions lies at $q \approx 1.0$Å^{-1}, 0.7Å^{-1}, 1.0Å^{-1}, 1.0Å^{-1} and 1.1Å^{-1}, respectively. In the present study the phonon frequencies of $Zr_{83}Ni_{17}$ glass, calculated using HB approach with the five screening functions are shown in Figure 3(e). It is noticed that both the phonon modes are very sensitive to the screening effect.

The first depth in the longitudinal branch for H, T, IU, F and S-local field correction functions is around at $q \approx 1.0$Å^{-1}, 0.7Å^{-1}, 1.0Å^{-1}, 1.0Å^{-1} and 1.1Å^{-1}, respectively. The present results are found to be in qualitative agreement with the available theoretical data [11, 12, 15, 17]. The phonon modes for longitudinal and transverse branches of $Zr_{17}Ni_{83}$, $Zr_{33}Ni_{67}$, $Zr_{50}Ni_{50,}$ $Zr_{67}Ni_{33}$ and $Zr_{83}Ni_{17}$ metallic glasses calculated using BS approach with the five screening functions are shown in figures 4(a)-(e). To study the screening influence on the phonon eigen frequencies, the longitudinal and transverse phonon modes obtained using BS approach are shown in Figure 4(a) for $Zr_{17}Ni_{83}$ glass. The enhancement in both the phonon branches is concluded due to the effect of exchange and correlation functions. The present results of PDC due to H, T and F are lying between those due to IU and S-screening. The first minimum in the longitudinal branch for H, T, IU, F and S-local field correction functions occurs at $q \approx 0.7$Å^{-1}, respectively.

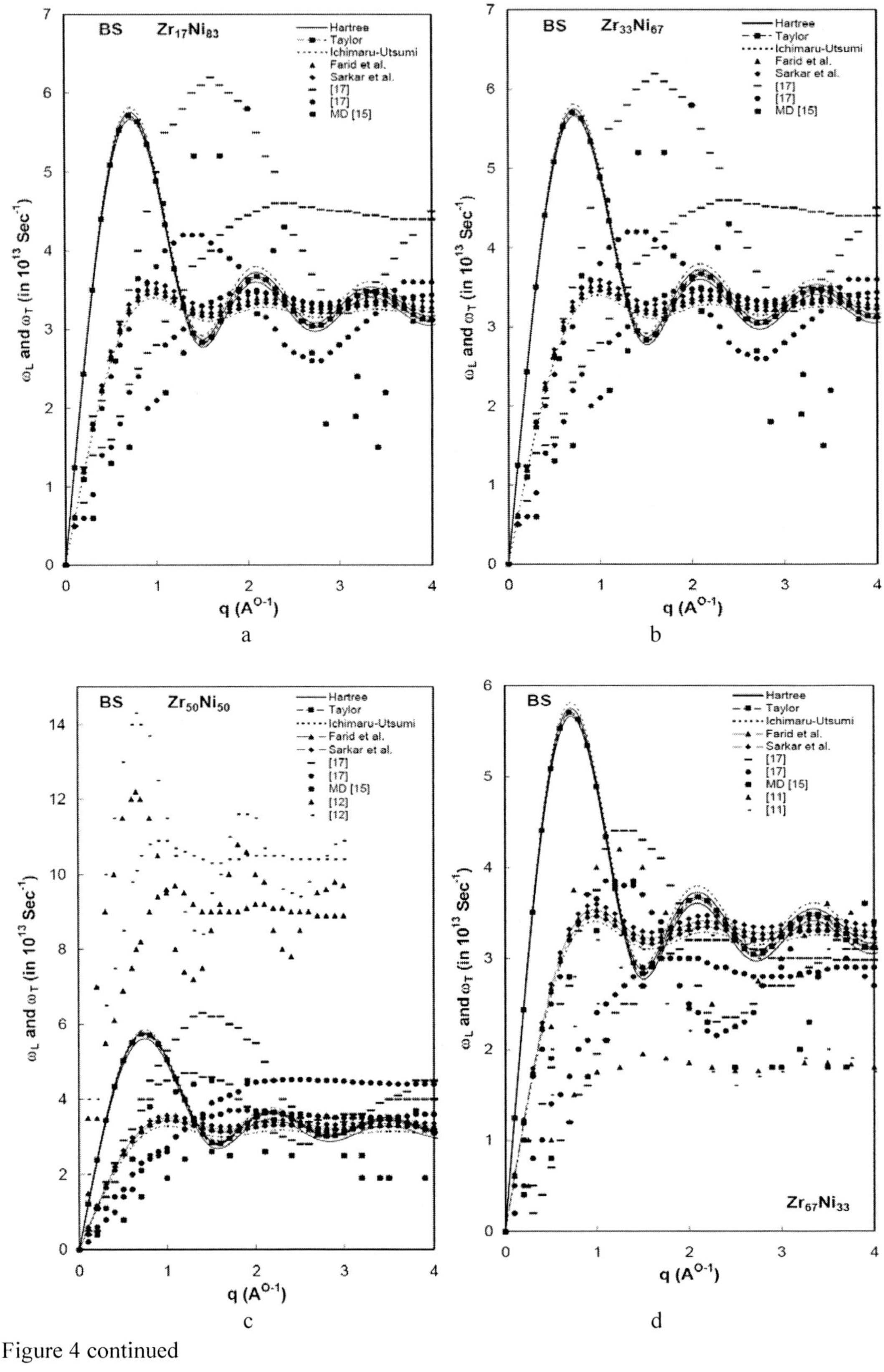

Figure 4 continued

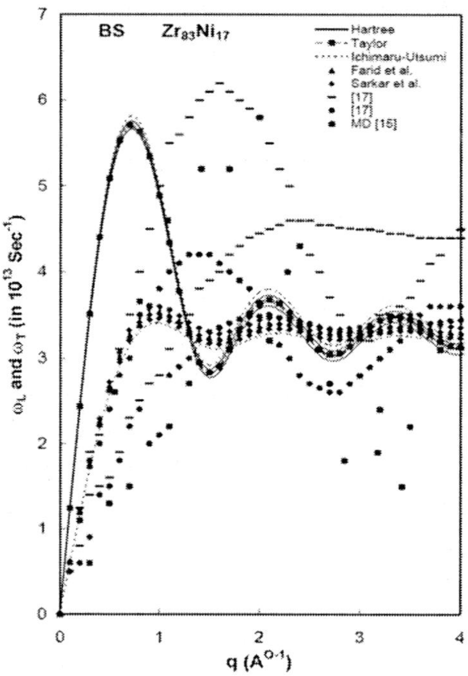

Figure 4. Screening dependence of the phonon dispersion curves for (a) $Zr_{17}Ni_{83}$, (b) $Zr_{33}Ni_{67}$, (c) $Zr_{50}Ni_{50}$, (d) $Zr_{67}Ni_{33}$ and (e) $Zr_{83}Ni_{17}$ metallic glasses using BS approach.

Figure 4(b) shows the phonon frequencies for $Zr_{33}Ni_{67}$ glass obtained for various local field correction functions. It is seen that the first minimum in the longitudinal branch for H, T, IU, F and S-local field correction functions occurs at $q \approx 0.7Å^{-1}$, respectively. The longitudinal and transverse phonon frequencies of $Zr_{50}Ni_{50}$ glass calculated using BS approach to study the screening influence is shown in Figure 4(c). It is seen that the influence of exchange and correlation effect lower down the phonon frequencies except for S-local field correction function. The first minimum in the longitudinal branch for H, T, IU, F and S-screening is occurs at $q \approx 0.7Å^{-1}$, respectively.

The screening influence on phonon frequencies of $Zr_{67}Ni_{33}$ glass is shown in Figure 4(d) where the results due to BS approach are included.

It is apparent that both the longitudinal and transverse phonon frequencies are lower down by the inclusion of exchange and correlation effect. The first minimum in the longitudinal branch for H, T, IU, F and S-local field correction functions lies at $q \approx 0.7Å^{-1}$, respectively. In the present study the phonon frequencies of $Zr_{83}Ni_{17}$ glass, calculated using HB approach with the five screening functions are shown in Figure 4(e). It is noticed that both the phonon modes are very sensitive to the screening effect. The first depth in the longitudinal branch for H, T, IU, F and S-local field correction functions is around at $q \approx 0.7Å^{-1}$, respectively. The present results are found to be in qualitative agreement with the available theoretical data [11, 12, 15, 17].

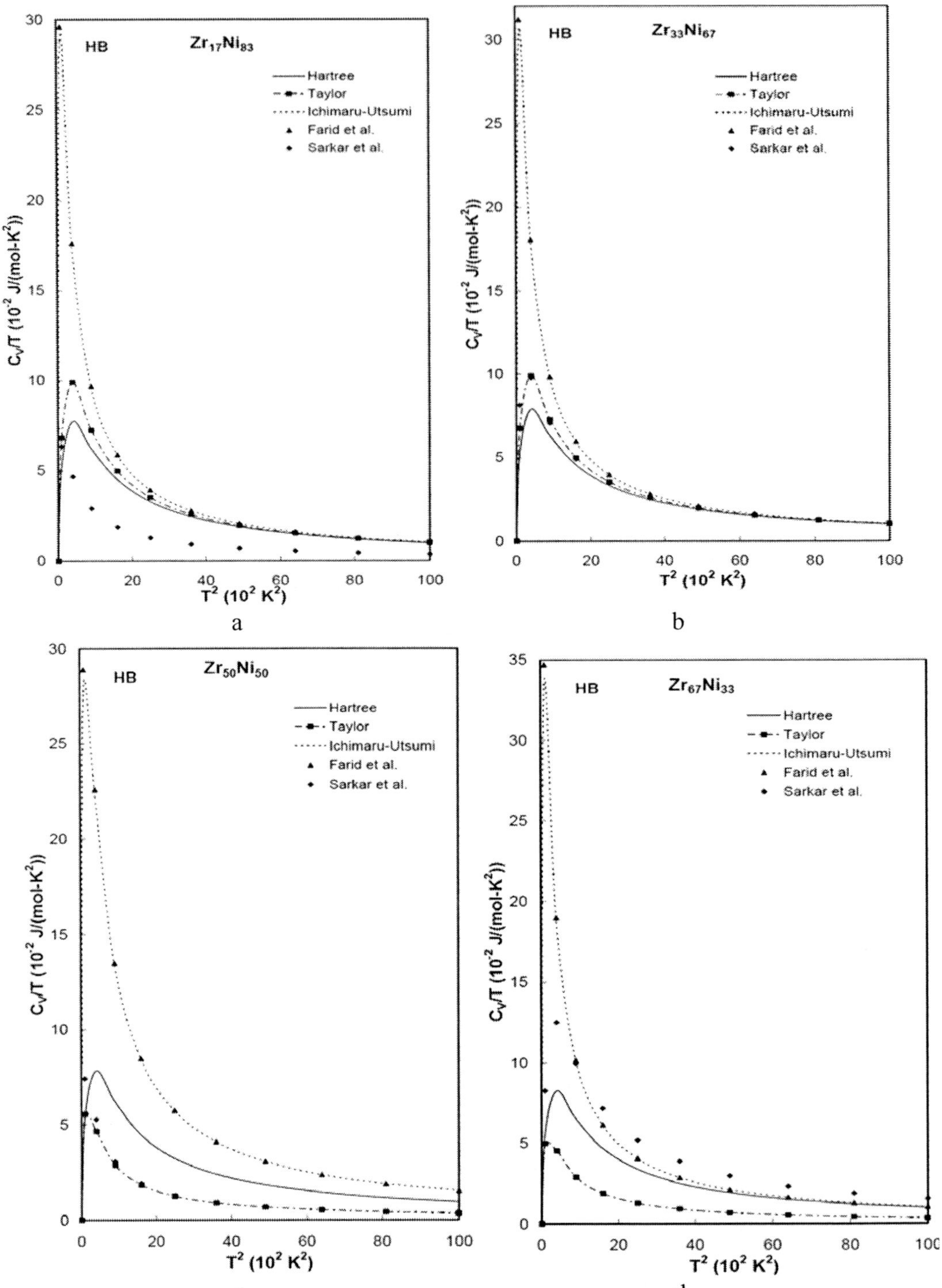

Figure 5. Continued on next page.

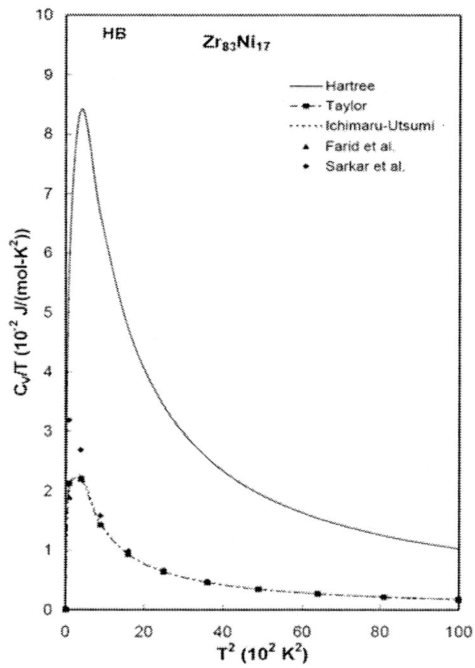

e

Figure 5. Screening dependence of the low temperature specific heat for (a) $Zr_{17}Ni_{83}$, (b) $Zr_{33}Ni_{67}$, (c) $Zr_{50}Ni_{50}$, (d) $Zr_{67}Ni_{33}$ and (e) $Zr_{83}Ni_{17}$ metallic glasses metallic glasses computed from HB approach.

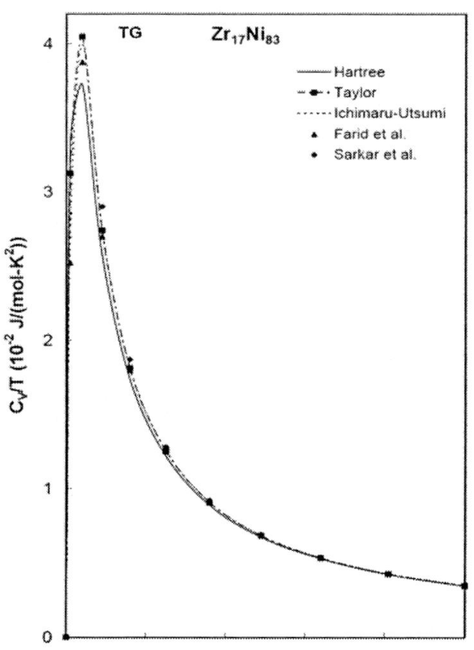

a

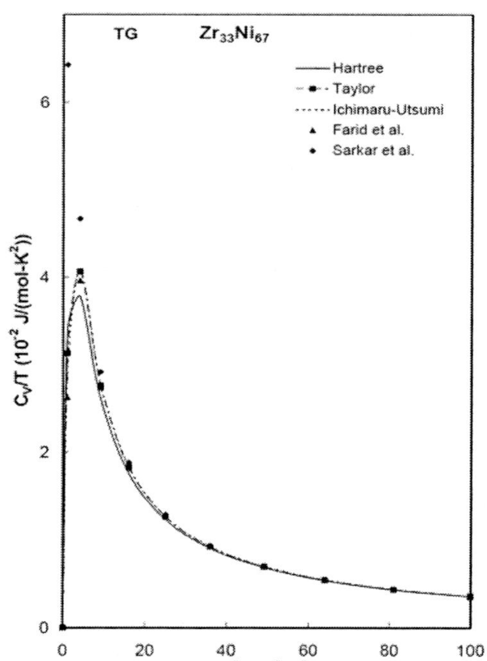

b

Figure 6. Continued on next page.

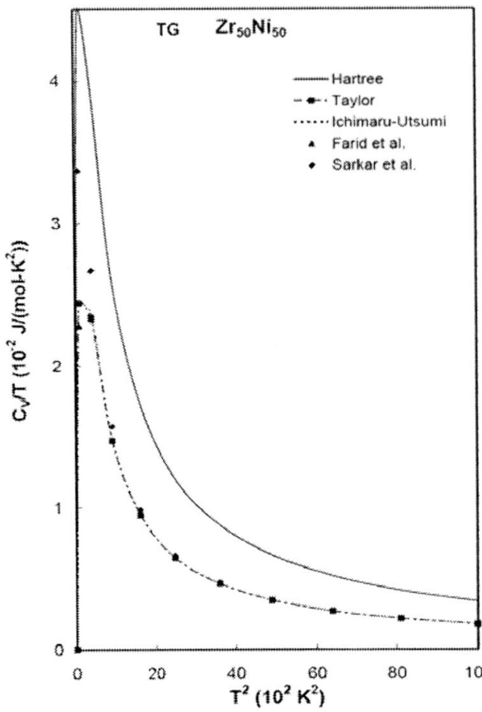

c

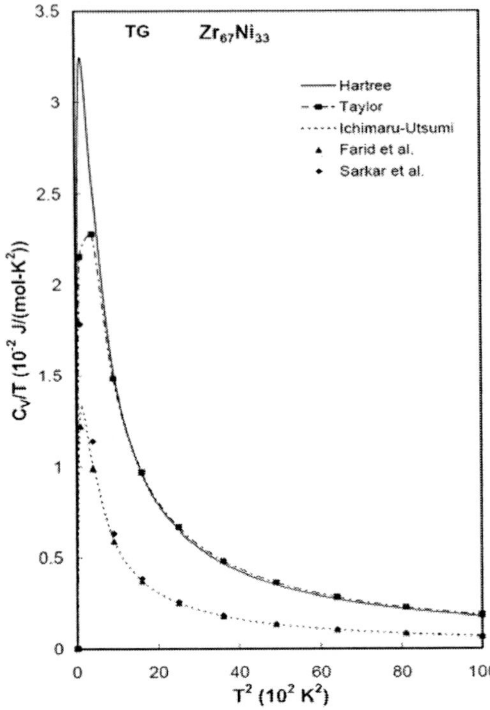

d

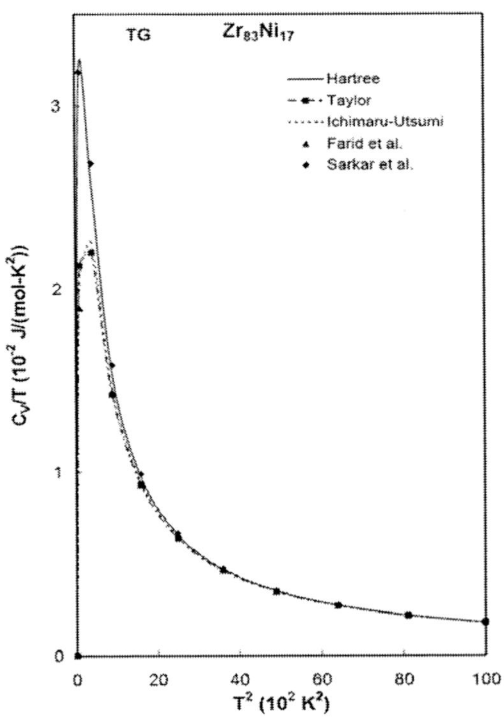

e

Figure 6. Screening dependence of the low temperature specific heat for (a) $Zr_{17}Ni_{83}$, (b) $Zr_{33}Ni_{67}$, (c) $Zr_{50}Ni_{50}$, (d) $Zr_{67}Ni_{33}$ and (e) $Zr_{83}Ni_{17}$ metallic glasses metallic glasses computed from TG approach.

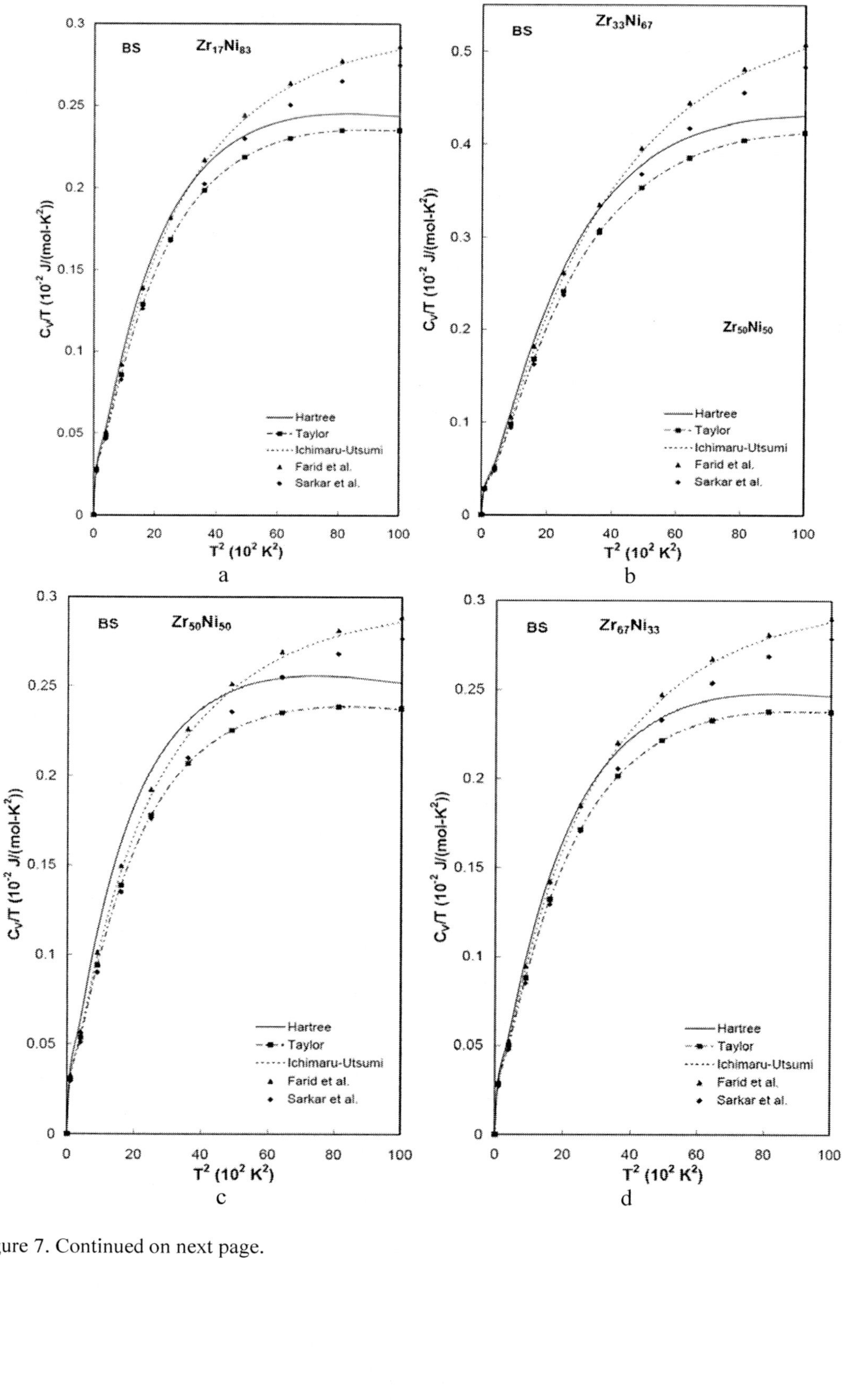

Figure 7. Continued on next page.

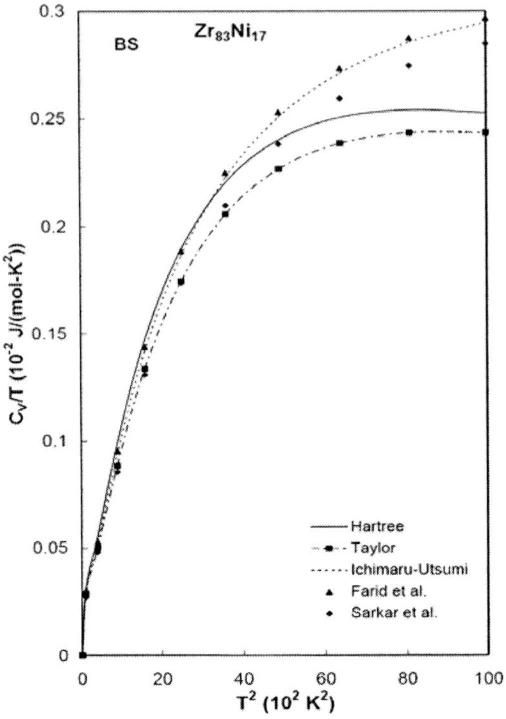

e

Figure 7. Screening dependence of the low temperature specific heat for (a) $Zr_{17}Ni_{83}$, (b) $Zr_{33}Ni_{67}$, (c) $Zr_{50}Ni_{50}$, (d) $Zr_{67}Ni_{33}$ and (e) $Zr_{83}Ni_{17}$ metallic glasses metallic glasses computed from BS approach.

It is also observed from the Figures 2-4 that, the oscillations are more prominent in the longitudinal phonon modes as compared to the transverse modes in all three approaches. This shows the existence of collective excitations at larger momentum transfer due to longitudinal phonons only and the instability of the transverse phonons due to the anharmonicity of the atomic vibrations in the metallic systems. Moreover, the present outcomes of both the phonon modes due to HB as well as TG approaches are more enhanced than the BS approach. Here in transverse branch, the frequencies increase with the wave vector q and then saturates at $\approx q = 2.0\text{Å}^{-1}$, which supports the well known Thorpe model [32] in which, it describes a glass like a solid containing finite liquid cluster. The transverse phonons are absorbed for frequencies larger than the smallest eigen frequencies of the largest cluster.

As shown in figures 5-7 for $Zr_{17}Ni_{83}$, $Zr_{33}Ni_{67}$, $Zr_{50}Ni_{50}$, $Zr_{67}Ni_{33}$ and $Zr_{83}Ni_{17}$ metallic glasses, the exchange and correlation functions also affected in the anomalous behaviour (i.e. deviation from the T^3 law), which is observed in the vibrational part of the low temperature specific heat C_V. The reason behind the anomalous behaviour may be due to the low frequency modes modify the generalized vibrational density of states of the glass with that of the polycrystal. These modes are mainly responsible for the difference in the temperature dependence of the vibrational part of the specific heat which departs from the normal behaviour. The existence of a portion of the spectrum with 'softer phonons' (resembling rotons in liquid helium) may be cause of anomalous behaviour of low temperature specific heat C_V. In the low temperature region, a contribution to the low temperature specific heat C_V is made by phonons of the initial part of the frequency spectrum.

Table 2. Thermodynamic and Elastic properties of $Zr_{17}Ni_{83}$ metallic glass

App.	SCR	v_L x 10^5 cm/sec	v_T x 10^5 cm/sec	B_T x 10^{11} dyne/cm^2	G x 10^{11} dyne/cm^2	σ	Y x 10^{11} dyne/cm^2	θ_D (K)
HB	H	1.8246	1.0534	1.3320	0.7992	0.2500	1.9979	131.59
	T	1.8337	1.0587	1.3453	0.8072	0.2500	2.0179	132.24
	IU	0.9279	0.5357	0.3445	0.2067	0.2500	0.5167	66.92
	F	0.9152	0.5284	0.3351	0.2011	0.2500	0.5027	66.00
	S	1.5197	0.8774	0.9240	0.5544	0.2500	1.3860	109.60
TG	H	2.7070	1.4173	3.3486	1.4466	0.3112	3.7935	178.37
	T	2.7802	1.6084	3.0825	1.863	0.2485	4.6519	200.87
	IU	2.7503	1.7147	2.6243	2.1174	0.1821	5.0058	212.58
	F	2.8032	1.7594	2.6866	2.2294	0.1750	5.2391	217.98
	S	2.1055	1.85700	0.1188	2.4835	1.2513	1.2481	216.78
BS	H	7.1769	3.4442	25.0234	8.3169	0.3504	22.4621	431.84
	T	7.2984	3.5679	25.4458	8.9254	0.343	23.9732	446.91
	IU	7.2508	3.5201	25.2777	8.6874	0.3458	23.3835	441.08
	F	7.2261	3.5027	25.1404	8.6017	0.3464	23.1634	438.94
	S	7.3643	3.6236	25.7486	9.2059	0.3403	24.6768	453.72
Others [17]		4.67, 4.36	2.59, 2.28	1.29, 9.0	4.99, 3.87	–	–	–
MD [17]		4.34	2.13	9.5	3.37	–	–	–

When the temperature reaches a value at which the energy of the thermal motions becomes comparable to the energy of 'softer phonons' minimum, an additional contribution appears to heat capacity from the roton portion of the phonon frequency $\omega(q)$. The theoretical or experimental data is not available for further comparisons.

Furthermore, the thermodynamic and elastic properties of Zr-based glassy alloys estimated from the elastic limit of the PDC are tabulated in Tables 2-6. The present results of thermodynamic and elastic properties of Zr-based glasses are found lower in line with other theoretical data [11, 12, 17]. But among the three approaches adopted, the outcome due to HB and TG approaches are lower than those due to BS approach. The screening influence on PDC is reflected on these properties also.

From the overall picture of the present study it is noticed that, the proposed model potential is successfully applicable to study the vibrational dynamics of some bulk metallic glass. The influences of various local field correction functions are also observed in the present study. The experimental or theoretical data of most of the bulk metallic glass are not available in the literature but the present study is very useful to form a set of theoretical data of particular bulk metallic glass.

Table 3. Thermodynamic and Elastic properties of $Zr_{33}Ni_{67}$ metallic glass

App.	SCR	v_L x 10^5 cm/sec	v_T x 10^5 cm/sec	B_T x 10^{11} dyne/cm^2	G x 10^{11} dyne/cm^2	σ	Y x 10^{11} dyne/cm^2	θ_D (K)
HB	H	1.8129	1.0467	1.3037	0.7822	0.2500	1.9556	130.37
	T	1.8416	1.0632	1.3453	0.8072	0.2500	2.0179	132.43
	IU	0.9081	0.5243	0.3271	0.1963	0.2500	0.4907	65.31
	F	0.8937	0.516	0.3168	0.1901	0.2500	0.4752	64.27
	S	1.5439	0.8914	0.9455	0.5673	0.2500	1.4182	111.02
TG	H	2.6991	1.4147	3.2963	1.4290	0.3106	3.7457	177.52
	T	2.7855	1.6094	3.0739	1.8495	0.2494	4.6216	200.45
	IU	2.7406	1.7105	2.5775	2.0890	0.1810	4.9340	211.43
	F	2.7933	1.7553	2.6378	2.1999	0.1737	5.1640	216.81
	S	2.0852	1.8523	0.1618	2.4498	1.3707	1.8164	215.22
BS	H	7.1769	3.4442	25.0234	8.3169	0.3504	22.4621	431.84
	T	7.2984	3.5679	25.4458	8.9254	0.343	23.9732	446.91
	IU	7.2508	3.5201	25.2777	8.6874	0.3458	23.3835	441.08
	F	7.2261	3.5027	25.1404	8.6017	0.3464	23.1634	438.94
	S	7.3643	3.6236	25.7486	9.2059	0.3403	24.6768	453.72
Others [17]		5.11, 4.95	2.68, 2.81	1.50, 10.2	5.29, 5.82	–	–	–
MD [17]		4.67	2.16	11.5	3.44	–	–	–

Table 4. Thermodynamic and Elastic properties of $Zr_{50}Ni_{50}$ metallic glass

App.	SCR	v_L x 10^5 cm/sec	v_T x 10^5 cm/sec	B_T x 10^{11} dyne/cm^2	G x 10^{11} dyne/cm^2	σ	Y x 10^{11} dyne/cm^2	θ_D (K)
HB	H	1.7809	1.0282	1.2926	0.7755	0.2500	1.9389	132.33
	T	1.7041	0.9838	1.1835	0.7101	0.2500	1.7752	126.62
	IU	1.0348	0.5974	0.4364	0.2618	0.2500	0.6546	76.89
	F	1.0174	0.5874	0.4219	0.2531	0.2499	0.6328	75.60
	S	1.5031	0.8678	0.9208	0.5525	0.2500	1.3812	111.69
TG	H	2.6120	1.2664	3.4363	1.1766	0.3463	3.1682	164.99
	T	2.6385	1.5173	2.8551	1.6889	0.2529	4.2322	195.35
	IU	2.5943	1.5407	2.6153	1.7414	0.2275	4.2753	197.78
	F	2.6396	1.5804	2.6682	1.8323	0.2206	4.4730	202.72
	S	2.3400	1.3482	2.2391	1.3334	0.2516	3.3376	173.54
BS	H	6.8930	3.2011	24.8332	7.5170	0.3625	20.4841	417.93
	T	7.0921	3.3970	25.6101	8.4655	0.3511	22.8760	442.83
	IU	7.0581	3.3556	25.5318	8.2602	0.3540	22.3683	437.60
	F	7.0258	3.3297	25.3670	8.1330	0.3552	22.0433	434.29
	S	7.1585	3.4599	25.8830	8.7818	0.3476	23.6686	450.82
Others [12, 17]		9.11, 9.53 4.62	6.58, 6.96, 2.71	1.74, 1.80, 8.30	5.26	–	–	370.50, 419.90, 443.77
MD [17]		4.65	2.11	11.1	3.19			

Table 5. Thermodynamic and Elastic properties of $Zr_{67}Ni_{33}$ metallic glass

App.	SCR	v_L x 10^5 cm/sec	v_T x 10^5 cm/sec	B_T x 10^{11} dyne/cm^2	G x 10^{11} dyne/cm^2	σ	Y x 10^{11} dyne/cm^2	θ_D (K)
HB	H	1.7890	1.0329	1.2466	0.7480	0.2499	1.8699	127.87
	T	1.8080	1.0438	1.2732	0.7639	0.2500	1.9099	129.23
	IU	0.8688	0.5016	0.2940	0.1764	0.2500	0.4410	62.10
	F	0.8507	0.4911	0.2819	0.1691	0.2500	0.4228	60.80
	S	1.5912	0.9187	0.9862	0.5917	0.2499	1.4793	113.73
TG	H	2.4117	1.3652	2.3357	1.3067	0.2642	3.3041	169.30
	T	2.8000	1.6136	3.0626	1.8256	0.2514	4.5689	199.80
	IU	2.7245	1.7035	2.4913	2.0347	0.1790	4.7979	209.26
	F	2.7767	1.7484	2.5479	2.1433	0.1715	5.0217	214.60
	S	2.4025	1.4765	2.0089	1.5284	0.1966	3.6576	181.70
BS	H	7.1769	3.4442	25.0234	8.3169	0.3504	22.4621	431.84
	T	7.2984	3.5679	25.4458	8.9254	0.3430	23.9732	446.91
	IU	7.2508	3.5201	25.2777	8.6874	0.3458	23.3835	441.08
	F	7.2261	3.5027	25.1404	8.6017	0.3464	23.1634	438.94
	S	7.3643	3.6236	25.7486	9.2059	0.3403	24.6768	453.72
Others [11, 17]		5.00, 4.19, 4.75, 4.46	2.50, 2.05, 1.95, 1.88	1.176, 0.846, 10.4	2.43	–	–	311, 256, 254±11, 372.53
MD [11]		4.10	2.02	7.8	2.80	–	–	–

Table 6. Thermodynamic and Elastic properties of $Zr_{83}Ni_{17}$ metallic glass

App.	SCR	v_L x 10^5 cm/sec	v_T x 10^5 cm/sec	B_T x 10^{11} dyne/cm^2	G x 10^{11} dyne/cm^2	σ	Y x 10^{11} dyne/cm^2	θ_D (K)
HB	H	1.7781	1.0266	1.2210	0.7326	0.2500	1.8314	126.73
	T	1.8173	1.0492	1.2755	0.7653	0.2500	1.9133	129.53
	IU	0.8516	0.4917	0.2801	0.168	0.2500	0.4201	60.70
	F	0.8317	0.4802	0.2671	0.1603	0.2500	0.4007	59.28
	S	1.6116	0.9304	1.0030	0.6018	0.2500	1.5045	114.86
TG	H	2.4182	1.3671	2.3329	1.2992	0.2651	3.2874	169.07
	T	2.8078	1.6162	3.0593	1.8159	0.2522	4.5478	199.57
	IU	2.7185	1.701	2.4557	2.0113	0.1783	4.7398	208.33
	F	2.7705	1.7458	2.5109	2.1188	0.1707	4.9610	213.66
	S	2.3911	1.4781	1.9496	1.5187	0.1908	3.6169	181.27
BS	H	7.1769	3.4442	25.0234	8.3169	0.3504	22.4621	431.84
	T	7.2984	3.5679	25.4458	8.9254	0.3430	23.9732	446.91
	IU	7.2508	3.5201	25.2777	8.6874	0.3458	23.3835	441.08
	F	7.2261	3.5027	25.1404	8.6017	0.3464	23.1634	438.94
	S	7.3643	3.6236	25.7486	9.2059	0.3403	24.6768	453.72
Others [17]		4.32	2.37	7.2	3.64	–	–	–
MD [17]		4.27	1.80	9.0	2.10	–	–	–

In all three approaches, it is very difficult to judge which approaches is best for computations of vibrational dynamics of binary metallic glass, because each approximation has its own identity. The HB approach is simplest and older one, which generating consistent results of the vibrational data of these bulk metallic glass, because the HB approaches needs minimum number of parameters. While TG approach is developed upon the quasi-crystalline approximation in which effective force constant depends upon the correlation function for the displacement of atoms and correlation function of displacement itself depends on the phonon frequencies. The BS approach is retained the interatomic interactions effective between the first nearest neighbours only hence, the disorderness of the atoms in the formation of metallic glasses is more which show deviation in magnitude of the PDC and their related properties. From the present study we are concluded that, all three approaches are suitable for studying the vibrational dynamics of the amorphous materials. Hence, successful application of the model potential with three approaches is observed from the present study.

The dielectric function plays an important role in the evaluation of potential due to the screening of the electron gas. For this purpose in the present investigations the local filed correction function due to H, T, IU, F and S are used. Reason for selecting these functions is that H-function does not include exchange and correlation effect and represents only static dielectric function, while T-function cover the overall features of the various local field correction functions proposed before 1972. While, IU, F and S functions are recent one among the existing functions and not exploited rigorously in such study. This helps us to study the relative effects of exchange and correlation in the aforesaid properties. Hence, the five different local field correction functions show variations up to an order of magnitude in the vibrational properties.

CONCLUSIONS

Lastly it is concluded that, in the study of phonon dynamics of metallic glasses, the pair potentials and its derivatives as well as pair correlation function plays an important role. In the present computation, the WH form is adopted to generate the pair potentials, which ignores the angular interaction due to partially filled d-bands in transition metals. Most recent model potential with WH model and HB approach produces consistent results of phonon dynamics for all metallic glasses. Hence, the present model potential is suitable for studying the phonon dynamics of five Zr-based metallic glasses, which confirms applicability of the model potential in the aforementioned study. Such study on phonon dynamics of other binary as well as ternary liquid alloys and metallic glasses is in progress, which will be communicated in near future.

REFERENCES

[1] M. Vora, Ph. D. Thesis, Sardar Patel University, INDIA (2004).

[2] A. M. Vora, M. H. Patel, P. N. Gajjar and A. R. Jani, *Solid State Phys*. 46, 315 (2003).

[3] P. N. Gajjar, A. M. Vora and A. R. Jani, In: *Proceedings of the 9th Asia Pacific Physics Conference Hanoi,* Vietnam The Gioi Publication (2006) 429.

[4] A. M. Vora, *Chinese Phys. Lett.* 23, 1872 (2006).

[5] A. M. Vora, *J. Mater. Sci.* 42, 935 (2007).
[6] A. M. Vora, *Acta Phys. Pol. A* 111, 859 (2007).
[7] A. M. Vora, *J. Non-Cryst. Sol.* 352, 3217 (2006).
[8] A. M. Vora, *Front. Mater. Sci.* 1, 366 (2007).
[9] A. M. Vora, *FIZIKA A* 16, 187 (2007).
[10] A. M. Vora, *Rom. J. Phys.* 53 (2008) 517.
[11] A. Gupta, A. Prasad, D. Bhandari, K. C. Jain and N. S. Saxena, *J. Phys. Chem. Sol.* 58, 33 (1997).
[12] N. Gupta, K. C. Jain and N. S. Saxena, *Phys. Stat. Sol.* (b) 176, 81 (1993).
[13] Ch. Hausleitner and J. Hafner, *Phys. Rev.* B42, 5683 (1990).
[14] T. Aihara Jr. and T. Masumoto, J. *Phys. Cond. Matt.* 7, 1525 (1995).
[15] T. Aihara Jr., Y. Kawazoe and T. Masumoto, *J. Non-Cryst. Sol.* 250-252, 645 (1996).
[16] T. Otomo, M. Arai, J.–B. Suck and S. M. Bennington, *J. Non-Cryst. Sol.* 312-314, 599 (2002).
[17] Kirit N. Lad and Arun Pratap, *Physica* B334, 135 (2003).
[18] J. Hubbard and J. L. Beeby, J. *Phys. C: Solid State Phys*. 2, 556 (1969).
[19] A. B. Bhatia and R. N. Singh, *Phys. Rev* B31, 4751 (1985).
[20] S. Takeno and M. Goda, Prog. Thero. Phys. 45, 331 (1971); *Prog. Thero. Phys*. 47, 790 (1972).
[21] J. M. Wills and W. A. Harrison, *Phys. Rev*. B28, 4363 (1983).
[22] M. M. Shukla and J. R. Campanha, *Acta Phys. Pol.* A94, 655 (1998).
[23] W. A. Harrison, Pseudopotentials in the Theory of Metals, W. A. Benjamin, New York, 1966.
[24] R. Taylor, J. Phys. F: *Met. Phys*. 8, 1699 (1978).
[25] S. Ichimaru and K. Utsumi, *Phys. Rev*. B24, 3220 (1981).
[26] B. Farid, V. Heine, G. Engel and I. J. Robertson, *Phys. Rev*. B48, 11602 (1993).
[27] A. Sarkar, D. Sen, H. Haldar and D. Roy, *Mod. Phys. Lett*. B12, 639 (1998).
[28] V. Heine, D. Weaire, *Solid State Physics*, Vol. 24, Eds. H. Ehrenreich *et al.*, Academic Press, New York, 1970, p. 249.
[29] J. Hafner, V. Heine, J. Phys. F: *Met. Phys*. 13, 2479 (1983).
[30] N. P. Kovalenko and Y. P. Krasny, *Physics* B162, 115 (1990).
[31] J. L. Bretonnet, A. Derouiche, *Phys. Rev*. B43, 8924 (1990).
[32] M. F. Thorpe, J. *Non-Cryst. Sol*. 57, 355 (1983).

In: New Trends in Chemical Physics Research
Editor: James N. Lapherty
ISBN: 978-1-61668-535-5

Chapter 16

WAY FROM PHYSICAL NONEQUILIBRIUM TO BIOLOGICAL EVOLUTION

Vladimir Kompanichenko*
Institute for Complex Analysis, 4, Sholom-Aleyhem Street,
Birobidzhan 679016, Russia

ABSTRACT

This chapter is devoted to consideration of the method of the transition from prebiotic organic microsystems to the primordial communities of prokaryotes. During the initial transformations prebiotic organic microsystems acquired the unstable bifurcate state, with its following relative stabilization by means of balanced oscillations around the bifurcation point. Then microsystems of the type converted into the primary living units – probionts, which were characterized with the ability to concentrate free energy and information already. Through the synergy effect probionts were organized into the communities that represented themselves minimum self-sufficient unit of life.

The appeared communities of probionts evolved to more complex communities of progenotes (characterized by C.Woese) and then prokaryotes. This earliest biological evolution proceeded through development of the initial nucleoprotein interaction, formation of RNA and then DNA, etc. Emergence of species had led to conversion of the initial communities of probionts and progenotes ("embryos" of ecosystems) into the real ecological systems, which were hierarchically organized of different microbial species.

INTRODUCTION

Investigation of the prehistory of initial microbial population on our planet is important to understand further evolution of the biosphere. The basic effort in this way was done by C. Woese (1987). He theoretically reconstructed *the progenote* – a type of the extinct organism that was considered as the universal ancestor for Archaea and Bacteria. However, the earlier

* kompanv@yandex.ru

prehistory of the microbial world goes into uncertainty because a method of the transformation of abiotic organic microsystems into simplest living units is unclear. The attempt to clarify this method is made in the framework of the systemic conception of the origin of life (Kompanichenko, 2004, 2008, 2009). The systemic difference between an assembly of organic matter (abiotic system) and a living unit (biotic system) is expressed in terms of the four key biological properties: a) the ability to concentrate free energy and information (by means of their extraction from the environment); b) the ability to exhibit an intensified counteraction to external influences; c) expedient behavior (or expedient character of interaction with the environment); d) regular self-renovation at different hierarchical levels (from molecular to ecosystemic), including self-reproduction. In accordance with the conception, these properties first appeared in the simplest living units – probionts. Further evolution of probionts up to prokaryotes was not reconstructed in the mentioned articles. The aim of this work is to consider briefly the entire way from organic microsystems to the present prokaryotes being at the root of the Phylogenetic Tree. The proposed theoretical reconstruction includes four intermediate units between them: (a) a prebiotic microsystem at the unstable (bifurcate) state; (b) pulsing (bistate) prebiotic microsystem; (c) probiont; (d) progenote.

1. Prebiotic Organic Microsystems

Several prebiotic organic microsystems were proposed by various researchers: proteinoide microspheres (Fox, Dose, 1977; Fox et al, 1996), RNA-World macromolecules (Gilbert, 1986; Joice and Orgel, 1993), liposomes (Deamer, 1986, 2004; Luisi, 2000), aromatic hydrocarbons (Ehrenfreund et al, 2006). The prebiotic models demonstrate some signs of activity relevant to the vital processes. For instance, a coacervate is able to grow and selectively extract substance from the environment, a catalytic activity peculiar to proteinoide microspheres, some RNA macromolecules are able to self-replicate. However, although these models displayed the ability for chemical complication during the conducted experiments, they have not been converted into simplest living unit by now. It seems that the principal problem consists in their inability to force the initial signs of activity up to the self-maintaining regime, giving rise to metabolism, regular self-replication and other vital dynamic processes. The all-round consideration of this problem in the framework of the systemic conception allowed formulate the following thesis: transformation of a prebiotic organic microsystem into a living unit is possible only at the state of nonequilibrium bifurcate transition and under oscillating conditions in the outside world (Kompanichenko, 2003, 2004, 2008, 2009).

Behaviour of the biologically important organic molecules and the suggested prebiotic models is still not experimentally explored under nonequilibrium oscillating conditions: almost all experiments on prebiotic chemistry have been carried out in stable conditions. So, it is unclear which of the models, or their combination, was the most suitable to be transformed into initial forms of life on the early Earth. Due to this reason, it is accepted in this chapter the conditional composition of a prebiotic microsystem: mixture of the biologically important organic molecules (aromatic hydrocarbons, lipids, amino acids, nucleic acids) and/or their precursors. The optimum ratio between them can be defined further in the course of the experiments on prebiotic chemistry under oscillating far-from-equilibrium conditions.

The minimum temperature in the geological cradle of life can be evaluated as 70°C at the time of the origin of life on Earth (3.8-4.4 Ga). This magnitude characterizes seawater temperature of the Hadean ocean. The oxygen isotopic record in marine cherts shows the temperature of the Hadean ocean +80°C at age 3.8 billion years ago (Ga) and +70°C (±15°C) at age 3.5 Ga (Knauth, 1992; Knauth and Lowe, 2003; Schwartzman, 1999). Similar result was obtained on the basis of the exploration of the silicon isotopic compositions (Chaussidon and Franquis, 2006). The oldest records of life are about 3.5 Ga, in accordance with the exploration of microfossils, and 3.85 Ga on the basis of biogeochemical data (Schidlowski, 1997; Schopf et al, 2002; Westall et al, 2006). The maximum temperature limit in the geological cradle of life is uncertain because a temperature of water in hydrothermal systems and their submarine discharges can be much higher than water boiling point, up to 300-400°C. The upper temperature limit for some present Archaea is 121°C. These data allow us to agree with the microbiologists, who place thermophilic or hyperthermophiic forms Archaea and Bacteria at the root of the Phylogenetic Tree (Stetter, 1995; Woese, 1987; Xue et al, 2003). Of course, emergence of some thermophilic species from mesophilic forms during the following biological evolution seems to be possible.

High-temperature and wet conditions are appropriate for the listed organic compounds or their assemblies to exist. In particular, we found some aromatic hydrocarbons (naphthalene, biphenyl, fluorene, phenathrene, etc.) in the sterile water-steam mixture with the temperature 175°C from the borehole in Kamchatka geothermal region. In this region Mukhin et al (1979) discovered glycine in the sterile water-steam mixture, and several other amino acids in the inhabited hot spring. Fatty acids, their precursors and liposomes are usual components of high temperature fluid. Upper temperature limit for the synthesis of nucleotides is 50-60°C, as the exception up to 90 °C. So, in principle, prebiotic organic microsystems could be composed of all biologically important organic compounds in the high temperature medium (70 °C and higher) (Holm and Andersson, 2005; Russell, 2003; Schwartzman, 1999).

The requirement of nonequilibrium conditions connects the origin of life process with the wide range of the specific events related with nonequilibrium bifurcate transitions and formation of dissipative structures. The broad class of dissipative structures embraces as some specific non-living systems (oscillating chemical reactions, Benard cells, etc.) as all living organizations (Nicolis and Prigogine, 1977; Prigogine and Stengers, 1984).

2. Prebiotic Microsystems at the Unstable Bifurcate State

There exist two principal mechanisms of evolution of natural systems. The first one consists in gradual accumulation of small changes in a system. Importance of such mechanism for the biological evolution was emphasized C. Darwin. The second mechanism is revolutionary and leads to the radical transformation of a system. This mechanism is a bifurcation. The mechanism acts through the bifurcate transition of a system under nonequilibrium conditions. A bifurcation occurs when a system cannot exist further in the current conditions due to changes that have occurred in the surroundings or inner mutations. The universal scheme of bifurcate transition of a system is the following: stable existence of a system → rise of instability through powerful fluctuations → the highest point of instability (point of bifurcation, or critical point), radical change of the system's structure → choice of new path of development → the next period of its stable existence (Figure 1 (a)). At the bifurcation point a system undergoes many accidental changes that may influence the choice of its further development path. This is the reason why potential paths of a system's development bifurcate at the moment of its highest instability.

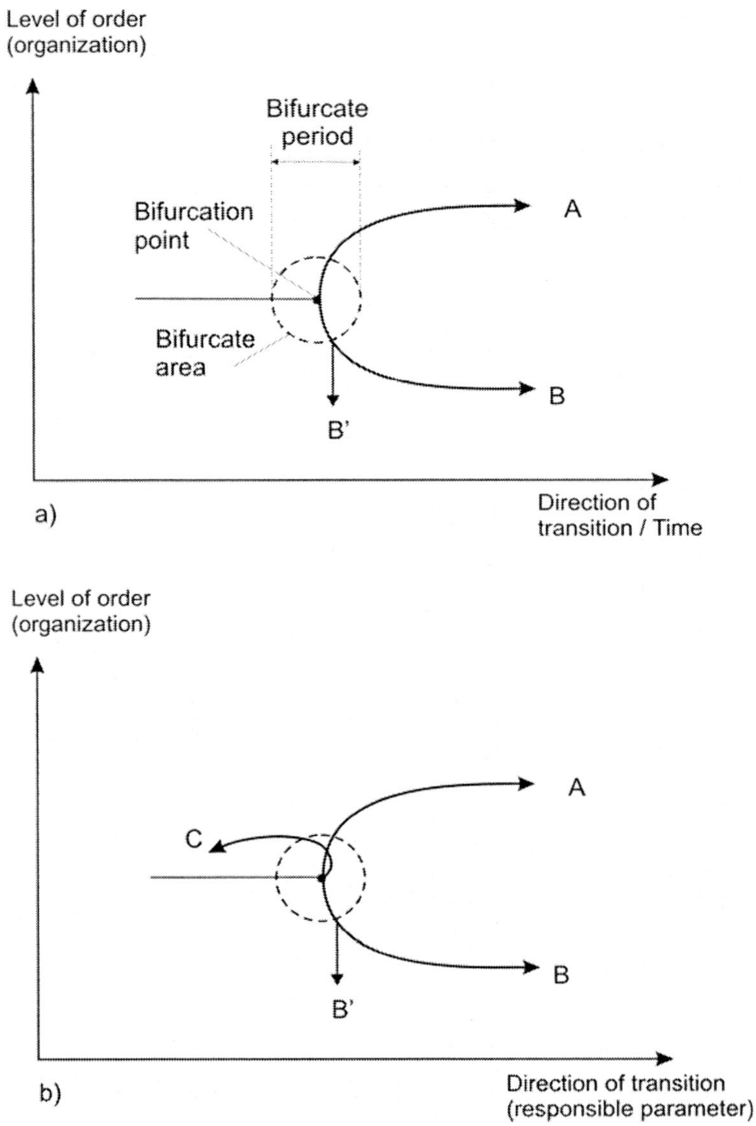

Figure 1. Principal scheme of bifurcate transition of a natural system under nonequilibrium conditions: a – direct transition from the initial stable state into one of permissible advanced stable states due to changed conditions in the outside world;. b – direct and reverse transitions (in case oscillating conditions in the outside world). A –trend to advanced higher-organized state; B – trend to advanced lower-organized state; B' – trend to complete destruction; C - reverse trend to the approximately initial state.

Finally, the system chooses one of many permissible ways, which can be united into two principal trends: (1) complication through self-organization (Trend A), (2) simplification and degradation (Trend B), up to full destruction (Trend B'). A reverse change of the external conditions initiates the back transition of a system into the initial state in accordance with the Trend C (Figure 1 (b)).

The mechanism of radical transformations of natural systems under far-from-equilibrium conditions is well-explored in the framework of the theory of dissipative structures and synergetics (Haken, 1978, 2003; Nicolis and Prigogine, 1977; Prigigine, 2003; Prigogine and Stengers, 1984). Behaviour of a chemical system under nonequilibrium conditions radically differs from its behaviour under conditions near equilibrium. Around the bifurcation point a chemical system acquires some specific properties, which are not peculiar to it during a period of the stable existence. The most essential of these bifurcate (or critical) properties are the following: (1) sharp heterogeneity with the intensive counter processes along and against the gradients; (2) continuous fluctuations (waves) and re-arrangement of molecules; (3) integrity through cooperative processes; (4) incessant exchange by matter and energy with the surroundings (open system). Let us consider the obvious analogies between these properties and the corresponding characteristics of living organisms. A comparison is drawn in the Table 1. Each of four main critical properties is distinctly reflected in the vital processes. Although the corresponding characteristics of living cells do not express the unique essence of life, they are in its background. Loss of any of these characteristics would make life impossible. Meanwhile, the critical properties are temporal characteristics of a chemical system: they appear at the beginning of the bifurcate transformation and disappear with the completed transition to the advanced stable state. On the basis of these facts the following thesis is formulated: *Life on Earth arose from the bifurcate state* that was inherent to still indeterminate kind of organic microsystems.

This thesis leads to the paradoxical situation: the bifurcate state should be considered as a starting-point for living systems, but the bifurcation point is in principle unstable. A system cannot be at the bifurcation point for a long time, because the occurrence of many accidental changes inevitably turns its development either into the initial state or into the advanced one. However, nature on Earth found a way to prolong a bifurcate state of the prebiotic microsystems about four billion years ago. The permissible opportunities for this to occur are investigated below.

3. Bistate (Pulsing) Prebiotic Microsystems

The period of a bifurcate transition is characterized by the appearance of two opposite forces in a system. They simultaneously strive to turn development of a system into the opposed initial and advanced states. Action of these forces dualizes the system's structure, maintains the internal tension, and keeps the bifurcate properties considered above. The bifurcation point designates peak of the bifurcate period. To prevent of the completed transition to the initial or advanced stable states that would lead to loss of the critical properties, an organic microsystem must oscillate around the bifurcation point staying inside the bifurcate area all the time.

Table 1. Correlation between the principal properties of a chemical system at the bifurcate state and the corresponding characteristics of a living organism

Main properties of a chemical system at the bifurcate state	The corresponding characteristics of a living cell
1. Sharp heterogeneity with the intensive counter processes along and against the gradients. Multitude high gradients (concentration, pressure, temperature, electric field, and other kinds) appear throughout in a chemical system that approaches the critical point of transition. Its structure becomes highly heterogeneous. Values of the gradients continuously decrease due to inevitable spontaneous processes (diffusion, heat conductivity, etc.), but at the same time they restore as a result of self-organization. This means the circulation of counter spontaneous processes (eroding the heterogeneity) and non-spontaneous ones (maintaining it) takes place around the point of bifurcation.	1. Heterogeneous structure and the counter processes of synthesis and destruction. Distinct heterogeneity of a cell is displayed on the ionic-molecular level as on the structural level. It is maintained by extraordinary complex circulation of the spontaneous processes (decreasing the gradients) and the non-spontaneous ones (increasing them) providing simultaneously synthesis and destruction (assimilation and dissimilation). Thus, spontaneous diffusion decreases concentration gradients, while non-spontaneous active transport increases them at the expense of energy expenditure. Permanent ionic gradients are maintained between a cell and the environment, their reversible changes support electric field gradients. Structural heterogeneity is more obvious in eucariotic cells: the nucleus, protoplasm, mitochondrions, endoplasmatic network, structures of Goldzhi complex are surrounded and permeated by plentiful membranes fixing gradients of concentrations, pressure, and electric field.
2. Continuous waves and re-arrangement of molecules. The particles are organized by fluctuations through cooperative events and synergism. A competition between the fluctuations continuously changes their radii of correlation. It follows that from time to time the particles transit from one temporal association to another. As a result, continuous re-arrangement of the particles takes place.	2. Continuous dynamic processes and re-arrangement of molecules/atoms. A cell is an assembly-line system in which manifold dynamic processes proceed uninterruptedly. Permanent chemical nonequilibrium between a cell and the outside world supports continuity of the processes. Re-arrangement of molecules takes place as due to their directed transfer as through their transformations during nucleo-protein interaction, ATP synthesis and decomposition, etc.
3. Integrity through cooperative processes. At the critical point the dominant fluctuation integrates all particles because its radius of correlation expands to the entire system. The particles are perceptive each other on macroscopic distances due to long-range links and the cooperative effect. The echo of any local event spreads over the whole system.	3. Integrity based on cooperative interaction. Integrity of a cell is maintained by the organization and regulation of cellular functions on different hierarchical levels. The organization of an entire cell appears in its organic process of division, coordinate movement, etc. The local structures integrate own parts and molecules as well, being simultaneously dirigible subsystems of the cell. Thus, mitochondrions may contain DNA and be able to self-reproduce, but their reproductive division is dependent on the nucleus.
4. Uninterrupted exchange by matter and energy with the surroundings (open system). Powerful fluctuations arising around the critical point open the system and maintain the exchange by matter and energy, as well as of simple structural information, with the surroundings. Through the exchange the system acquires extraordinary sensitivity to occurrences in the outside world.	4. Continuous exchange of a cell of matter, energy and information with the environment (open system). Existence of any organism is maintained through the various exchange processes with the outside world. An organism is highly sensitive to external changes and responds to them on different hierarchical levels. Thus, ions and molecules move through the plasmatic membrane both inside and outside of a cell. Large molecules and particles can be actively extracted from the environment by means of the specific membranes' properties. Unicellular organisms may respond by using complex taxis.

Such oscillation in the microsystem might be supported by the optimum oscillations in the surroundings, because any system at the critical state is open and very sensitive to the external changes. The type of a system that continuously oscillates around the bifurcation point between two polar stable states is called a *bi-state system,* or, in short, a *bisystem* (Kompanichenko, 2004, 2008).

In fact, a system of the type is pulsing. The biforked in this manner system simultaneously exists in two autonomous and interrelated states, which become dominating in turn due to oscillations around the central bifurcation point.

A bistate organic microsystem may appear under oscillating conditions in the maternal medium, when the regular inversions of the responsible parameters maintain the relative equilibrium between the competing nonequilibrium states. Once formed, the bistate system resists its own destruction through adequate response to external actions. This opportunity follows the Le Chatelier principle: any action to a chemical system due to change of the external conditions initiates the counteraction.

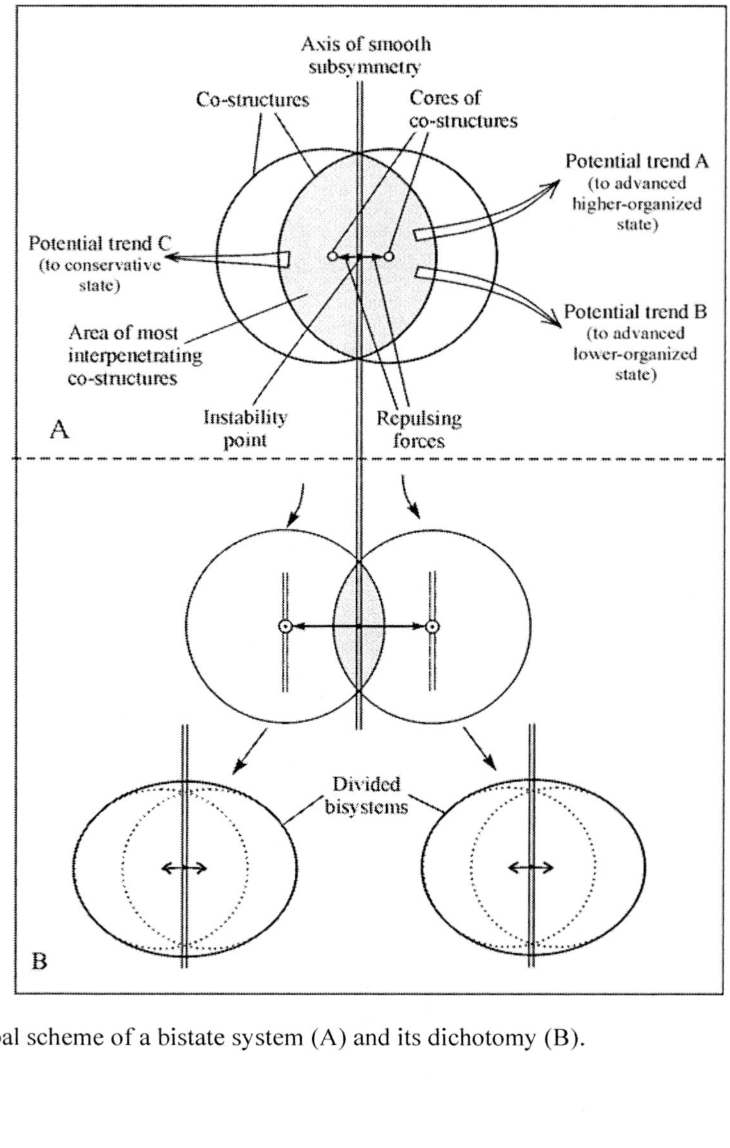

Figure 2. Principal scheme of a bistate system (A) and its dichotomy (B).

The principal scheme of a bistate system is shown on Figure 2. The point of instability repulses the co-structures to the opposite sides from the axis of smooth subsymmetry (Figure 2 (a)). A bisystem simultaneously strives to evolve in accordance with three main potential trends peculiar to a system during the bifurcate transition: to the conservative state (Trend C), to the advanced higher-organized state (A) and to the advanced lower-organized (B) state (Figure 1). Maintenance of the relative equilibrium between the competing co-structures works in the direction to their balanced separation that is finally displayed in the organic division of a bisystem into two sub-identical components (Figure 2 (b)).

According to the systemic conception, first cells derived of the bistate prebiotic microsystems and converted their method of organization. This thesis can be confirmed by the obvious correlation between the theoretically distinguished properties of a bistate prebiotic system and the real characteristics of a living organism (Table 2). The table shows that the intermediate position of a bistate system between "life and non-life" allows in principle to explain the fission of a cell, inner biophysical and biochemical rhythms, motion of unicellular organisms by means of oscillations of flagellums, accidental mutations in the genetic structures, morphological subsymmetry (dissymmetry) of living beings, their ability to correct own behaviour and modify functions, etc. The theoretically substantiated process of the dichotomy of a bistate system correlates with the real methods of dichotomy of simple and complex cells - amitosis, mitosis, and meiosis (Figure 3).

Scheme of dichotomous division of a cell (A - amitosis, B - mitosis and meyosis)

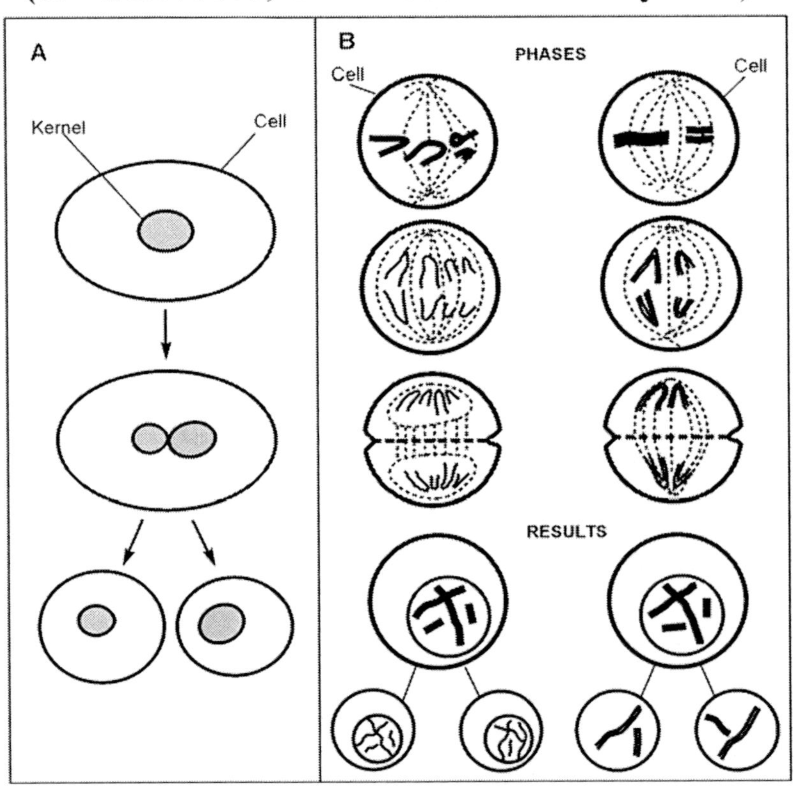

Figure 3. Scheme of dichotomy of a cell (A – amitosis, B – mitosis and meiosis).

Table 2. Correlation between the theoretically based properties of a bistate system and the corresponding real characteristics of a living organism

Main properties of a bistate system	The corresponding characteristics of a living organism
1. Latent biforked structure consisting of two autonomous co-structures divided by the plane of smooth subsymmetry. The plane is maintained by relative equilibrium between the opposite forces. Because the co-structures develop in different directions they are not absolutely identical. That means a bistate system is characterized with the paradoxical combination of the tendency to symmetry and asymmetry that results in its specific dissymmetric method of organization.	1. The biforked structure of a simple procariotic or eucariotic cell is latent and displays during the final fission into two sub-identical units. The biforked structure of advanced multicellular organisms is evident: they possess pair limbs, eyes, ears, cerebral hemispheres, etc., which are developed into different sides from the coincident planes of smooth subsymmetry. Disymmetry is a peculiar property of living organisms.
2. Dichotomy at the end of normal cycle of the existence. This property follows from the paradoxical combination of instability and integrity in a bistate system. Maintenance of the relative balance on a background of the repulsive forces should inevitably lead to the final division of the prebiotic bisystem into two sub-identical components. Figuratively speaking, normal cycle of its existence represents the gradual 'sawing' combined with the tendency to form sub-identical dichotomous units.	2. Fission of a cell at the end of normal cycle of the existence. The division begins with self-replication and divergence of the chromosomes in the core of a cell.
3. Oscillating character of existence related to periodic change of the dominant states. Oscillations of all inner processes support stability of a bisystem and sustain the internal tension that is a source of incessant rearrangement of molecules. Ceasing of the oscillations constrains a bisystem to leave the bifurcate area with a following irreversible transition into one of the stable equilibrium states.	3. Various inner processes in a living organism have oscillating character, for instance, oscillations of flagellums in independent eucariotic cells or hierarchical biophysical rhythms in advanced forms of life.
4. Display of mutability and heredity maintaining the ability to self-evolve. According to the theory of dissipative structures, each forward transition over the bifurcation point (from the initial into advanced state) brings some accidental changes in the chemical system. During the back transition a system strives to return into the initial state that is altered already because of the changes that have occurred. So, a reverse transition can be considered as a way towards conservation of the previous state of a chemical system. In this context, regular oscillations around the bifurcation point occurring in a bistate chemical system work step by step towards accumulation of the accidental changes that provides its ability to evolve. The forward transitions initiate new transformations in a bisystem (mutability), while the reverse ones retain its state (heredity).	4. Mutability and heredity are well-known processes investigated first of all in the framework of evolutional biology and genetics. Periodic mutations reorganize genetic structures at different hierarchical levels (in a gene, chromosome and entire genome).

4. Emergence of Living Probionts and Their Communities

As it was based, a prebiotic bistate microsystem possesses many properties that are essential for life. However, it is still not characterized with the following key unique properties of biological systems listed in the introduction. The essence of the key biological properties could be boiled down to the following thesis: a viable organism concentrates free energy (F) and information (I) through the ability to intensify and make expedient reaction to external changes. Rewording, it is able to keep the positive F and I gradients with respect to the environment. By means of the positive free energy gradient a viable living system is able to transform the environment more efficiently than the environment transforms it (this demonstrates the ability for intensified counteraction). Due to the positive information gradient, a viable biological system knows about the environment more than the environment knows about it (this is a basis for its expedient behaviour). To keep this path of existence, the inflow of free energy F (providing the ability to do a work) into a biological system must be higher than inflow of entropy S (devaluing the ability to carry out a work). The entropy contribution is connected with the total input of the spontaneous (basic) processes in a natural system, while the free energy contribution is related with the total input of non-spontaneous (coupled) processes. That means the ratio 'free energy contribution to entropy contribution' can be used to fix the approximate balance between these counter universal processes. The universal basic and coupled processes embrace all processes that proceed in a system. All non-living natural systems are characterized with negative F/S ratio, while viable living systems possess positive value of the ratio. The permissible method of the inversion of the ratio F/S contribution is theoretically outlined below.

Being a non-living system, a prebiotic bistate microsystem is characterized by a negative total F/S ratio. However, because of the oscillating character of existence the F/S ratio in such a system changes all the time. Besides, sometimes amplitude of the swing of the ratio may rise extraordinary due to a radical constructive or destructive transformation initiated by the accumulation of accidental changes. In the case high amplitude, from time to time the free energy contribution exceeds the entropy contribution although the average long-term F/S balance is still negative (Figure 4-I).

In the changeable conditions prebiotic bistate microsystems undergo periodic stress. The stress affecting a bistate prebiotic microsystem for the period of negative F/S balance leads to its degradation because of the prevailing entropy contribution (Figure 4-III). However, in the case of the period of positive F/S ratio a stress stimulates intensive reorganization of the microsystem possessing an excess of free energy. The surplus of free energy can be used for constructive transformations in the bistate microsystem as a respond to stress, with the following rise of free energy production due to more efficient organization of the network of chemical reactions. The enhanced inflow of free energy (as well as information) in the microsystem increases the positive F/S balance further and facilitates new constructive transformations. Several successive actions from the outside world having an influence for the favorable period may result in an increasing "avalanche-like" constructive conversion of the prebiotic bisystem which turn the *average F/S balance* onto the positive branch (Figure 4-II). In this way probionts on the early Earth acquired the *first unique biological property* – the ability to concentrate free energy and information.

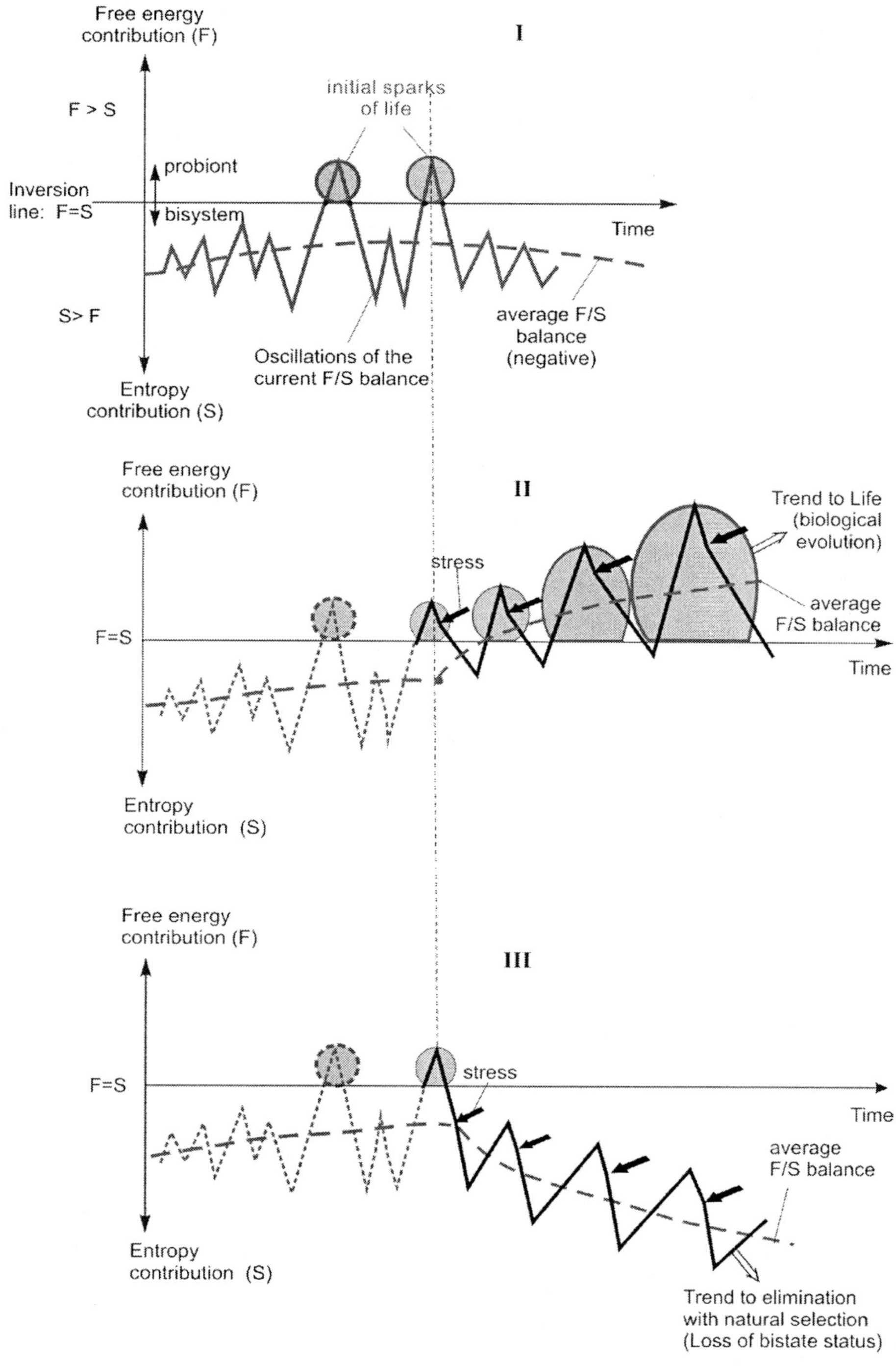

Figure 4. Oscillations of the balance 'Free energy contribution/Entropy contribution' in a prebiotic bistate (micro)system. The different trends of change of the balance: (I) Initial sparks of life in the bistate microsystem that appear because of the oscillating character of its existence; (II) Transformation of a bistate prebiotic microsystem into a primary living unit (probiont) through efficient responses to changes in the outside world (stresses) executed at the periods of the positive F/S balance; (III) Degradation of the bistate microsystem due to the changes influenced at the periods of the negative F/S balance.

The process of free energy concentration was maintained through the second *unique biological property* - the ability to exhibit an intensified counteraction to external influences. Like a free energy, information concentrated in a probiont faster than decomposed. This inverted direction of the information exchange gave rise the formation of *bioinformation*. New quality of information of this kind consisted in the appearance of the functional sequence complexity and programming (Abel, Trevors, 2005). The primary bioinformation that was involved into the network of biochemical reactions allowed a probiont to execute expedient counteraction to external influences (the *third unique biological property).* The incessant self-renovating processes in a viable probiont combined with its tendency to dichotomy (that appeared at the previous stage - formation of the prebiotic bistate microsystem) represent the *forth unique biological property* – regular self-renovation at different levels (from the continuous molecular reorganization up to self-reproduction).

The inversion of the F/S ratio leads to remarkable changes in the network of chemical reactions. This radical transformation gives rise to metabolism of an initial biological organism and transition of chemical reactions from *in vitro* into *in vivo*. Once formed, a probiont has opportunities to support existence because of the Le Chatelier principle: under regular external actions a probiont persistently develops a counteraction directed to maintain its own structural and functional stability, i.e. the biological method of organization (Kompanichenko, 2008a).

The proposed mechanism of probionts formation is statistical: only part of prebiotic bistate microsystems can be transformed into probionts while the rest ones should loss their bistate status due to natural selection at this prebiotic stage. So, the statistically sufficient quantity of the bistate microsystems is probably necessary to produce small number of probionts. The mechanism itself is based on the synergy effect as a capital factor of the maintenance of a system. On this basis the following thesis can be formulated: just a community of probionts, not a single probiont, should be considered as a minimum self-sufficient unit of the primary life.

Optimum fluctuations of the parameters in the geological cradle of life were necessary as to keep stability of prebiotic bistate microsystems (oscillating around the point of bifurcation) as to realize the ability of probionts to exhibit the intensified and expedient reaction to external changes. Such conditions are most peculiar to hydrothermal medium where stable regimes of existence interchange with powerful fluctuations that are generated by tectonic-magmatic activity of our planet. In this context, initial communities of probionts probably appeared around the hot springs in the shallow or deep zones of the Hadean ocean. Besides, the primary communities might originate around discharges of terrestrial hydrothermal systems. These communities, which probably were initially isolated of each other, respond to the embryonic stage of the evolution of ecological systems.

5. Biological Evolution from Probionts to Progenotes

The progenote that was theoretically reconstructed by C. Woese (1987) can be considered as an intermediate form of life between probionts and prokaryotes. It is correlated with the Last Common Universal Ancestor (LUCA) in a structure of the extinct part of the Phylogenetic Tree. Organisms of this type could have had a genotype and phenotype (i.e., information stored in a quiescent [replicative] form in one class of molecules that was also manifested in an active [functional] form in another), but their genes would for the most part have been physically separate units; they would not be organized into large contiguous linear arrays. Proteins would have been small or nonunique sequence or both. It follows of the reason that proteins of normal size could not have been synthesized without introducing (many) errors. As a consequence, enzymes would not be as accurate and specific as their modern counterparts. The progenote reasonably had error rates two or three orders of magnitude greater then found in cells today. To keep existence of minimally functional cells, genes should be disjoint, and they could have existed in high copy numbers. Probably, the informational macromolecule at this stage was RNA, the functional form of nucleic acid today, not DNA.

RNA is a complex macromolecule therefore the initial nucleoprotein interaction had to appear at the earlier stage of the biological evolution – in a probiont. At that period chemical reactions in a probiont were already involved into the organized network of metabolic processes *in Vivo*. The new biological organization facilitated an increase of thermal stability of nucleic acids. High average temperature of the Hadean ocean (70-80°C that could be higher around black smokers) probably didn't allow even short chains of nucleic acids to exist in abiotic bistate microsystems, at the exception of short periods of life sparks in them during the temporal positive F/S ratio. Bulk of the information in prebiotic bisystems was probably contained in the autocatalytic network (existence of a system possessed such level of information retention is supposed by C.Woese too). New opportunities for the synthesis of polynucleotide chains appeared immediately a prebiotic bisystem had transformed into probiont. Thermal stability of amino acids is higher than nucleic acids. In a general case, the stability of amino acids raises when the degree of their polymerization grows. Therefore, a probiont, inheriting from the previous bisystem oscillating character of existence and incessant circulation of matter/energy, and acquiring the reverse direction of the inner processes (F>S), should inevitably involve into the process of own "self-building" differently organized chains of amino acids and nucleic acids, which functionally complemented each other. Lipids (that play an important role in high temperature transformations of organic matter) and some other compounds should be involved in this process as well.

6. Biological Evolution from Progenotes to Prokaryotes

Evolution of progenotes into prokaryotic cells proceeded through significant rise of a level of the translation accuracy, elongation of strings of the both classes of biopolymers, the formation of DNA and a genome (Woese, 1987). This process led to an extraordinary raise of the structural-functional complexity of living beings and their communities. In the ecosystemic aspect, this process was characterized with the emergence of new species and appearance of parallel ways in the earliest biological evolution. Existence of manifold cross branches in the Phylogenetic Tree of prokaryotes was emphasized, in particular, by Doolittle (2000). According to the wide-spread opinion among microbial ecologists, (hyper)thermophilic Archaea and Bacteria are at the root of the Phylogenetic Tree (Stetter, 1995; Woese, 1987; Xue et al, 2003), although some of them could evolve from mesophilic forms at later stages of biological evolution. The opinion coincides with the above evaluations of temperature conditions in the hydrosphere of the early Earth. Emergence of species had led to conversion of the initial communities of probionts and progenotes ("embryos" of ecosystems) into the real ecological systems, which were hierarchically organized of different microbial species. Due to their expansion life spread over the entire Earth's surface film. It was a starting point of the microbial biosphere development.

Conclusion

The proposed method of the transition from abiotic organic microsystems to living cells needs further specification and experimental proof. Some relevant kinds of experimental research can be suggested. First, various prebiotic models should be explored at the bifurcate state and under oscillating conditions in the experimental chamber. Exploration of behaviour of organic assemblies under such far-from-equilibrium conditions allows us to concentrate attention just on the moment of the emergence of the key biological properties. Then, it seems necessary to determine the regime of fluctuations (amplitudes, frequencies) of thermodynamic and physico-chemical parameters in the hydrothermal medium that is the most optimal for the transition. The oscillations can be explored in the course of natural experimental research with the following using of the obtained data to conduct laboratory experiments. The investigation of the extinct types of microorganisms at the root of the Phylogenetic Tree should be brought to the probiont's level. Through these and other relevant scientific research the fundamental regularities of the biosphere functioning that were founded at the time of its origin can be clarified. Later on these regularities might serve as a scientific basis that would be helpful to control the global ecological changes on our planet.

References

Abel, D. L. and Trevors, J. T. (2005). Three subsets of sequence complexity and their relevance to biopolymeric information. *Theoretical Biology and Medical Modeling*, 2:29doi:10.1186/1742-4682-2-29.

Chaussidon, M. and Franquis, R. (2006). Constrains from silicon isotopic compositions on seawater temperature in the Archaean. Abstracts *of 6th European Workshop on Astrobiology*, Lyon, October 2006.

Deamer, D. W. (1986). Role of amphiphilic compounds in the evolution of membrane structure on early Earth. *Origins Life,* 17, 3-25.

Deamer, D. W. (2004). Prebiotic amphiphilic compounds. In: J. Seckbach (Ed.). *Origins.* (pp. 75-89). Netherlands, ST: Kluwer.

Doolittle, W. F. (2000). Uprooting the Tree of Life. *Sci. Am.,* February, 90-95.

Ehrenfreund, P; Rasmussen, S; Cleaves, J; Chen, L. (2006). Experimentally tracing the key steps in the origin of life: the aromatic world. *Astrobiology, 6*(3), 490-520.

Fox, S. and Dose, K. (1977). *Molecular evolution and the origin of life*. New York, ST: Dekker.

Fox, S; Balin, P; Pappelis, A; Yu, B. (1996). Experimental retracement of terrestrial origin of an excitable cell: was it predictable? In: Chela-Flores, J and. Raulin, F (Eds.) *Chemical evolution: Physics of the origin and evolution of life* (pp. 21-32). Netherlands, ST: Kluwer.

Gilbert, W. (1986). The RNA World. *Nature*, 319, 618.

Haken, H. (1978). *Synergetics. Berlin*, New York, ST: Springer-Verlag.

Haken, H. (2003). Special issue: Nonlinear Phenomena in Complex Systems, 5 (4).

Holm, N. and Andersson, E. (2005). Hydrothermal simulation experiments as a tool for studies for the origin of life on Earth and other terrestrial planets: a review. *Astrobiology*, 5(4), 444-460.

Joice, G. F and Orgel, L. E. (1993). Prospects for understanding the origin of the RNA world. In: *The RNA World* (pp. 1-25). New York, ST: Gold Spring Harbor laboratory Press.

Knauth, L. P. (1992). Origin and diagenesis of cherts: an isotopic perspective. In. N. Clanet and S. Chaudhuri (Eds.). Isotopic signatures and sedimentary records, *Lecture notes in Earth Sciences,* No 43 (pp. 123-152). Heidelberg, New York, ST: Springer-Verlag.

Knauth, L. P. and Lowe, D. R. (2003). High Archaean climatic temperature inferred from oxygen isotope geochemistry of cherts in the 3.5 Ga Swaziland Supergroup, South Africa. *Geol. Soc. Amer. Bull.*, 115(5), 566-580.

Kompanichenko, V. N. (2003). Distinctive properties of biological systems: the all-round comparison with other natural systems. *Frontier Perspectives*, 12(1), 23-35.

Kompanichenko, V. N. (2004). Systemic approach to the origin of life. *Frontier Perspectives*, 13(1), 22-40.

Kompanichenko, V. N. (2008). Three stages of the origin-of-life process: bifurcation, stabilization and inversion. *Int. J. Astrobiology*, 7(1), 27-46.

Kompanichenko, V. N. (2009). Changeable hydrothermal media as a potential cradle of life on a planet. *Planetary and Space Science*, 57, 468-476.

Luisi, P. L. (2000). The relevance of supramolecular chemistry for the origin of life. *Advances in Supramolecular Chemistry*, 6, 287-307.

Mukhin, L. M; Bondarev, V. B; Kalinichenko, V.I; Vakin, E.A. (1979). *Amino acids in hydrotherms of Southern Kamchatka*. Doklady AN USSR, 244(4), 974-977.

Nicolis, G. and Prigogine, I. (1977). *Self-organization in nonequilibrium systems*. New York, ST: Wiley.

Prigogine, I. (1997). *End of Certainty*. The Free Press.

Prigogine, I. and Stengers, I. (1984). *Order out of chaos*. New York, ST: Bantam.

Russell, M. J. (2003). The importance of being alkaline. *Science,* 302, 580-581.

Schidlowski, M. (1997). Carbon isotopes and the oldest record of life: potential and limits. *SPIE*, 3111, 462-471.

Schopf, J. W; Kudryavtsev, A. B; Argesti, D. G; Wdowiak, T. G; Czaja, A. D. (2002). Laser-Raman imagery of Earth's earliest fossils. *Nature*, 416, 73-76.

Schwartzman, D. (1999). *Life, temperature, and the Earth.* New York, ST: Columbia University Press.

Stetter, K. O. (1995). Microbial life in hyperthermal environments. *ASM News*, 61 (6), 328-340.

Westall, F; de Vries, T; Nijaman, W. (2006). The 3.466 Ga "Kitty's gap Chert" an early Archaean microbial ecosystem. In: W. U. Reymond and R. L. Gibson (Eds.) *Processes on the early Earth* (pp.105-131). Geological Society of America, 405.

Woese, C. R. (1987). Bacterial evolution. *Microbiol. Rev.*, 51(2), 221-271.

Xue, H; Tong, K; Marck, C; Grosjean, H; Wong, J. T-F. (2003). Transfer RNA paralogs: evidence for genetic code-amino acid biosynthesis coevolution and an archaean root of life. *Gene*, 310, 59-66.

In: New Trends in Chemical Physics Research
Editor: James N. Lapherty
ISBN: 978-1-61668-535-5

Chapter 17

THE STRUCTURE OF PHOSPHATE, ALKALI PHOSPHATE, BORATE AND ALKALI BORATE GLASSES

Mehdi Mirzayi[*]
Faculty of Basic Science,
Islamic Azad University Miyaneh branch,
Miyaneh, Iran

ABSTRACT

Glassy materials are crucial to a number of major technological devices notably in domains of electronics, energy storage, and converters and environmental monitoring such as batteries, fuel cells and sensor technology. In this paper the structure of vitreous phosphate and borate glasses have been investigated. Also, effects of addition of alkali ions on the structure of vitreous phosphate and borate have been reviewed. It is observed that the structure of glassy phosphoric oxide is based on tetrahedral building block which termed as ultraphosphate. By Addition of metal oxides to phosphoric oxide, other structures are possible such as meta, pyro, and ortho phosphate which can be made depending on the [O]/[P] ratio. Investigations have been demonstrated that the structure of vitreous B_2O_3 is formed by boroxol rings which joined together by a common oxygen atom. Addition of alkali ions to B_2O_3 glass converts two BO_3 (triangle) units into two BO_4 (tetrahedron) units.

INTRODUCTION

Amorphous materials are known by disorder arrangement of their atoms. Thus appear it is difficult to give definition of amorphous structure. In the amorphous materials unlike the crystalline materials, no single structure exists for any given glass composition, or even spatially with in a given sample. Instead, we must discuss an idealized structure, which is

[*] Corresponding author: Email; m_mirzayi2003@yahoo.com

known to vary within certain bonds throughout the specimen. These structural models will prove to be much more useful if they are always based on a desire to explain all of the observations [1].

The structural properties of amorphous semiconductors have been investigated by means of X-ray, electron and neutron diffraction [2], transmission electron micro-scope (TEM) and scanning electron microscope (SEM) [3], extended X-ray absorption fine structure (EXAFS), small-angle X-ray scattering (SAXS), Raman scattering, infrared absorption (IR) and nuclear magnetic resonance (NMR) [4,5,6]. These modern spectroscopic methods now allow researchers to propose models for new glasses. It should be recognized that, these methods are best suited for examining the local structure about an ion, with some extension to the intermediate range of order, and in general are quiet incapable of application to materials where the properties are controlled by microstructure rather than short to intermediate range atomistic order.

The coordination number of all cations is very important parameter which provide basic element of model for glass structure. Since the building blocks are usually well defined structure such as tetrahedral or triangles, they exhibit order at the level of several associated atoms or ions. It is estimated by radius ratio or by consideration of the coordination state usually observed for those cations in other materials. Thus we can state the silicon in silicate glass has tetrahedral coordination and boron can potentially exist in either three or four-fold coordination. The advent of magic angle spinning nuclear magnetic resonance (MAS-NMR) has virtually removed the element of speculation form models for the coordination state of Si^{4+}, B^{3+}, P^{5+} and other ions in glass. It should be mentioned experimental determination of bond angle distribution is very difficult and has been carried out for very few glasses.

Description of glass structure also address the number and arrangement of bridging and non-bridging bonds which link each of the building blocks to their neighbors. Most models for glasses only consider the non-bridging oxygens (NBO) distribution for explanation the structure. Spectral techniques such as IR and MAS-NMR have proven to be invaluable in establishing the distribution of non-bridging oxygens[1]. For example in silicate glasses, oxygen atoms bonded to only one Si atom are usually termed "Single-bonded" or "non-bridging oxygen". Those linked to two silicons are termed "bridging oxygen". The same terminology is applicable to glasses based on the other glass-forming oxides such as borate and phosphate [7].

For modeling the structure of glasses we need to dimensionality of network. The coordination number only considers the average number of bridging and non-bridging bonds per network unit and does not consider how they are arranged. For example, boric oxide glass consists of a network of boron-oxygen triangles, while silica glass consists of a network of silicon-oxygen tetrahedra. Boric oxide glasses can be considered to have a planar structure, existing in three dimension in much the same way a drawing on a piece of paper becomes three dimensional when the paper is crumpled into a ball. On the other hand, silicate glasses are a true three dimensional network. Other glasses may even exist as a tangled structure of long chain polymers, which basically have a dimensionality of one and occupy three dimensions in the same way as does a ball of yarn.

Field strength of both network-forming and network-modifier cations are very important factors in structure modeling. Since many modern studies include variation in the identity of the anions which link the structure, their field strength and ionic radii are also important. The atomic radii of mobile cations, anions, and atoms and molecules must be considered in structural models [1].

The detailed properties of structural defects have been elucidated by means of optical spectroscopies such as electron microscopy, local chemical analysis, infrared and dark field microscopy, small angle scattering, and resonance methods such as electron spin resonance (ESR) [6]. In this paper the structure of phosphate, alkali phosphate, borate and alkali borate glasses have been reviewed.

THE STRUCTURE OF PHOSPHATE AND ALKALI PHOSPHATE GLASSES

Phosphate-based glasses have been evaluated as materials for soft and hard tissue engineering, and are of particular interest in the biomedical field due to their unique properties [8,9,10]. These glasses are particularly attractive because they can accommodate large concentrations of active ions without losing the useful properties of the material [11].

Phosphate glasses usually melt at temperatures less than 1000°C, have thermal expansion coefficients in the range of 90 to 250×10^{-7} /°C, have glass transition temperatures (T_g) less than below 350°C and softening temperature (T_d) below 400°C. Thus, they are of increasing interest for many applications, e.g. glass to metal seals, thick film paste, the molding of optical elements, low temperature enamels for metals [12,13].

The main advantage of a phosphate glass over other oxide glasses (e.g. silicate and borate) is its ability to accommodate high concentrations of transition metal ions and remain amorphous. In addition, phosphate glasses enjoy a range of compositional and structural possibilities (ultra, meta, pyro, and ortho) that facilitate tailoring chemical and physical properties of interest for specific technological applications [14,15,16]. The structure of glassy phosphoric oxide is based on tetrahedral building block [17]. Since phosphorus is pentavalent ion, the formation of a phosphorus/oxygen tetrahedron with four bridging oxygen would result in a unit with a net positive charge of +1. A charge balance tetrahedron can be created if one of the oxygens forms a double bond with the P^{+5} ion, while the other three oxygens form bridging oxygen (BO) with adjacent tetrahedral. The two dimension network formed by the connection of these tetrahedral at three corners. This network is easily disrupted, resulting in a very low glass transformation temperature for vitreous phosphoric oxide [1,18]. PO_4, tetrahedral units are described by the Q^n notion where n represents the number of bridging oxygens that link neighboring P-tetrahedra. Vitreous P_2O_5 structure has Q^3 P-tetrahedra with three bridging oxygens and one double bonded oxygen P=O, forming a three-dimensional cross-linked network. Ultraphosphate glasses have both Q^3 and Q^2 P-tetrahedra structural units. Metaphosphate glasses have Q^2 P-tetrahedra with two bridging oxygens and two terminal oxygens that are shared with the modifying cation polyhedra [13,19]. The Q^2 tetrahedra link to form long chains (or rings). Phosphate glasses based on pyro (Q^1) and orthophosphate (Q^0) units are also possible [20]. Figure 1 shows the possible phosphate tetrahedral units.

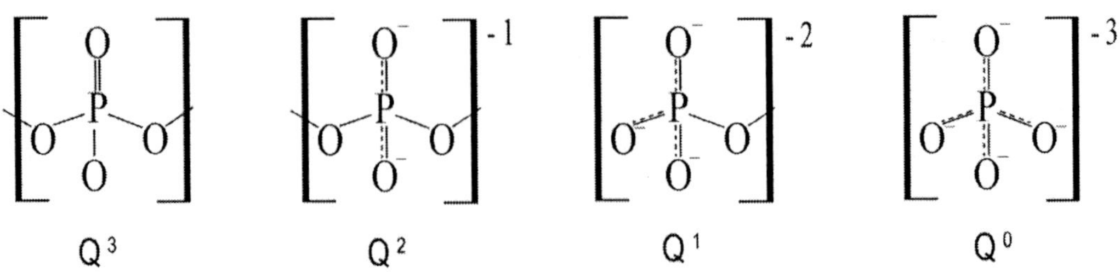

Figure 1. Phosphate tetrahedral sites that can exist in phosphate glasses.

Thus, a range of structures can be made depending on the [O]/[P] ratio as set by glass Composition. The term ultraphosphate refers to compositions where O/P < 3.0 [17].

Incorporating a metal oxide into the PO_4 network depolymerizes the network structure by converting bridging oxygen atoms to non-bridging oxygen atoms by breaking the P–O–P links, which produces Q^2 groups. In the ultraphosphate region both units Q3 and Q^2 are present in the glass network, whereas in polyphosphate glasses there are no Q^3 units. With increasing metal oxide content, the average chain length becomes progressively shorter. The structure of polyphosphate glasses with [O]/[P] > 3 is based on Q^2 chains terminated by Q^1 units. At pyrophosphate stoichiometry ([O]/[P] = 3.5), phosphate dimers, two Q^1 tetrahedra linked by a bridging oxygen, dominate the structure. Finally, isolated orthophosphate units (Q^0 groups) are present in glasses for which [O]/[P] = 4 [21].

Poor chemical durability of phosphate glasses limit their use in many applications [22,12]. One approach to increasing the chemical durability is to add various oxides to phosphate glasses, such as SnO, PbO, ZnO, CuO, Al_2O_3 and Fe_2O_3, etc. These oxides lead to the formation of Sn–O–P, Pb–O–P, Zn–O–P,Cu-O-P, Al–O–P and Fe–O–P bonds which replace the easily hydrolysable P–O–P bonds and improve dramatically the chemical durability of the modified phosphate glasses [23,24].

A second alternative of improving the chemical durability of phosphate glasses is to replace part of oxygen atoms by nitrogen [25]. It is down by exposure of the melt to ammonia, which decomposes to release nitrogen ions. Exposure of melts to N_2 molecules results in very little reaction. Nitrogen replaces oxygens in the network, allowing three tetrahedral to share a single corner. Since oxygen ions only allow sharing of two corners of the network units, replacement of oxygen by nitrogen increases the connectivity of the structure and improving the chemical durability [1,26,27].

Addition of alkali and alkaline earth oxide to phosphoric oxide lead to break of rings and the conversion of the network to a system of entangled linear chains of phosphorus-oxygen tetrahedral cross-linked by the monovalant or diavalant ions [1,28]. In other word, with increasing the alkali oxide content, average phosphate chain length decreases and lead to formation of short phosphate chains [29,30]. According to general reaction, we can write [13];

$$2Q^n + R_2O \rightarrow 2Q^{n-1} \quad (1)$$

For binary $xR_2O.(1-x)P_2O_5$ glasses, the concentrations of Q^n tetrahedral generally follow simple predictions based on glass composition. In the ultraphosphate region ($0 \le x \le 0.5$), the fraction, $f(Q^n)$, of Q^2 and Q^3 tetrahedra are given by [17];

$$f(Q^2) = \frac{x}{1-x} \text{ and } f(Q^3) = \frac{1-2x}{1-x} \quad (2)$$

For meta and pyro-phosphate region ($0.5 \le x \le 0.667$), $f(Q^2)$ and $f(Q^1)$ are obtained by;

$$f(Q^1) = \frac{2x-1}{1-x} \text{ and } f(Q^2) = \frac{2-3x}{1-x} \quad (3)$$

And finally for pyro and ortho-phosphate region ($0.667 \le x \le 0.75$) the relative concentration is given by [31];

$$f(Q^0) = \frac{3x-2}{1-x} \text{ and } f(Q^1) = \frac{3-4x}{1-x} \quad (4)$$

Hoppe et al. described the role that modifier coordination plays in determining the properties and structures of phosphate glasses in diffraction studies of binary ultraphosphate glasses [13,32]. Ultraphosphate glasses are based on Q^2 and Q^3 tetrahedra structure. He proposed that the structure (and properties) of ultraphosphate glasses would be dependent on the number of terminal oxygens available to coordinate the modifier ions ($R^{\upsilon+}$, where υ is the ion valence). Such cations will have a coordination number (CN_R) similar to that found in the appropriate crystalline phases. For glasses with the stoichiometry $xR_{2/\upsilon}O.(1-x)P_2O_5$, the number of terminal oxygens available per modifying ion is;

$$M_{TO} = \nu(\frac{1}{x}) \quad (5)$$

Hoppe et al. predict that there are two structurally distinct compositional ranges: when $M_{TO} > CN_R$ (region I) and when $M_{TO} < CN_R$ (region II). For region I, sufficient individual terminal oxygens are available for coordination with each $R^{+\upsilon}$ ion and so these ions can exist as isolated coordination polyhedra within the phosphate network and there are no R-O-R bonds. In region II, there are not enough individual terminal oxygens satisfy the coordination environments of every ion, $R^{+\upsilon}$ and so these ions must share the available terminal oxygens by forming R-O-R linkages. As a result, in region II, $R^{+\upsilon}$ coordination polyhedra share corners and edges, and act as bridges between neighboring Q^2 polyhedra [13,17]. Figure 2 shows the alkali coordination environments in $M_{TO} > CN_R$ and $M_{TO} < CN_R$ regions.

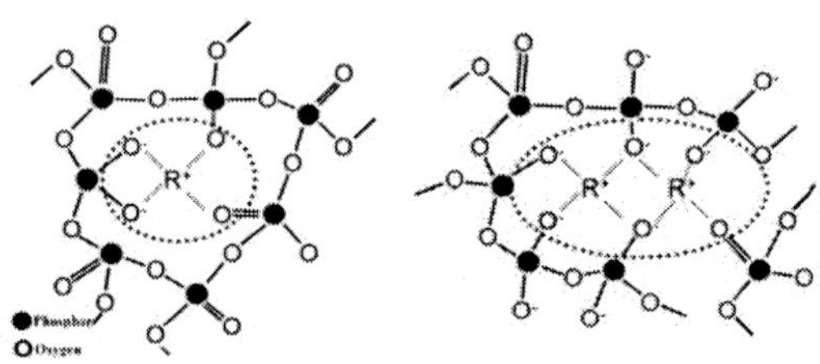

Figure 2. Schematic representation of alkali bonding In ultra-phosphate glasses when $M_{TO}>CN_{Me}$ (left) and $M_{TO}<CN_{Me}$ (right).

BORATE AND ALKALI BORATE GLASSES

B_2O_3 is one of the best glass formers known and is present in almost all commercially important glasses [33]. Borate base glasses are often used as dielectric and insulating material and it is known that borate is a good shield against infrared radiation. Also these glasses are used as electro-optic modulators, electro-optic switches, solid-state laser materials, non-linear optical parametric converters, thermoluminescence dosimetry devices and electrolytes for lithium batteries [34,35,36]. Alkali borate glasses have been used in diverse technologically important glasses and are one of the essential components of amorphous optical materials [37]. One of the primary drawbacks to the use of borate-based glasses is their relatively poor chemical durability [34].

Numerous investigations such as X-ray, neutron diffraction, infrared, Raman and NMR spectroscopies [38] have been demonstrated that the structure of vitreous B_2O_3 is formed by boroxol rings. Each boroxol ring consists of three [BO_3] triangles joined together by a common oxygen atom. One oxygen atom linking two adjacent boroxol rings leads to the formation of weakened intermolecular bonds between them [39,40]. Since the basic building block of this network is planner [41] rather than three dimensional, the three-dimensional linkage which occurs in a network of tetrahedra does not exist for vitreous boric oxide. A three-dimensional structure is developed by crumpling of the network, in much the same way that a two dimensional drawing on a sheet of paper develops a third dimension when the paper is crumpled to a ball[1].

Bonds in third dimension are Van der Waals and very weak. Thus the structure is easily disrupted and the melting temperature of boron oxide (450°C) is substantially lower than that of SiO_2 (1728°C). Note that, in this case, the strength of the B–O bond (496 kJ/mol) is relatively high and comparable to the strength of the Si–O bond (444 kJ/mol)[39].

The structure of vitreous boron oxide is shown in Figure 3. Vitreous boron oxide has a structure with a chain-layered motif in which boron atoms are in the threefold-coordinated state. The composition of the structural unit in the chain-layered structure corresponds to the B_6O_9 stoichiometry. The structure can be represented as a superposition of $2B_2O_3$ repetitive units. Moreover, it is quite probable that donor- acceptor oxygen bridges are formed in the structure and firmly cross-link the aforementioned chain-layered structural fragments [39].

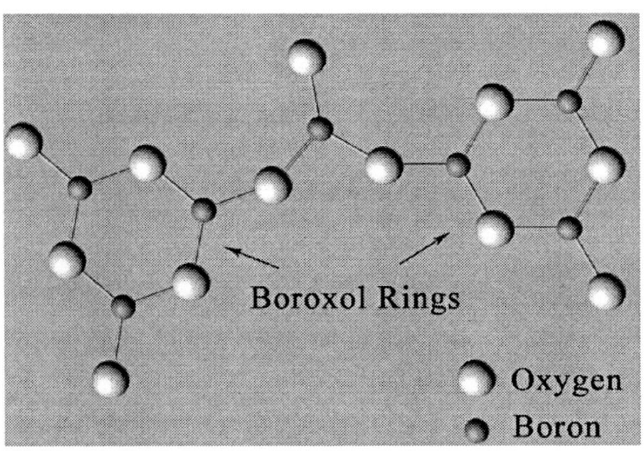

Figure 3. Structure of vitreous boric oxide glass.

The structure of vitreous boric oxide is also believed to contain a large concentration of boroxol group or boroxol ring. Monte-Carlo simulation, X-Ray diffraction and NMR Investigations have been showed that 60-80% of the boron atoms are constituents of boroxol rings [42,43]. Arguments supporting the existence of boroxol groups are based on observation of vary sharp band at 808 cm^{-1} in the Raman spectra of vitreous boric oxide [1]. Also this band has been observed in the vitreous boric oxide melt at 805 cm^{-1} and we conclude the boroxol group do exist in the melt [42,44].

The properties of borate glasses modified with alkali oxide showed a non-linear behavior when the alkali oxide was gradually increased. This departure from linearity was termed borate anomaly. They also claimed that Li^+ cations are easily inserted in the network interstices without substantial changes of the structure. At the same time, the rigidity increases as a result of the structure transformation [45].

Addition of alkali oxides (R_2O) into the borate glass leads to stable ionic glass and a considerable structural transformation due to the presence of fourfold-coordinated boron, i.e., the transformation of boron triangles into tetrahedral [39,46]. Also the introduction of a definite amount of alkali into borate glasses consolidates their structure and increases their density [4].

According to the Raman and NMR studies, the addition of one molecule of Na_2O or K_2O to B_2O_3 glass converts two BO_3 (triangle) units into two BO_4 (tetrahedron) units, as shown in Figure 4.

The fraction of the three-and four-coordinated boron, N_3 and N_4, respectively, can be represented by [48,49];

$$N_3 = \frac{1-2x}{1-x} \quad \text{and} \quad N_4 = \frac{1}{1-x} \tag{6}$$

where, x is the mole fraction of alkali metal oxide and up to x~0.3[50]. Accordingly, the coordination number of O atoms around a reference B atom is given by [51];

$$N_{B-O} = 3N_3 + 4N_4 = 3 + \frac{x}{1-x} \tag{7}$$

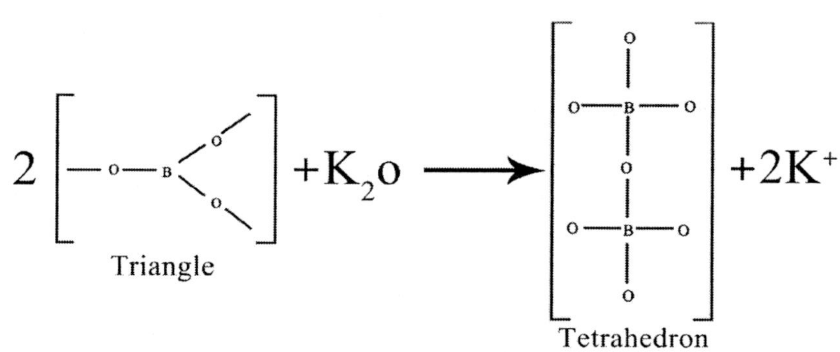

Figure 4. Conversion of triangle unit to tetrahedron unit by addition of alkali oxide in borate glass.

As mentioned, the addition of alkali oxide to B_2O_3 improves the stability of the continuous random network (CRN) by cross-linking BO_3 and BO_4 units. The concentration of non-bridging oxygen atoms is negligible up to about 30 mol% alkali oxide content. Beyond this concentration, alkalis begin to depolymerize the network with the formation of a significant amount of non-bridging oxygens [46,52]. Since the glass network consists of the BO_3 and BO_4 units joined together in a random way, an increase in the modifier concentration results, first, in the formation of tetrahedra and, then, in the appearance of non-bridging oxygen atoms in both the BO_3 triangles and the BO_4 tetrahedra [48].

According to NMR and XAFS investigations, BO_3 and BO_4 units are present in alkali borate glasses [50] and the four-coordinate boron population is independent of alkali up to about $R = 0.48$, where R is the nominal ratio of alkali oxide to diboron trioxide. Beyond this alkali loading, the heavier-alkali borate glasses preferentially generate trigonal borons with non-bridging oxygens, whereas lithium borates continue to form a higher proportion of tetrahedral boron in response to increasing modifier content. In fact, the data at $R = 0.7$ clearly indicate that the balance of tetrahedral boron and non-bridging oxygens is strongly dependent on the nature of the alkali modifier at high-alkali content. Heavier alkali ions appear to stabilize non-bridging oxygens, whereas lighter alkalis – particularly lithium – appear to stabilize tetrahedral boron. One possible explanation for this observation is related to charge balance in high-alkali glasses, where linkages between anionic tetrahedral borons become unavoidable in the absence of non- bridging oxygens. The significant negative charge present at sites of adjacent boron tetrahedra can be more easily balanced by multiple small lithium ions than by the larger alkali ions which repel each other due to their size. Consequently, adjacent tetrahedral borons are more energetically unfavourable when only large ions are available for charge balance, and non-bridging oxygens are preferentially formed at lower alkali content than in lithium borates [53,54].

Investigations on the rare-earth borate glasses have been showed that the rare-earth ions affect the structure of borate glass, very similar alkali ions. Presence of rare- earth ions causes to stability of glass and conversion of trigonal BO_3 units to tetrahedra BO_4 units. However, degree of both BO_3 and BO_4 units depends on rare earth occurrence and its concentration in glass composition [55,56].

The observed absorption bands in the FTIR spectra of rare-earth borate glasses is evidence to coexistence of trigonal BO_3 and tetrahedral BO_4 groups in these glasses. Also, equilibrium of the structural conversion between BO_3 and BO_4 units in the borate network depends on the chemical composition, the kind of modifiers and measurement conditions (temperature, irradiation)[57].

Figure 5 shows the Li cation in the tetraborate cluster. For all alkali borate glasses, the alkali metal cation M^+ is predominantly located near the edges of the BO_4 tetrahedron on the outside of the B-O-B comer fragments. Note that the positions, in which the cation can form additional bonds with oxygen atoms of the nearest environment, appear to be more stable (with energies less by 4-5 kcal/mol). Owing to these interactions, the cation can go away from the anion by one O-B-O unit without a considerable increase in the energy. Consequently, the cation can move only along such regions of chains formed by the boron and oxygen atoms that correspond to the outer parts of the B-O-B comer fragments. The migration to the opposite half-plane of this chain region, in actual fact, is forbidden. The cation can change the direction of motion and go to another chain in the pentaborate and tetraborate centers, which is improbable in the triborate centers [46,48].

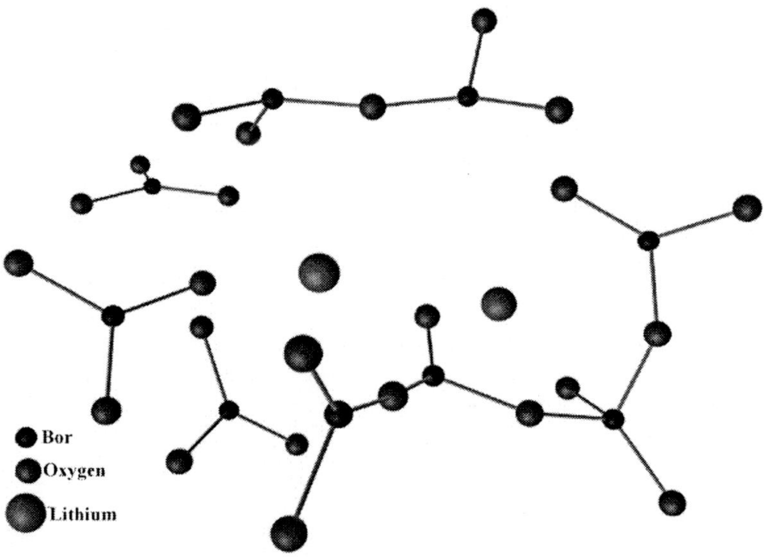

Figure 5. Tetraborate cluster in alkali borate glass.

As mentioned, in the alkali borate glasses, the alkali cations are not rigidly fixed in specific positions and can migrate over the continuous random network. According to experimental data, the conductivity sharply increases in the sample bulk beginning with a certain concentration of the Li_20 modifier. This is associated with a decrease in the activation energy and an increase in the mobility of modifier cations. It was demonstrated that the conductivity is hopping in character and the physical properties of ionic glasses depend on the nature and distribution of the cation-hosting sites. The conductivity can be explained by an increase in the number of negatively charged centers, which results in an increase in the electrostatic interaction between negative states, and by an increase in the cluster sizes, which rise the probability of the cation migration from center to center. As the Li_20 concentration increases, the number of carriers also increases, and their small size (as compared to other metals) favors the migration of cations over the bulk [48].

CONCLUSION

The structure of glassy phosphoric oxide is based on tetrahedral building block and has a range of compositional and structural possibilities (ultra, meta, pyro, and ortho) which can be made depending on the [O]/[P] ratio as set by glass composition. Also these glasses have three dimensional network structure. The term ultraphosphate refers to compositions where $O/P < 3.0$ is the main structure of vitreous phosphoric oxide. Incorporating a metal oxide into the phosphate network depolymerizes the network structure through converting bridging oxygen atoms to non-bridging oxygen atoms by breaking the P–O–P links. The polyphosphate, pyrophosphate and orthophosphate units produced for $[O]/[P] > 3$, $[O]/[P] = 3.5$ and $[O]/[P] = 4$ ratio respectively.

Modifier coordination determines the properties and structures of phosphate glasses. The structure (and properties) of phosphate glasses would be dependent on the number of terminal oxygens available to coordinate the modifier ions. When the number of terminal oxygens was bigger than coordination number of cations, there was sufficient individual terminal oxygens available for coordination so these ions can exist as isolated coordination polyhedra within the phosphate network and there are no R-O-R bonds in which R represents the modifier atom. But when the number of terminal oxygens was smaller than coordination number of cations, the modifier coordination polyhedra shared corners and edges, and acted as bridges between neighboring polyhedra.

The structure of vitreous B_2O_3 is formed by boroxol rings in which boroxol rings joined together by a common oxygen atom. an oxygen atom linking two adjacent boroxol rings together leads to the formation of weakened intermolecular bonds between them and the basic building block of this network is planner rather than three dimensional.

Addition of alkali oxides (R_2O) into the borate glass leads to stable ionic glass and a considerable structural transformation due to the presence of fourfold-coordinated boron, i.e., the transformation of boron triangles into tetrahedral.

REFERENCES

[1] J.E.Shelby, *Introduction to glass science and technology*, 0-85404-533-3, Royal society chemistry, Cambridge, 1997.

[2] G.J.Cuello, M.J.Ruiz, M.A.Gonzalez. *J. Non-Cryst Solids* 2007, *Vol 353*, 724–728.

[3] L.A.Bendersky, F.W.Gayle. *J. Res. Natl Inst Stand Technol* 2001, *Vol 106*, 997–1012.

[4] M.Karabulut, G.K.Marasinghe, C.S.Ray, D.E.Day, G.D.Waddill, C.H.Booth, P.G. Allen, J.J.Bucher, D.L.Caulder, D.K.Shuh *J Non-Cryst Solids* 2002, *Vol 306*, 182–192.

[5] J.W.Wiench, M.Pruski, B.Tischendorf, J.U.Otaigbe, B.C.Sales *J Non-Cryst Solids* 2000,*Vol 263and264*, 101-110.

[6] P.G.Le.Comber, J.Mort, *Electronic and Structural Properties of Amorphous Semiconductors*, 0-12-440550-9 , Academic press, London, 1973.

[7] M.H.Hekmat-Shoar, Ph.D thesis, Brunel University, 1979, UK. J.C.Knowles *J. Mater Chem* 2003, *Vol 13* , 2395.

[8] P.Hartmann, J.Vogel, U.Friedrich, C.Jager *J. Non-Cryst Solids* 2000,*Vol 263and264* , 94-100.

[9] M.Karabulut, E.Melnik, R.Stefan, G.K.Marasingh, C.S.Ray , C.R.Kurkjian, D.E.Day *J non-crystalline solids* 2001, *Vol288* , 8-17.

[10] R.V.S.S.N.Ravikumar, A.V.Chandrasekhar, L.Ramamoorthy, J.Reddy, Y.P. Reddy, Jun Yamauchi, P.S. Rao. *J. Alloy Compounds* 2004, *Vol 364*, 176–179.

[11] 12.P.Y.Shih, S.W.Yung, T.S.Chin *J Non-Cryst Solids* 1999,*Vol 244*, 211-222.

[12] R.K.Brow. *J. Non-Cryst Solids* 2000, *Vol 263and264* , 1-28.

[13] G.D.Khattak, E.E.Khawaja, L.E.Wenger, D.J.Thompson, M.A.Salid, A.B.Hallak, M.A.Daous, *J. Non-Cryst Solids* 1996, Vol 194, 1-12.

[14] A.V.Chandrasekhar, R.V.S.S.Ravikumar, B.J.Reddy, Y.P.Reddy, P.S.Rao, *Glass Technol.* 2002 , *Vol 43*, 32-39.

[15] E.Metwalli, M.Karabulut, D.L.Sidebottom, M.M.Morsi, R.K.Brow *J. Non-Cryst Solids* 2004, *Vol 344*, 128–134.

[16] R.K.Brow, C.A.Click , T.M.Alam. *J. Non-Cryst Solids* 2000, *Vol 274*, 9-16.

[17] B.C.Sales, L.A.Boatner, J.O.Ramey *J. Non-Cryst Solids* 2000, *Vol 263and264*, 155-166.
[18] E.Metwalli, R.K.Brow *J. Non-Cryst Solids* 2001, *Vol 289* , 113-122.
[19] J.E.stanworth *J soc glass technol* 1946 , *Vol 30* , 54-59.
[20] G.Walter, G.Goerigk,C.Russel. *J. Non-Cryst Solids* 2006, *Vol 352*, 4051–4061.
[21] S.T.Reis, A.M.Milankovic, V.Licina, J.B.Yang, M.Karabulut, D.E.Day , R.K.Brow *J Non-Cryst Solids* 2007, *Vol 353*, 151–158.
[22] S.T. Reis, M. Karabulut, D.E.Day, *J Non-Cryst Solids* 2001, *Vol 292* ,150-157.
[23] M.Karabulut, E.Metwalli, D.E.Day, R.K.Brow *J. Non-Cryst Solids* 2003, *Vol 328* , 199–206.
[24] A.L.Sauze, R.Marchand *J. Non-Cryst Solids* 2000 ,*Vol 263and264* , 285-292.
[25] M.R.Reidmeyer, D.E.Day, *J. Non-Cryst Solids*1995, *Vol 181* , 201-214.
[26] A.L.Sauze, L.Montagne,G.Palavit, F.Fayon ,R.Marchand. *J. Non-Cryst Solids* 2000, *Vol 263and264* , 139-145.
[27] R.Ciceo Lucacel, A.O.Hulpus , V.Simon, I.Ardelean. *J. Non-Cryst Solids* 2009, *Vol 355*, 425–429.
[28] G.Walter, J.Vogel, U.Hoppe, P.Hartmann *J. Non-Cryst Solids* 2001, *Vol296* , 212-223.
[29] P.Y.Shih, J.Y.Ding, S.Y.Lee *Mater Chem Phys* 2003, *Vol 80* ,391–396.
[30] J.W.Wiench, M.Pruski, B.Tischendorf, J.U.Otaigbe, B.C.Sales *J. Non-Cryst Solids* 2000,*Vol 263and264*, 101-110.
[31] M.Karabulut, E.Metwalli, A.K.Wittenauer, R.K.Brow , G.K.Marasinghe ,C.H.Booth , J.J.Bucher, D.K.Shuh. *J. Non-Cryst Solids* 2005, *Vol 351*, 795–801.
[32] T.Yano, N.Kunimine, S.Shibata, M.Yamane *J. Non-Cryst Solids* 2003 , *Vol 321*, 147–156.
[33] M.Bengisu, R.K.Brow, E.Yilmaz, A.M.Milankovic, S.T.Reis. *J. Non-Cryst Solids* 2006, *Vol 352* , 3668–3676.
[34] R.V.S.S.N.Ravikumar, R.Komatsu, K.Ikeda, A.V.Chandrasekhar, B.J.Reddy, Y.P. Reddy, P.S.Rao. *J. Phys Chem Solids* 2003 ,*Vol 64* , 261–264.
[35] Y.B.Saddeeka, H.A.Afifi, N.S.AbdEl-Aal *Physica B* 2007, *Vol 398*, 1–7.
[36] S.Keun Lee, P.J.Eng, H.k.Mao,Y.Meng, J.Shu. *Phys Rev lett* 2007 , *Vol 98* , 105502.1-105502.4.
[37] P.Hudon, D.R.Baker *J. Non-Cryst Solids* 2002 , *Vol 303*, 354–371.
[38] N.M.Bobkova *Glass Phys Chem* 2003,*Vol 29*, 501–507.
[39] N.A.Minakova, A.V.Zaichuk, Y.I.Belyi, *Glass Ceram* 2008, *Vol 65*, 70-73.
[40] M.M.Islam, T.Bredow, Ch.Minot *Chem Phys Lette* 2006, *Vol 418* , 565–568.
[41] J.Sakowski, G.Herms *J. Non-Cryst Solids* 2001, *Vol 293-295*, 304-311.
[42] J.Deppe, M.Balkanski, R.F.Wallis *Phys Rev B* 1990 , *Vol 41*, 7767-7773.
[43] T.Yano, N.Kunimine, S.Shibata, M.Yamane *J. Non-Cryst Solids* 2003 , *Vol 321*, 157–168.
[44] Y.B.Saddeek *Physica B* 2004, *Vol 344* , 163–175.
[45] C.P.E.Varsamis, A.Vegiri, E.I.Kamitsos *Phys. Rev. B* 2002, *Vol 65*, 104203.1-104203.14.
[46] I.Kashif, A.A.Soliman, H.Farouk, M.El-Shorpagy, A.M.Sanad *Physica B* 2008 , *Vol 403*, 3903–3906.
[47] O.A.Kondakova, A.S.Zyubin, S.A.Dembovsky. *Glass Phys Chem* , *Vol 26*, 418-424.

[48] H.Ushida, Y.Iwadate, T.Hattori, S.Nishiyama ,K.Fukushima , Y.Ikeda M.Yamaguchi, M.Misawa,T.Fukunaga, T.Nakazawa , S.Jitsukawa *J. Alloy Compound* 2004, *Vol 377* , 167–173.

[49] N.Ohtori, K.Takase, I.Akiyyama, Y.Suzuki, K.Handa, I.Sakai, Y.Iwadate, T.Fukunaga, N.Umesaki *J. Non-Cryst Solids* 2001, *Vol 293-295*, 136-145.

[50] K.Handa, Y.Kita, S.Kohara, K.Suzuya, T.Fukunagae, M.Misawa, T.Iida, H.Iwasaki, N.Umesaki. *J. Phys Chem Solids* 1999, *Vol 60* , 1465–1471.

[51] L.Cormier, G.Calas, B.Beuneu *J. Non-Cryst Solids* 2007, *Vol 353* , 1779–1784.

[52] V.K.Michaelis, P.M.Aguiar, S.Kroeker *J. Non-Cryst Solids* 2007, *Vol 353*, 2582–2590.

[53] P.M.Aguiar, S.Kroeker *J. Non-Cryst Solids* 2007, *Vol 353*, 1834–1839.

[54] W.A.Pisarski, T.Goryczka , B.W.Dus, M.Plonska, J.Pisarska *Mater Sci Engine B* 2005, *Vol 122* , 94–99.

[55] A.E.Burns, D.W.Winslow, W.J.Clarida, M.Affatigato, S.A.Feller, R.K.Brow. *J. Non-Cryst Solids* 2006, *Vol 352* , 2364–2366.

[56] W.A.Pisarski, J.Pisarska, W.R.Romanowski. *J. Mole Struc* 2005, *Vol 744–747* , 515–520.

In: New Trends in Chemical Physics Research
Editor: James N. Lapherty
ISBN: 978-1-61668-535-5

Chapter 18

CORRELATIONS BETWEEN ELECTRICAL PROPERTIES AND FLASH POINT WITH METHANOL CONTENT IN BIODIESEL

S. D. Romano[1,2] ***and P. A. Sorichetti***[3]

[1]Grupo de Energías Renovables, Facultad de Ingeniería,
Universidad de Buenos Aires, Av. Paseo Colón 850 (1063) Buenos Aires, Argentina
[2]CONICET (Consejo Nacional de Investigaciones Científicas y Técnicas),
Av. Rivadavia 1917 (1033) Buenos Aires, Argentina
[3]Laboratorio de Sistemas Líquidos, Facultad de Ingeniería,
Universidad de Buenos Aires, Av. Paseo Colón 850 (1063)
Buenos Aires, Argentina

ABSTRACT

This article presents a systematic exploration of the direct relation existing between electrical properties (permittivity and conductivity) and free methanol content (MC) in biodiesel (BD), as a function of temperature, and also between flash point (FP) and MC.

Precisely known amounts of methanol (up to 2.5% V/V) were added to BD, and electrical properties were determined between 20Hz and 2 MHz from room temperature to 81°C. Measurements show a clear dependence between MC and the permittivity and conductivity of the samples. Particularly, the value of the real part of the permittivity extrapolated to 0°C shows a linear dependence on MC. Moreover, since MC is closely related to FP of biodiesel, the correlation between both properties was also determined. An excellent fitting to a potential function was found.

The results presented in this article show that electrical properties (EP) and flash point measurements are an interesting alternative for the verification of methanol content in biodiesel.

Introduction

Biodiesel is a liquid biofuel that can replace petroleum-based Diesel fuel (DF) or blended with it, since they can be mixed in any ratio. ASTM (formerly known as the American Society for Testing and Materials) defines biodiesel as a fuel comprised of mono-alkyl esters of long chain fatty acids derived from vegetable oils or animal fats, designated B100, and meeting the requirements of ASTM D 6751 [1].

In many countries, biodiesel in blends with DF in a low proportion is routinely used in Diesel engines without any modification. In consequence, world production and use of biodiesel has significantly increased in recent years, since it is a renewable fuel that offers an alternative to fossil fuels in automotive applications.

Biodiesel [2], [3] is obtained by transesterification of vegetable oils or animal fats with short linear - chain alcohol (methanol or ethanol). The reaction must take place under well defined conditions. Raw materials for BD production may include new or used, edible or non-edible vegetal oils and animal fats. Therefore, biodiesel may be produced from oil from soybean, sunflower, corn, peanut, cotton, palm tung, castor seed, jatropha, rapeseed, micro-algae, and animal fats from cows, pigs, chicken, fish, etc.

Measurement of electric properties is a widely used technique in liquid samples [4], [5]. Among other applications, electrical techniques make possible to characterize mixtures and detect contaminants independently from the color or turbidity of the liquid phases.

Quality control of fuel during production and distribution is essential to ensure the quality of the product delivered to users and the good performance of engines. To establish the good quality of the liquid biofuel, several properties must be measured, including the free methanol content and flash point.

Acceptance ranges for BD properties are included in the standards set by different organizations (ASTM D 6751, EN 14214, IRAM 6515, DIN V 51606, ÖN C 1191, UNI 10635, etc.) [1], [6], [7].

Since it is an impurity, standards require that the free methanol content (MC) in BD be lower than 0.2 %. Therefore, the excess of alcohol must be removed from the mix of methyl esters by an adequate purification process. Also, flash point (FP) is an important parameter concerning the safety in BD distribution and storage, since higher values lead to a safer handling of the fuel. The minimum value required by standards is usually 100°C or higher.

This article presents the results of a systematic study of the correlations between electrical properties and MC in biodiesel, as a function of temperature, and also between FP and MC.

Experimental

I. Preparation of Biodiesel Samples

The samples of biodiesel studied in this chapter were prepared at the Grupo de Energías Renovables -GER- (Renewable Energy Group) at the Facultad de Ingeniería (Engineering School) of the University of Buenos Aires (UBA). Biofuel was made by the basic transesterification of new oil from soybeans and methanol, under constant stirring during one hour at a temperature of 60°C [4], [5]. Sodium hydroxide was used as a catalyst, and a 1:4 (in volume) alcohol to oil ratio. The mixture of methyl esters obtained as a product of the reaction was separated from glycerin and then purified by three washing steps. The first washing step was carried out with water acidified with hydrochloric acid and the remaining two with deionized water. Finally, samples were dried, in order to comply with the maximum water content value allowed by the standards [8].

In this article, the final product that complies with standards after the purification process is designated as "BD".

II. Characterization of Biodiesel According to International Standards

The samples of BD studied in this article were characterized by the following properties: viscosity, density, acid number, flash point (FP), cloud point, pour point, copper strip corrosion, free methanol content (MC) , free and total glycerin content and water content. All the measurement results were within the allowable ranges as required by international standards.

III. Preparation of Biodiesel Samples with Known Methanol Content

To obtain the value of EP and FP as a function of MC, biodiesel samples prepared by the process described in the previous paragraphs were mixed with known volumes of methanol (analytical grade).

IV. Measurement of Permittivity and Conductivity in Liquids

Broadly speaking, the relative permittivity of a substance characterizes its dielectric polarization, that is, the macroscopic (average) response of the permanent and induced dipole moments at the atomic and molecular scales, when an external electric field is applied. Similarly, conductivity determines the macroscopic current originated by transport processes of free charges (electrons and ions), when the substance is subjected to an external electric field [9].

The conductivity and permittivity of the sample as a function of the excitation angular frequency, ω, are usually jointly described by the complex relative permittivity $\varepsilon_r(\omega)$, defined as the ratio of the capacitance of the cell containing the sample, $C_s(\omega)$, to the capacitance of the empty cell, C_0.

$$\varepsilon_r(\omega) = \frac{C_s(\omega)}{C_0} \tag{1}$$

where the imaginary part of $\varepsilon_r(\omega)$ is written with a negative sign:

$\varepsilon_r(\omega)=\varepsilon_r'(\omega)- i\varepsilon_r''(\omega)$ (2)

As follows from (1) the real part of $\varepsilon_r(\omega)$ is proportional to the electrostatic energy stored as dielectric polarization, whereas the imaginary part $\varepsilon_r''(\omega)$ represents the energy lost through dissipative processes. Usually, the imaginary part of the permittivity is often written as the sum of two terms:

$$\varepsilon_r''(\omega) = \varepsilon_{r\,pol}''(\omega) + \frac{\sigma}{\varepsilon_0 \omega} \tag{3}$$

The first term, $\varepsilon_{r\,pol}''(\omega)$ represents the dissipative processes (relaxation) associated to the dielectric polarization. Therefore, it is related to the same molecular and electronic processes that give rise to the dipole moments described by the real part of the permittivity, $\varepsilon_r'(\omega)$. The second term describes the energy loss associated to the electrical conduction due to free charge carriers (ions and electrons). It is proportional to the conductivity parameter, σ, and inversely proportional to the angular frequency of the excitation, ω. Clearly, at low frequencies ($\omega \to 0$) conductivity often is the most important dissipative process and therefore dominates the imaginary part of complex permittivity. The constant ε_0 is the permittivity of free space (8.854 10^{-12} F/m).

The phase angle (with a minus sign) of the complex permittivity as a function of ω is customarily designated as $\delta(\omega)$. It is also usual in technological applications to define the dissipation factor $D_f(\omega)$ as the tangent of $\delta(\omega)$, that is, as the ratio (with a minus sign) of the imaginary to the real part of the complex permittivity $\varepsilon_r(\omega)$:

$$D_f(\omega) = tg\,\delta(\omega) = \frac{\varepsilon''_r(\omega)}{\varepsilon'_r(\omega)} \tag{4}$$

It is easy to see from equations. (1) and (2) that, under steady - state time - harmonic excitation, the dissipation factor is proportional to the ratio of the energy dissipated to the energy stored per cycle in the dielectric.

The results of complex permittivity measurements are usually presented as the real and imaginary parts, $\varepsilon_r'(\omega)$ and $\varepsilon_r''(\omega)$.

Measurements presented in this article were carried out by an automated dielectric system [10], [11] consisting of a GW - 830 Signal Generator, a custom - design Dielectric Measuring Interface and a Tektronix TDS - 210 Digital Oscilloscope. The sample cell is of the usual type used for electrochemical work, with platinized platinum electrodes, and the temperature was controlled by a Lauda thermostat to within +/- 0.1 °C .

V. Determination of Flash Point

Flash point measurements were carried out using the Pensky-Martens Closed Cup Tester according to ASTM D 93 [12].

It should be noted that, according to most standards, FP of biodiesel should be higher than 100°C, and usually it is above 130°C. This is a significant difference with the FP of Diesel fuel, which is around 50°C, and it is one of the advantages for the transport and storage of BD.

RESULTS AND DISCUSSION

I. Determination of the Temperature Dependence of Electrical Properties of Biodiesel

In the frequency range of interest in this article, from about 20 Hz to 2 MHz, no dielectric relaxation processes are observed in biodiesel; therefore, ε_r' may be considered as independent of ω and $\varepsilon_{r\ pol}''$ is negligible. On the other hand, the presence of free charge carriers due to the presence of contaminants and thermal effects in BD originate dissipative effects and the conductivity term σ/ω must be included in the imaginary part of the complex permittivity. Moreover, since thermal agitation tends to oppose the polarization associated to molecular orientation, ε_r', may be expected to decrease with temperature.

In summary, in this article the dependence of electrical properties of BD with the absolute temperature T and the angular frequency of the excitation, ω, will be fitted to the following expression:

$$\varepsilon_r(\omega, T) = \varepsilon_r'(T) - \mathrm{i}\,\frac{\sigma(T)}{\varepsilon_0\ \omega} \qquad (5)$$

The real part $\varepsilon_r'(T)$ will be assumed to have a linear dependence on temperature

$$\varepsilon_r'(T) = a - b\ T \qquad (6)$$

where the parameters a and b depend on the BD composition and also are affected by the presence of contaminants.

The parameter a in equation (6) corresponds to the value of $\varepsilon_r'(T)$ extrapolated to 0°C. It must be remarked that the direct measurement of $\varepsilon_r'(T)$ in biodiesel at low temperatures is difficult, since the cloud point and, sometimes, the pour point of could be reached.

On the contrary, the conductivity of pure BD is very low but increases exponentially with temperature [5]; moreover, the presence of contaminants changes significantly the room temperature value of conductivity and its dependence with the temperature.

II. Permittivity and Conductivity in Biodiesel Containing Methanol

Regarding electrical properties, experimental data show that biodiesel containing methanol has a higher permittivity at room temperature than pure BD [4], [5].

As indicated in equation (6), the dependence of the real part of the complex permittivity, $\varepsilon_r'(T)$, may be fitted to a linear function, where the parameters *a* and *b* depend on methanol concentration. From the measurement results plotted in Figure 1, the fitting parameters are summarized in Table I.

From Figure 1 it may be seen that the rate of decrease of permittivity with temperature increases with methanol content. Also, the fitting of permittivity data as a linear function of temperature is better at lower methanol concentrations (c.f. Table I). At higher temperatures $\varepsilon_r'(T)$ tends to the value of pure BD. This is not surprising since methanol evaporates at 65°C.

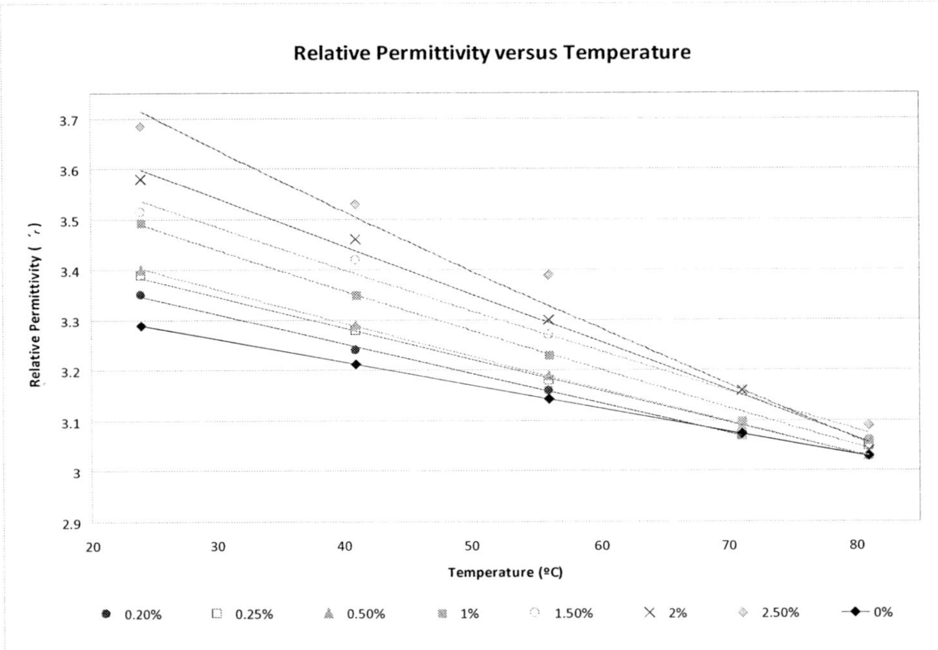

Figure 1. Relativity Permittivity versus Temperature as a function of methanol content in Biodiesel.

Table I. Parameters of the linear fit and correlation coefficient for permittivity versus temperature at different values of methanol content in biodiesel

Methanol concentration in Biodiesel (%V/V)	a	b	R^2
0.00	3.398	0,0046	0,999
0.20	3.490	0,0059	0,998
0.25	3.534	0,0062	0,988
0.50	3.544	0,0061	0,998
1.00	3.645	0,0074	0,994
1.50	3.747	0,0084	0,996
2.00	3.828	0,0095	0,993
2.50	3.997	0,0117	0,949

In Figure 2 it may be seen that the dependence of relative permittivity extrapolated to 0° C on methanol contents fits very well to a linear function:

$$z = 3.436 + 0.213\,x \tag{7}$$

where x is the methanol content in biodiesel (% V/V) and z is the relative permittivity extrapolated to 0°C. The correlation coefficient (R^2) is 0.98.

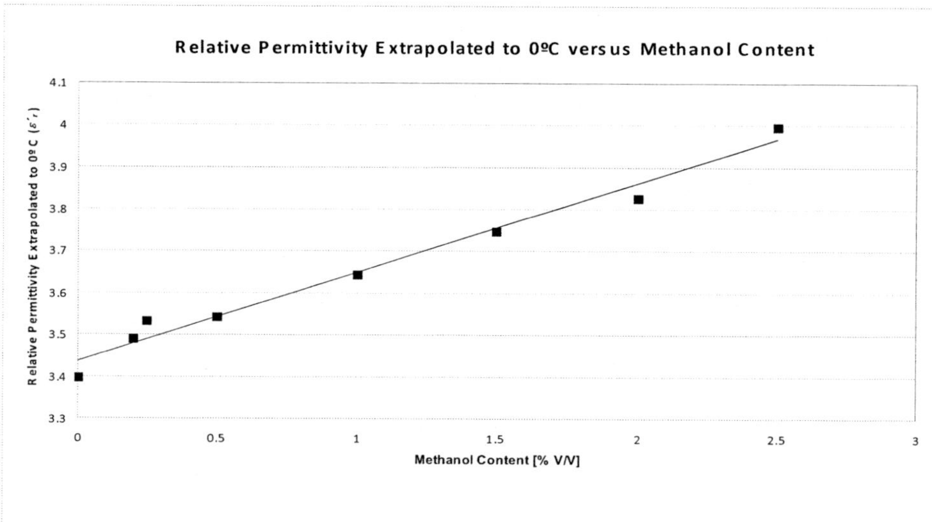

Figure 2. Relativity Permittivity extrapolated to 0°C versus methanol content in biodiesel.

From Figure 3, below 65°C the conductivity of biodiesel increases with higher methanol content.

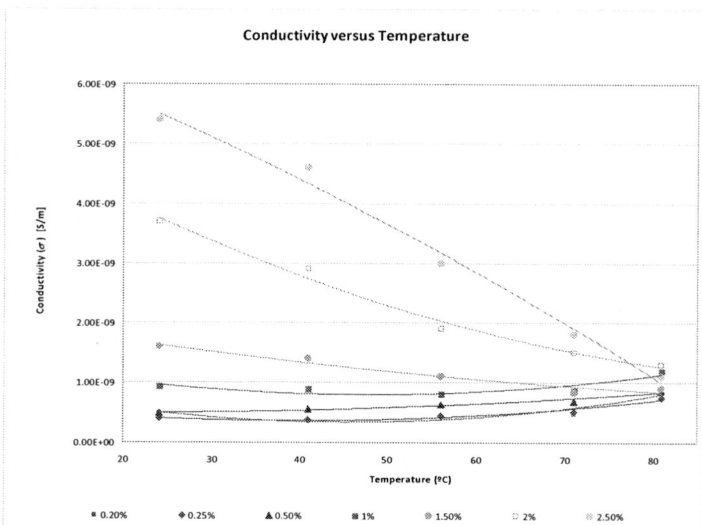

Figure 3. Conductivity versus Temperature as a function of methanol content in Biodiesel.

III. Correlation between Flash Point and Methanol Content in Biodiesel

Measurements show that there is a strong correlation between flash point and methanol content (Figure 4). The following expression fits very well the experimental data, with a correlation coefficient (R^2) greater than 0.99:

$$y = 38 \quad x^{-0.6} \tag{8}$$

where x is the methanol content in biodiesel (% V/V) and y is the flash point of the sample (°C), with $x \geq 0.2$. FP of BD samples with no methanol added always exceeded the maximum FP value that could be measured with the equipment available at the GER (> 180°C).

It is important to remark that the measured flash point value at the maximum methanol content set by international standards (0.2%) corresponds to the minimum allowable flash point (100°C) according to the standards. Therefore, compliance with the flash point requirement is also a good indicator of compliance with maximum methanol content. As it is easy to see in Figure 4, a small excess in methanol content originates a sharp decline in flash point of biodiesel. It is very important to take into account that free methanol content and flash point are closely related. Since the FP of pure methanol is 11°C, it is not surprising that FP of biodiesel will be lower than 100°C if the MC exceeds 0.2%.

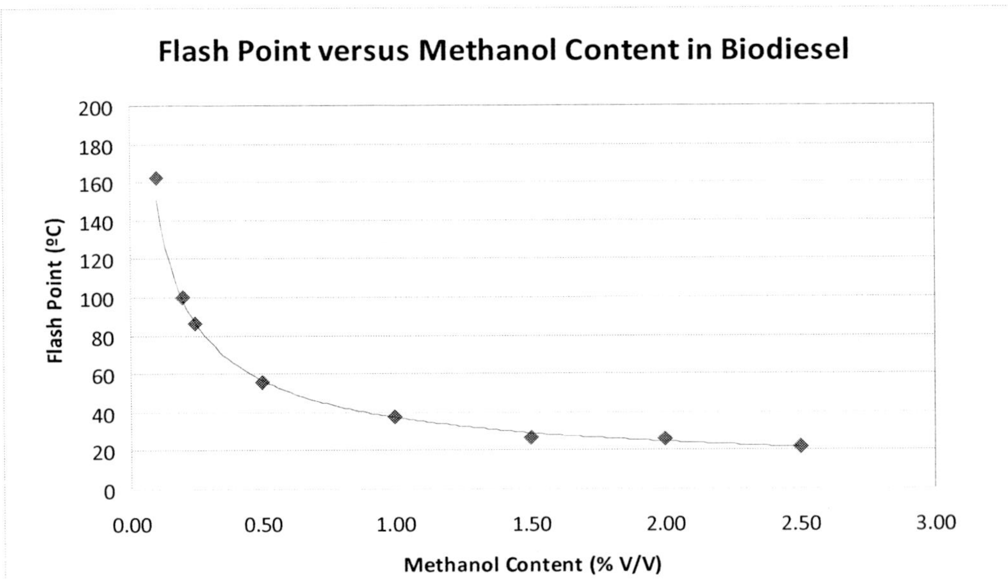

Figure 4. Flash Point versus Methanol Content in Biodiesel.

CONCLUSION

Experimental results show that there is a strong correlation between electrical properties and methanol content of biodiesel. Permittivity measurements as a function of temperature have a linear dependence on MC. This makes possible to determine accurately the methanol content in biodiesel within a wide range of concentrations, including the maximum value set by the standards (0.2%). In particular, the value of the real part of the permittivity, $\varepsilon_r'(T)$, extrapolated to 0°C shows a linear dependence on MC.

The conductivity of biodiesel below 65°C increases with higher methanol content. Therefore, measurements of biodiesel conductivity at room temperature (25°C) may be used for a preliminary check of methanol concentrations above 0.5%. This is very convenient in industrial applications.

Measurement data of flash point as a function of methanol content show clearly that small increases of alcohol concentration originate a steep decline in FP. This result is important for production, distribution and utilization of BD. In particular, excess of methanol must be eliminated with an adequate purification process.

From the strong correlation between EP and FP with MC, it is also possible to use these measurements for the preliminary verification (prior to certification) of the compliance with the methanol content requirements of BD standards. This is relevant since EP and FP techniques are less expensive than the procedure for determination of methanol content recommended by standards (chromatographic analysis) [13].

ACKNOWLEDGMENTS

The authors thank Mr. Ignacio Buesa Pueyo who carried out the measurements at the Renewable Energy Group (GER) of the University of Buenos.

The work in this article was funded by Project UBACYT I409 of the University of Buenos Aires: "Optimization of process variables for biodiesel production in an automated pilot plant, from different alternative raw materials".

REFERENCES

[1] ASTM D 6751- 03 a (American Standard for Biodiesel): Standard Specification for Biodiesel Fuel Blend Stock (B100) for Middle Distillate Fuels, 2003.

[2] Knothe G, Van Gerpen J, and Krahl J. *The Biodiesel Handbook*. AOCS Press, 2005.

[3] Romano, S. D., González Suárez, E., and Laborde, M. A. *Combustibles Alternativos*. Ediciones Cooperativas, 2nd Ed. 2006. Chapter I: Biodiesel, S. D. Romano, pp. 11 – 88.

[4] Sorichetti, P. A.; Romano, S. D. *Physics and Chemistry of Liquids*, 2005, *43 (1)*, 37 – 48.

[5] González Prieto, L. E.; Sorichetti, P. A.; Romano, S. D. *International Journal of Hydrogen Energy*, 2008, *33*, 3531 – 3537.

[6] EN 14214 (European Standard for Biodiesel): Automotive Fuels, *Fatty Acid Methyl Esters (FAME) for Diesel Engines*, Requirements and Test Methods.

[7] IRAM 6515-1/02 (Argentine Standard for Biodiesel): Calidad de Combustibles. Combustibles líquidos para uso automotor. *Biodiesel*, 2002.

[8] ASTM D 4928 – 00: Standard Test Methods for Water in Crude Oils by Coulometric Karl Fischer Titration, 2000.

[9] von Hippel, A. R.. *Dielectric and Waves*, John Wiley and Sons, New York, 1st Edition, 1954, Chapter 1.

[10] Schenkel, C. D.; Sorichetti, P. A.; Romano, S. D. *Anales de la Asociación Física Argentina,* 2005, *17*, 283 - 287.

[11] Sorichetti, P. A.; Matteo, C. L. *Measurement*, 2007, *40 (4)*, 437 - 449.

[12] ASTM D 93 – 02 a: Standard Test Methods for Flash Point by Pensky-Martens Closed Cup Tester, 2002.

[13] EN-14110: *Fat and oil derivatives – Fatty Acid Methyl Esters (FAME)* – Determination of methanol content.

INDEX

A

B

C

D

E

H

I

N

O

P

Q

R

S

T

U

V

W

X

Y

Z